# FUNDAMENTALS OF
# FRACTURE MECHANICS

# FUNDAMENTALS OF FRACTURE MECHANICS

TRIBIKRAM KUNDU

CRC Press is an imprint of the
Taylor & Francis Group, an **informa** business

CRC Press
Taylor & Francis Group
6000 Broken Sound Parkway NW, Suite 300
Boca Raton, FL 33487-2742


Printed in the United States of America on acid-free paper
10 9 8 7 6 5 4 3 2 1

International Standard Book Number-13: 978-0-8493-8432-5 (Hardcover)



**Library of Congress Cataloging-in-Publication Data**

Kundu, T. (Tribikram)
Fundamentals of fracture mechanics / Tribikram Kundu.
p. cm.
Includes bibliographical references and index.
ISBN 978-0-8493-8432-5 (alk. paper)
1. Fracture mechanics. I. Title.

TA409.K86 2008
620.1'126--dc22 2007038845

**Visit the Taylor & Francis Web site at**
**http://www.taylorandfrancis.com**

**and the CRC Press Web site at**
**http://www.crcpress.com**

# Dedication

*To my wife, Nupur, our daughters, Auni and Ina,*

*and our parents, Makhan Lal Kundu, Sandhya Rani Kundu,*

*Jyotirmoy Naha, and Rubi Naha*

# Contents

# *Preface*

My students motivated me to write this book. Every time I teach the course on fracture mechanics my students love it and ask me to write a book on this subject, stating that my class notes are much more organized and easy to understand than the available textbooks. They say I should simply put together my class notes in the same order I teach so that any entry level graduate student or senior undergraduate student can learn fracture mechanics through self-study. Because of their encouragement and enthusiasm, I have undertaken this project.

When I teach this course I start my lectures reviewing the fundamentals of continuum mechanics and the theory of elasticity relevant to fracture mechanics. Chapter 1 of the book does this. Students lacking a continuum mechanics background should first go through this chapter, solve the exercise problems, and then start reading the other chapters. The materials in this book have been carefully selected and only the topics important enough to be covered in the first course on fracture mechanics have been included. Except for the last chapter, no advanced topics have been covered in this book. Therefore, instructors of elementary fracture mechanics courses should have a much easier time covering the entire book in a three-unit graduate level course; they will not have to spend too much time picking and choosing appropriate topics for the course from the vast knowledge presented in most fracture mechanics books available today.

A professor who has never taught fracture mechanics can easily adopt this book as the official textbook for his or her course and simply follow the book chapters and sections in the same order in which they are presented. A number of exercise problems that can be assigned as homework problems or test problems are also provided. At the end of the semester, if time permits, the instructor can cover some advanced topics presented in the last chapter or topics of his or her interest related to fracture mechanics.

From over 20 years of my teaching experience I can state with confidence that if the course is taught in this manner, the students will love it. My teaching evaluation score in fracture mechanics has always been very high and often it was perfect when I taught the course in this manner. Since many students of different backgrounds over the last two decades have loved the organization of the fracture mechanics course presented in this book, I am confident that any professor who follows this book closely will be liked by his or her students.

The book is titled *Fundamentals of Fracture Mechanics* because only the essential topics of fracture mechanics are covered here. Because I was motivated by my students, my main objective in writing this book has been to keep the materials and explanations very clear and simple for the benefit of students and first-time instructors. Almost all books on fracture mechanics available in the market today cover the majority of the topics presented in this book and often much more. These books are great as reference books but not necessarily as textbooks because the materials covered are not necessarily presented in the same order as most instructors present them in their lectures. Over half of the materials presented in any currently available fracture mechanics book is not covered in an introductory fracture mechanics course. For this reason, the course instructors always need to go through several fracture mechanics books' contents carefully and select appropriate topics to cover in their classes. It makes these books expensive and difficult for self-study. Often, instructors find that some important topics may be missing or explained in a complex manner in the fracture mechanics books currently available. For this reason, they are forced to follow several books in their course or provide supplementary class notes for clearer explanations of difficult topics. *Fundamentals of Fracture Mechanics* overcomes this shortcoming. Since it only covers the essential topics for an introductory fracture mechanics course, it is the right book for first-time learners, students, and instructors.

**Tribikram Kundu**

# The Author

**Tribikram Kundu,** Ph.D., is a professor in the Department of Civil Engineering and Engineering Mechanics and the Aerospace and Mechanical Engineering Department at the University of Arizona, Tucson. He is the winner of the Humboldt Research Prize (senior scientist award) and Humboldt Fellowship from Germany. He has been an invited professor in France, Sweden, Denmark, Russia, and Switzerland. Dr. Kundu is the editor of 14 books and 3 research monographs, as well as author or coauthor of 2 textbooks and over 200 scientific papers, 100 of which have been published in refereed scientific journals; 3 have received "best paper" awards. Among his noteworthy recognitions are receipt of the President of India gold medal for ranking first in his graduating class in IIT Kharagpur and the regents' fellowship and outstanding MS graduate award from UCLA. He is a fellow of ASME, ASCE, and SPIE.

# 1

# *Fundamentals of the Theory of Elasticity*

## 1.1 Introduction

It is necessary to have a good knowledge of the fundamentals of continuum mechanics and the theory of elasticity to understand fracture mechanics. This chapter is written with this in mind. The first part of the chapter (section 1.2) is devoted to the derivation of the basic equations of elasticity; in the second part (section 1.3), these basic equations are used to solve some classical boundary value problems of the theory of elasticity. It is very important to comprehend the first chapter fully before trying to understand the rest of the book.

## 1.2 Fundamentals of Continuum Mechanics and the Theory of Elasticity

Relations among the displacement, strain, and stress in an elastic body are derived in this section.

### 1.2.1 Deformation and Strain Tensor

Figure 1.1 shows the reference state R and the current deformed state D of a body in the Cartesian $x_1x_2x_3$ coordinate system. Deformation of the body and displacement of individual particles in the body are defined with respect to this reference state. As different points of the body move, due to applied force or change in temperature, the configuration of the body changes from the reference state to the current deformed state. After reaching equilibrium in one deformed state, if the applied force or temperature changes again, the deformed state also changes. The current deformed state of the body is the equilibrium position under current state of loads. Typically, the stress-free configuration of the body is considered as the reference state, but it is not necessary for the reference state to always be stress free. Any possible configuration of the body can be considered as the reference state. For simplicity, if it is not stated otherwise, the initial stress-free configuration of the body, before applying any external disturbance (force, temperature, etc.), will be considered as its reference state.

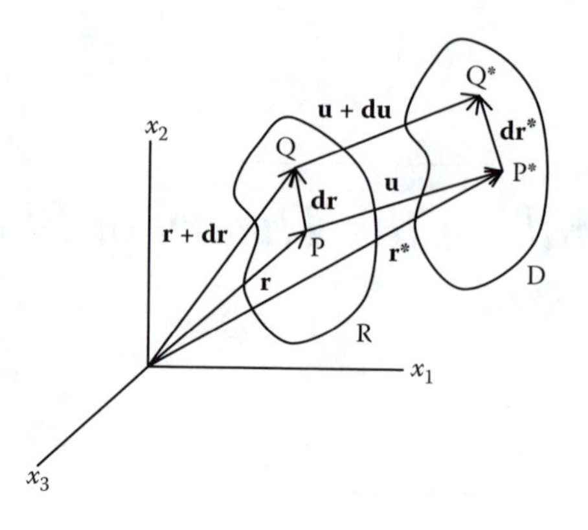


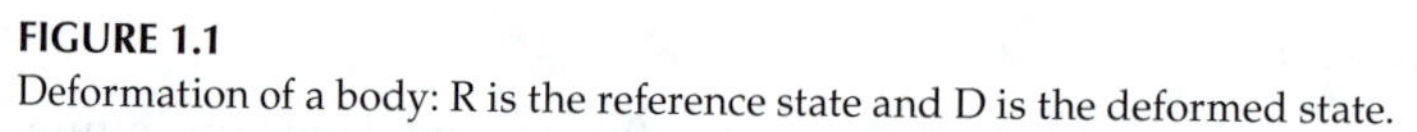

**FIGURE 1.1**
Deformation of a body: R is the reference state and D is the deformed state.

Consider two points P and Q in the reference state of the body. They move to P* and Q* positions after deformation. Displacement of points P and Q is denoted by vectors **u** and **u** + **du**, respectively. (Note: Here and in subsequent derivations, vector quantities will be denoted by boldface letters.) Position vectors of P, Q, P*, and Q* are **r**, **r** + **dr**, **r***, and **r*** + **dr***, respectively. Clearly, displacement and position vectors are related in the following manner:

$$\begin{aligned} \mathbf{r}^* &= \mathbf{r} + \mathbf{u} \\ \mathbf{r}^* + \mathbf{dr}^* &= \mathbf{r} + \mathbf{dr} + \mathbf{u} + \mathbf{du} \\ \therefore \mathbf{dr}^* &= \mathbf{dr} + \mathbf{du} \end{aligned} \tag{1.1}$$

In terms of the three Cartesian components, the preceding equation can be written as:

$$(dx_1^*\mathbf{e}_1 + dx_2^*\mathbf{e}_2 + dx_3^*\mathbf{e}_3) = (dx_1\mathbf{e}_1 + dx_2\mathbf{e}_2 + dx_3\mathbf{e}_3) + (du_1\mathbf{e}_1 + du_2\mathbf{e}_2 + du_3\mathbf{e}_3) \tag{1.2}$$

where $\mathbf{e}_1$, $\mathbf{e}_2$, and $\mathbf{e}_3$ are unit vectors in $x_1$, $x_2$, and $x_3$ directions, respectively.

In index or tensorial notation, equation (1.2) can be written as

$$dx_i^* = dx_i + du_i \tag{1.3}$$

where the free index $i$ can take values 1, 2, or 3.

Applying the chain rule, equation (1.3) can be written as

$$\begin{aligned} dx_i^* &= dx_i + \frac{\partial u_i}{\partial x_1}dx_1 + \frac{\partial u_i}{\partial x_2}dx_2 + \frac{\partial u_i}{\partial x_3}dx \\ \therefore dx_i^* &= dx_i + \sum_{j=1}^{3}\frac{\partial u_i}{\partial x_j}dx_j = dx_i + u_{i,j}dx_j \end{aligned} \tag{1.4}$$

In the preceding equation, the comma (,) means "derivative" and the summation convention (repeated dummy index means summation over 1, 2, and 3) has been adopted.

Equation (1.4) can also be written in matrix notation in the following form:

$$\begin{Bmatrix} dx_1^* \\ dx_2^* \\ dx_3^* \end{Bmatrix} = \begin{Bmatrix} dx_1 \\ dx_2 \\ dx_3 \end{Bmatrix} + \begin{bmatrix} \frac{\partial u_1}{\partial x_1} & \frac{\partial u_1}{\partial x_2} & \frac{\partial u_1}{\partial x_3} \\ \frac{\partial u_2}{\partial x_1} & \frac{\partial u_2}{\partial x_2} & \frac{\partial u_2}{\partial x_3} \\ \frac{\partial u_3}{\partial x_1} & \frac{\partial u_3}{\partial x_2} & \frac{\partial u_3}{\partial x_3} \end{bmatrix} \begin{Bmatrix} dx_1 \\ dx_2 \\ dx_3 \end{Bmatrix} \tag{1.5}$$

In short form, equation (1.5) can be written as

$$\{\mathbf{dr}^*\} = \{\mathbf{dr}\} + [\nabla \mathbf{u}]^{\mathrm{T}} \{\mathbf{dr}\} \tag{1.6}$$

If one defines

$$\varepsilon_{ij} = \frac{1}{2}(u_{i,j} + u_{j,i}) \tag{1.7a}$$

and

$$\omega_{ij} = \frac{1}{2}(u_{i,j} - u_{j,i}) \tag{1.7b}$$

then equation (1.6) takes the following form:

$$\{\mathbf{dr}^*\} = \{\mathbf{dr}\} + [\varepsilon]\{\mathbf{dr}\} + [\omega]\{\mathbf{dr}\} \tag{1.7c}$$

#### 1.2.1.1 Interpretation of $\varepsilon_{ij}$ and $\omega_{ij}$ for Small Displacement Gradient

Consider the special case when $\mathbf{dr} = dx_1\mathbf{e}_1$. Then, after deformation, three components of $\mathbf{dr}^*$ can be computed from equation (1.5):

$$\begin{aligned} dx_1^* &= dx_1 + \frac{\partial u_1}{\partial x_1} dx_1 = (1 + \varepsilon_{11}) dx_1 \\ dx_2^* &= \frac{\partial u_2}{\partial x_1} dx_1 = (\varepsilon_{21} + \omega_{21}) dx_1 \\ dx_3^* &= \frac{\partial u_3}{\partial x_1} dx_1 = (\varepsilon_{31} + \omega_{31}) dx_1 \end{aligned} \tag{1.8}$$

In this case, the initial length of the element PQ is $dS = dx_1$, and the final length of the element P*Q* after deformation is

$$dS^* = \left[(dx_1^*)^2 + (dx_2^*)^2 + (dx_3^*)^2\right]^{\frac{1}{2}} = dx_1\left[(1+\varepsilon_{11})^2 + (\varepsilon_{21}+\omega_{21})^2 + (\varepsilon_{31}+\omega_{31})^2\right]^{\frac{1}{2}}$$
$$\approx dx_1\left[1+2\varepsilon_{11}\right]^{\frac{1}{2}} = dx_1(1+\varepsilon_{11}) \tag{1.9}$$

In equation (1.9) we have assumed that the displacement gradients $u_{i,j}$ are small. Hence, $\varepsilon_{ij}$ and $\omega_{ij}$ are small. Therefore, the second-order terms involving $\varepsilon_{ij}$ and $\omega_{ij}$ can be ignored.

From its definition, engineering normal strain ($E_{11}$) in $x_1$ direction can be written as

$$E_{11} = \frac{dS^* - dS}{dS} = \frac{dx_1(1+\varepsilon_{11}) - dx_1}{dx_1} = \varepsilon_{11} \tag{1.10}$$

Similarly one can show that $\varepsilon_{22}$ and $\varepsilon_{33}$ are engineering normal strains in $x_2$ and $x_3$ directions, respectively.

To interpret $\varepsilon_{12}$ and $\omega_{12}$, consider two mutually perpendicular elements PQ and PR in the reference state. In the deformed state these elements are moved to P*Q* and P*R* positions, respectively, as shown in Figure 1.2.

Let the vectors PQ and PR be $(\mathbf{dr})_{PQ} = dx_1\mathbf{e}_1$ and $(\mathbf{dr})_{PR} = dx_2\mathbf{e}_2$, respectively. Then, after deformation, three components of $(\mathbf{dr}^*)_{PQ}$ and $(\mathbf{dr}^*)_{PR}$ can be written in the forms of equations (1.11) and (1.12), respectively:

$$(dx_1^*)_{PQ} = dx_1 + \frac{\partial u_1}{\partial x_1}dx_1 = (1+\varepsilon_{11})dx_1$$
$$(dx_2^*)_{PQ} = \frac{\partial u_2}{\partial x_1}dx_1 = (\varepsilon_{21}+\omega_{21})dx_1 \tag{1.11}$$
$$(dx_3^*)_{PQ} = \frac{\partial u_3}{\partial x_1}dx_1 = (\varepsilon_{31}+\omega_{31})dx_1$$

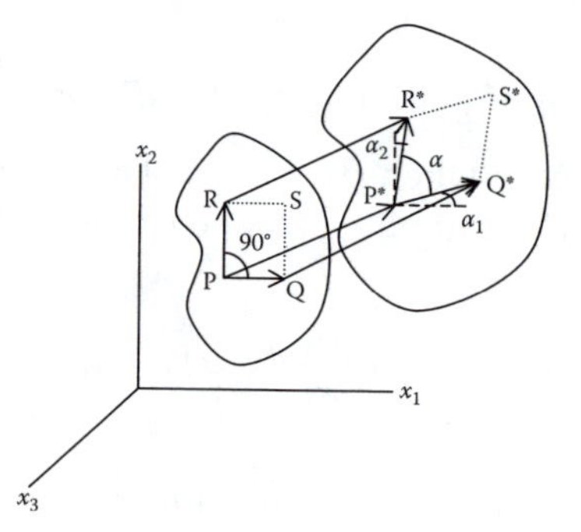


**FIGURE 1.2**
Two elements, PQ and PR, that are mutually perpendicular before deformation are no longer perpendicular after deformation.

$$(dx_1^*)_{PR} = \frac{\partial u_1}{\partial x_2} dx_2 = (\varepsilon_{12} + \omega_{12}) dx_1$$
$$(dx_2^*)_{PR} = \frac{\partial u_2}{\partial x_2} dx_2 = (1 + \varepsilon_{22}) dx_2 \quad (1.12)$$
$$(dx_3^*)_{PR} = \frac{\partial u_3}{\partial x_2} dx_2 = (\varepsilon_{32} + \omega_{32}) dx_1$$

Let $\alpha_1$ be the angle between P*Q* and the horizontal axis, and $\alpha_2$ the angle between P*R* and the vertical axis as shown in Figure 1.2. Note that $\alpha + \alpha_1 + \alpha_2 = 90°$. From equations (1.11) and (1.12), one can show that

$$\tan\alpha_1 = \frac{(\varepsilon_{21} + \omega_{21})dx_1}{(1 + \varepsilon_{11})dx_1} \approx \varepsilon_{21} + \omega_{21} = \varepsilon_{12} + \omega_{21}$$
$$\tan\alpha_2 = \frac{(\varepsilon_{12} + \omega_{12})dx_2}{(1 + \varepsilon_{22})dx_2} \approx \varepsilon_{12} - \omega_{21} \quad (1.13)$$

In the preceding equation, we have assumed a small displacement gradient and therefore $1 + \varepsilon_{ij} \approx 1$. For a small displacement gradient, $\tan\alpha_i \approx \alpha_i$ and one can write:

$$\alpha_1 = \varepsilon_{12} + \omega_{21}$$
$$\alpha_2 = \varepsilon_{12} - \omega_{21} \quad (1.14)$$
$$\therefore \varepsilon_{12} = \frac{1}{2}(\alpha_1 + \alpha_2) \quad \& \quad \omega_{21} = \frac{1}{2}(\alpha_1 - \alpha_2)$$

From equation (1.14) it is concluded that $2\varepsilon_{12}$ is the change in the angle between the elements PQ and PR after deformation. In other words, it is the engineering shear strain and $\omega_{21}$ is the rotation of the diagonal PS (see Figure 1.2) or the average rotation of the rectangular element PQSR about the $x_3$ axis after deformation.

In summary, $\varepsilon_{ij}$ and $\omega_{ij}$ are strain tensor and rotation tensor, respectively, for small displacement gradients.

**Example 1.1**

Prove that the strain tensor satisfies the relation $\varepsilon_{ij,k\ell} + \varepsilon_{k\ell,ij} = \varepsilon_{ik,j\ell} + \varepsilon_{j\ell,ik}$.

This relation is known as the compatibility condition.

*Solution*

$$\text{Left-hand side} = \varepsilon_{ij,k\ell} + \varepsilon_{k\ell,ij} = \frac{1}{2}(u_{i,jk\ell} + u_{j,ik\ell} + u_{k,\ell ij} + u_{\ell,kij})$$

$$\text{Right-hand side} = \varepsilon_{ik,j\ell} + \varepsilon_{j\ell,ik} = \frac{1}{2}(u_{i,kj\ell} + u_{k,ij\ell} + u_{j,\ell ik} + u_{\ell,jik})$$

Since the sequence of derivative should not make any difference, $u_{i,jk\ell} = u_{i,kj\ell}$; similarly, the other three terms in the two expressions can be shown as equal. Thus, the two sides of the equation are proved to be identical.

**Example 1.2**
Check if the following strain state is possible for an elasticity problem:

$$\varepsilon_{11} = k\left(x_1^2 + x_2^2\right), \qquad \varepsilon_{22} = k\left(x_2^2 + x_3^2\right), \qquad \varepsilon_{12} = kx_1x_2x_3, \qquad \varepsilon_{13} = \varepsilon_{23} = \varepsilon_{33} = 0$$

*Solution*
From the compatibility condition, $\varepsilon_{ij,k\ell} + \varepsilon_{k\ell,ij} = \varepsilon_{ik,j\ell} + \varepsilon_{j\ell,ik}$, given in example 1.1, one can write

$\varepsilon_{11,22} + \varepsilon_{22,11} = 2\varepsilon_{12,12}$ by substituting $i = 1, j = 1, k = 2, \ell = 2$.

$$\varepsilon_{11,22} + \varepsilon_{22,11} = 2k + 0 = 2k$$

$$2\varepsilon_{12,12} = 2kx_3$$

Since the two sides of the compatibility equation are not equal, the given strain state is not a possible strain state.

### 1.2.2 Traction and Stress Tensor

Force per unit area on a surface is called traction. To define traction at a point $P$ (see Figure 1.3), one needs to state on which surface, going through that point, the traction is defined. The traction value at point $P$ changes if the orientation of the surface on which the traction is defined is changed.

Figure 1.3 shows a body in equilibrium under the action of some external forces. If it is cut into two halves by a plane going through point $P$, in general, to keep each half of the body in equilibrium, some force will exist at the cut plane. Force per unit area in the neighborhood of point $P$ is defined as the traction at point $P$. If the cut plane is changed, then the traction at the same point will change. Therefore, to define traction at a point, its three components must be given and the plane on which it is defined must be identified. Thus, the traction can be denoted as $\mathbf{T}^{(\mathbf{n})}$, where the superscript $\mathbf{n}$ denotes the unit

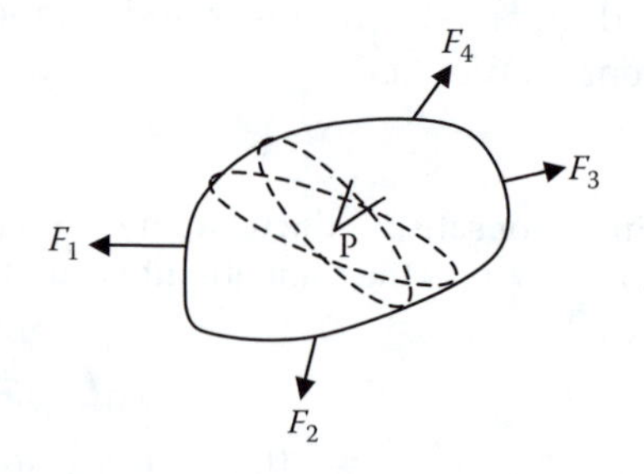


**FIGURE 1.3**
A body in equilibrium can be cut into two halves by an infinite number of planes going through a specific point $P$. Two such planes are shown in the figure.

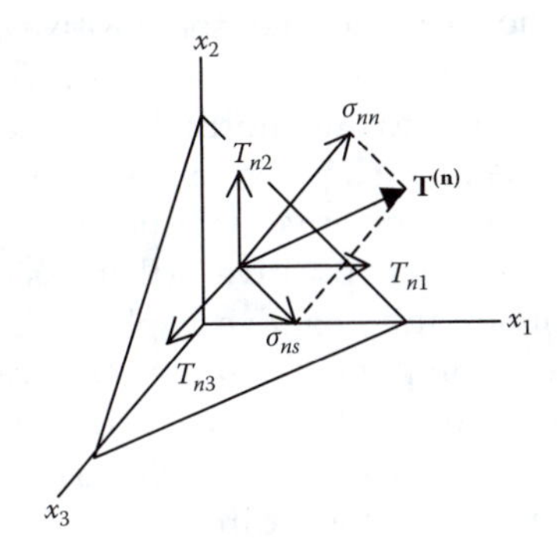


**FIGURE 1.4**
Traction $\mathbf{T}^{(\mathbf{n})}$ on an inclined plane can be decomposed into its three components, $T_{ni}$, or into two components: normal and shear stress components ($\sigma_{nn}$ and $\sigma_{ns}$).

vector normal to the plane on which the traction is defined and where $\mathbf{T}^{(\mathbf{n})}$ has three components that correspond to the force per unit area in $x_1$, $x_2$, and $x_3$ directions, respectively.

Stress is similar to traction; both are defined as force per unit area. The only difference is that the stress components are always defined normal or parallel to a surface, while traction components are not necessarily normal or parallel to the surface. A traction $\mathbf{T}^{(\mathbf{n})}$ on an inclined plane is shown in Figure 1.4. Note that neither $\mathbf{T}^{(\mathbf{n})}$ nor its three components $T_{ni}$ are necessarily normal or parallel to the inclined surface. However, its two components $\sigma_{nn}$ and $\sigma_{ns}$ are perpendicular and parallel to the inclined surface and are called normal and shear stress components, respectively.

Stress components are described by two subscripts. The first subscript indicates the plane (or normal to the plane) on which the stress component is defined and the second subscript indicates the direction of the force per unit area or stress value. Following this convention, different stress components in the $x_1x_2x_3$ coordinate system are defined in Figure 1.5.

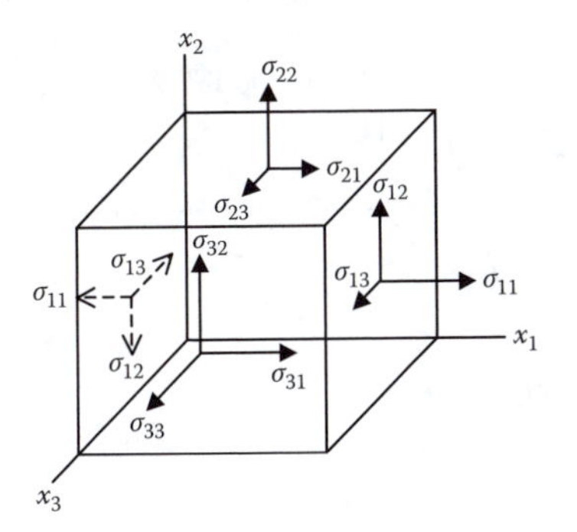


**FIGURE 1.5**
Different stress components in the $x_1x_2x_3$ coordinate system.

Note that on each of the six planes (i.e., the positive and negative $x_1$, $x_2$, and $x_3$ planes), three stress components (one normal and two shear stress components) are defined. If the outward normal to the plane is in the positive direction, then we call the plane a positive plane; otherwise, it is a negative plane. If the force direction is positive on a positive plane or negative on a negative plane, then the stress is positive. All stress components shown on positive $x_1$, $x_2$, and $x_3$ planes and negative $x_1$ plane in Figure 1.5 are positive stress components. Stress components on the other two negative planes are not shown to keep the Figure simple. Dashed arrows show three of the stress components on the negative $x_1$ plane while solid arrows show the stress components on positive planes. If the force direction and the plane direction have different signs, one positive and one negative, then the corresponding stress component is negative. Therefore, in Figure 1.5, if we change the direction of the arrow of any stress component, then that stress component becomes negative.

### 1.2.3 Traction–Stress Relation

Let us take a tetrahedron OABC from a continuum body in equilibrium (see Figure 1.6). Forces (per unit area) acting in the $x_1$ direction on the four surfaces of OABC are shown in Figure 1.6. From its equilibrium in the $x_1$ direction one can write

$$\sum F_1 = T_{n1}A - \sigma_{11}A_1 - \sigma_{21}A_2 - \sigma_{31}A_3 + f_1V = 0 \tag{1.15}$$

where $A$ is the area of the surface ABC; $A_1$, $A_2$, and $A_3$ are the areas of the other three surfaces OBC, OAC, and OAB, respectively; and $f_1$ is the body force per unit volume in the $x_1$ direction.

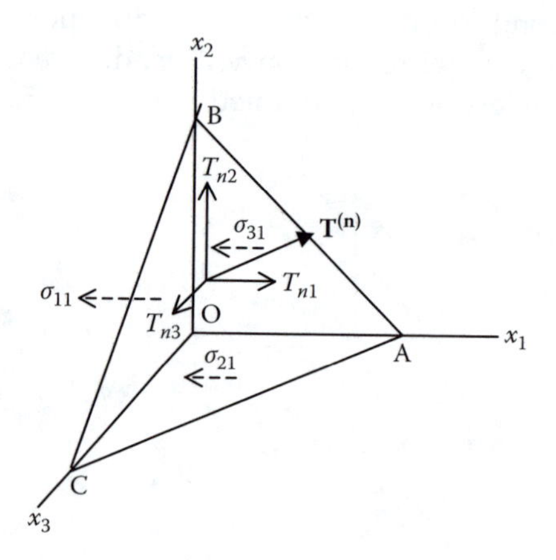


**FIGURE 1.6**
A tetrahedron showing traction components on plane ABC and $x_1$ direction stress components on planes AOC, BOC, and AOB.

If $n_j$ is the $j$th component of the unit vector $\mathbf{n}$ that is normal to the plane ABC, then one can write $A_j = n_j A$ and $V = (Ah)/3$, where $h$ is the height of the tetrahedron measured from the apex O. Thus, equation (1.15) is simplified to

$$T_{n1} - \sigma_{11}n_1 - \sigma_{21}n_2 - \sigma_{31}n_3 + f_1 \frac{h}{3} = 0 \tag{1.16}$$

In the limiting case when the plane ABC passes through point O, the tetrahedron height $h$ vanishes and equation (1.16) is simplified to

$$T_{n1} = \sigma_{11}n_1 + \sigma_{21}n_2 + \sigma_{31}n_3 = \sigma_{j1}n_j \tag{1.17}$$

In this equation the summation convention (repeated index means summation) has been used.

Similarly, from the force equilibrium in $x_2$ and $x_3$ directions, one can write

$$\begin{aligned} T_{n2} &= \sigma_{j2}n_j \\ T_{n3} &= \sigma_{j3}n_j \end{aligned} \tag{1.18}$$

Combining equations (1.17) and (1.18), the traction–stress relation is obtained in index notation:

$$T_{ni} = \sigma_{ji}n_j \tag{1.19}$$

where the free index $i$ takes values 1, 2, and 3 to generate three equations and the dummy index $j$ takes values 1, 2, and 3 and is added in each equation.

For simplicity, the subscript $n$ of $T_{ni}$ is omitted and $T_{ni}$ is written as $T_i$. It is implied that the unit normal vector to the surface on which the traction is defined is $\mathbf{n}$. With this change, equation (1.19) is simplified to

$$T_i = \sigma_{ji}n_j \tag{1.19a}$$

### 1.2.4 Equilibrium Equations

If a body is in equilibrium, then the resultant force and moment on that body must be equal to zero.

#### *1.2.4.1 Force Equilibrium*

The resultant forces in the $x_1$, $x_2$, and $x_3$ directions are equated to zero to obtain the governing equilibrium equations. First, $x_1$ direction equilibrium is studied. Figure 1.7 shows all forces acting in the $x_1$ direction on an elemental volume.

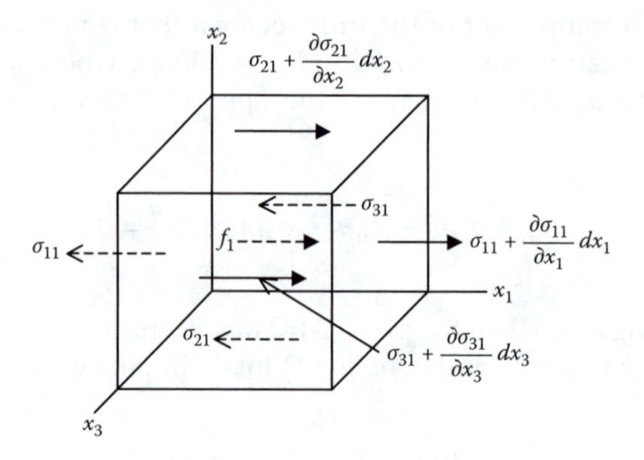


**FIGURE 1.7**
Forces acting in the $x_1$ direction on an elemental volume.

Thus, the zero resultant force in the $x_1$ direction gives

$$-\sigma_{11}dx_2dx_3 + \left(\sigma_{11} + \frac{\partial \sigma_{11}}{\partial x_1}dx_1\right)dx_2dx_3 - \sigma_{21}dx_1dx_3 + \left(\sigma_{21} + \frac{\partial \sigma_{21}}{\partial x_2}dx_2\right)dx_1dx_3$$

$$-\sigma_{31}dx_2dx_1 + \left(\sigma_{31} + \frac{\partial \sigma_{31}}{\partial x_3}dx_3\right)dx_1dx_2 + f_1\,dx_1dx_2dx_3 = 0$$

or

$$\left(\frac{\partial \sigma_{11}}{\partial x_1}dx_1\right)dx_2dx_3 + \left(\frac{\partial \sigma_{21}}{\partial x_2}dx_2\right)dx_1dx_3 + \left(\frac{\partial \sigma_{31}}{\partial x_3}dx_3\right)dx_1dx_2 + f_1\,dx_1dx_2dx_3 = 0$$

or

$$\frac{\partial \sigma_{11}}{\partial x_1} + \frac{\partial \sigma_{21}}{\partial x_2} + \frac{\partial \sigma_{31}}{\partial x_3} + f_1 = 0$$

or

$$\frac{\partial \sigma_{j1}}{\partial x_j} + f_1 = 0 \tag{1.20}$$

In equation (1.20) repeated index $j$ indicates summation.

Similarly, equilibrium in $x_2$ and $x_3$ directions gives

$$\begin{aligned}\frac{\partial \sigma_{j2}}{\partial x_j} + f_2 = 0 \\ \frac{\partial \sigma_{j3}}{\partial x_j} + f_3 = 0\end{aligned} \tag{1.21}$$

The three equations in (1.20) and (1.21) can be combined in the following form:

$$\frac{\partial \sigma_{ji}}{\partial x_j} + f_i = \sigma_{ji,j} + f_i = 0 \tag{1.22}$$

The force equilibrium equations given in equation (1.22) are written in index notation, where the free index $i$ takes three values—1, 2, and 3—and corresponds to three equilibrium equations, and the comma (,) indicates derivative.

### 1.2.4.2 Moment Equilibrium

Let us now compute the resultant moment in the $x_3$ direction (or, in other words, moment about the $x_3$ axis) for the elemental volume shown in Figure 1.8.

If we calculate the moment about an axis parallel to the $x_3$ axis and passing through the centroid of the elemental volume shown in Figure 1.8, then only four shear stresses shown on the four sides of the volume can produce moment. Body forces in $x_1$ and $x_2$ directions do not produce any moment because the resultant body force passes through the centroid of the volume. Since the resultant moment about this axis should be zero, one can write

$$\left(\sigma_{12} + \frac{\partial \sigma_{12}}{\partial x_1} dx_1\right) dx_2 dx_3 \frac{dx_1}{2} + (\sigma_{12}) dx_2 dx_3 \frac{dx_1}{2} - \left(\sigma_{21} + \frac{\partial \sigma_{21}}{\partial x_2} dx_2\right) dx_1 dx_3 \frac{dx_2}{2}$$

$$-(\sigma_{12}) dx_1 dx_3 \frac{dx_2}{2} = 0$$

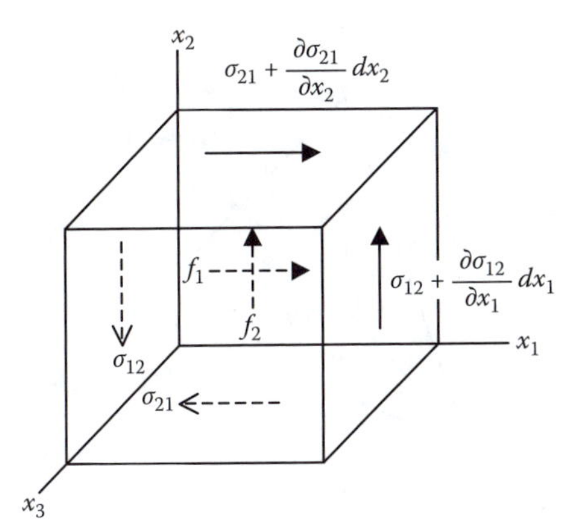


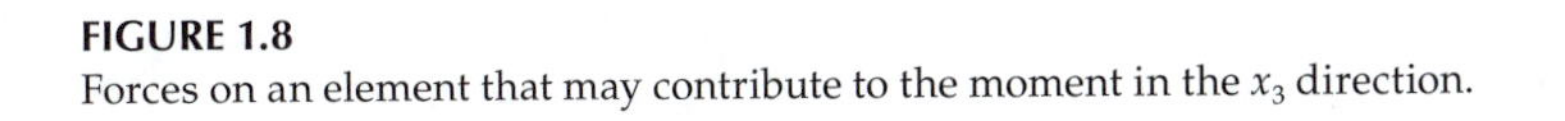

**FIGURE 1.8**
Forces on an element that may contribute to the moment in the $x_3$ direction.

Ignoring the higher order terms, one gets

$$2(\sigma_{21})dx_2dx_3\frac{dx_1}{2}-2(\sigma_{21})dx_1dx_3\frac{dx_2}{2}=0$$

or $\sigma_{12}=\sigma_{21}$.

Similarly, applying moment equilibrium about the other two axes, one can show that $\sigma_{13}=\sigma_{31}$ and $\sigma_{32}=\sigma_{23}$. In index notation,

$$\sigma_{ij}=\sigma_{ji} \tag{1.23}$$

Thus, the stress tensor is symmetric. It should be noted here that if the body has internal body couple (or body moment per unit volume), then the stress tensor will not be symmetric.

Because of the symmetry of the stress tensor, equations (1.19a) and (1.22) can be written in the following form as well:

$$\begin{aligned} T_i &= \sigma_{ij}n_j \\ \sigma_{ij,j}+f_i &= 0 \end{aligned} \tag{1.24}$$

### 1.2.5 Stress Transformation

Let us now investigate how the stress components in two Cartesian coordinate systems are related.

Figure 1.9 shows an inclined plane ABC whose normal is in the $x_{1'}$ direction; thus, the $x_{2'}x_{3'}$ plane is parallel to the ABC plane. Traction $\mathbf{T}^{(1')}$ is acting on this plane. Three components of this traction in $x_{1'}$, $x_{2'}$, and $x_{3'}$ directions are the three stress components $\sigma_{1'1'}$, $\sigma_{1'2'}$, and $\sigma_{1'3'}$, respectively. Note that the

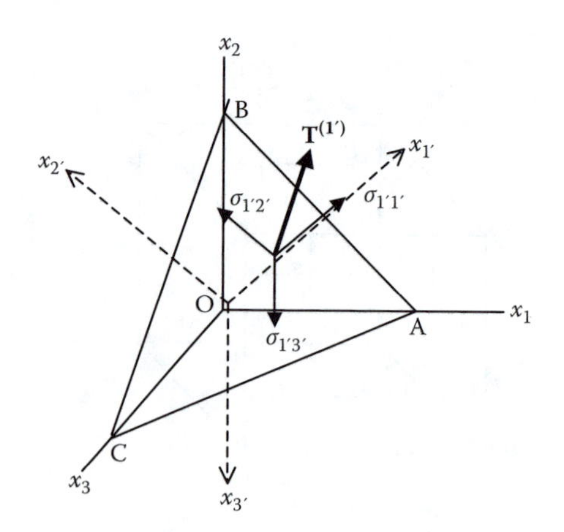


**FIGURE 1.9**
Stress components in $x_{1'}x_{2'}x_{3'}$ coordinate system.

first subscript indicates the plane on which the stress is acting and the second subscript gives the stress direction.

From equation (1.19) one can write

$$T_{1'i} = \sigma_{ji} n_j^{(1')} = \sigma_{ji} \ell_{1'j} \tag{1.25}$$

where $n_j^{(1')} = \ell_{1'j}$ is the $j$th component of the unit normal vector on plane ABC or, in other words, the direction cosines of the $x_{1'}$ axis.

Note that the dot product between $\mathbf{T}^{(1')}$ and the unit vector $\mathbf{n}^{(1')}$ gives the stress component $\sigma_{1'1'}$; therefore,

$$\sigma_{1'1'} = T_{1'i} \ell_{1'i} = \sigma_{ji} \ell_{1'j} \ell_{1'i} \tag{1.26}$$

Similarly, the dot product between $\mathbf{T}^{(1')}$ and the unit vector $\mathbf{n}^{(2')}$ gives $\sigma_{1'2'}$ and the dot product between $\mathbf{T}^{(1')}$ and the unit vector $\mathbf{n}^{(3')}$ gives $\sigma_{1'3'}$. Thus, we get

$$\begin{aligned} \sigma_{1'2'} &= T_{1'i} \ell_{2'i} = \sigma_{ji} \ell_{1'j} \ell_{2'i} \\ \sigma_{1'3'} &= T_{1'i} \ell_{3'i} = \sigma_{ji} \ell_{1'j} \ell_{3'i} \end{aligned} \tag{1.27}$$

Equations (1.26) and (1.27) can be written in index notation in the following form:

$$\sigma_{1'm'} = \ell_{1'j} \sigma_{ji} \ell_{m'i} \tag{1.28}$$

In this equation, the free index $m'$ can take values 1′, 2′, or 3′.

Similarly, from the traction vector $\mathbf{T}^{(2')}$ on a plane whose normal is in the $x_{2'}$ direction, one can show that

$$\sigma_{2'm'} = \ell_{2'j} \sigma_{ji} \ell_{m'i} \tag{1.29}$$

From the traction vector $\mathbf{T}^{(3')}$ on the $x_{3'}$ plane, one can derive

$$\sigma_{3'm'} = \ell_{3'j} \sigma_{ji} \ell_{m'i} \tag{1.30}$$

Equations (1.28) to (1.30) can be combined to obtain the following equation in index notation:

$$\sigma_{n'm'} = \ell_{n'j} \sigma_{ji} \ell_{m'i}$$

Note that in the preceding equation, $i$, $j$, $m'$, and $n'$ are all dummy indices and can be interchanged to obtain

$$\sigma_{m'n'} = \ell_{m'i} \sigma_{ij} \ell_{n'j} = \ell_{m'i} \ell_{n'j} \sigma_{ij} \tag{1.31}$$

#### 1.2.5.1 Kronecker Delta Symbol ($\delta_{ij}$) and Permutation Symbol ($\varepsilon_{ijk}$)

In index notation the Kronecker delta symbol ($\delta_{ij}$) and permutation symbol ($\varepsilon_{ijk}$, also known as the Levi–Civita symbol and alternating symbol) are often used. They are defined in the following manner:

$$\delta_{ij} = 1 \quad \text{for} \quad i = j$$

$$\delta_{ij} = 0 \quad \text{for} \quad i \neq j$$

and

$\varepsilon_{ijk} = 1$ for $i, j, k$ having values 1, 2, and 3; or 2, 3, and 1; or 3, 1, and 2.
$\varepsilon_{ijk} = -1$ for $i, j, k$ having values 3, 2, and 1; or 1, 3, and 2; or 2, 1, and 3.
$\varepsilon_{ijk} = 0$ for $i, j, k$ not having three distinct values.

#### 1.2.5.2 Examples of the Application of $\delta_{ij}$ and $\varepsilon_{ijk}$

Note that

$$\frac{\partial x_i}{\partial x_j} = \delta_{ij}; \qquad \mathbf{e_i} \cdot \mathbf{e_j} = \delta_{ij}$$

$$Det \begin{vmatrix} a_{11} & a_{12} & a_{13} \\ a_{21} & a_{22} & a_{23} \\ a_{31} & a_{32} & a_{33} \end{vmatrix} = \varepsilon_{ijk} a_{1i} a_{2j} a_{3k}; \qquad \mathbf{b} \times \mathbf{c} = \varepsilon_{ijk} b_j c_k \mathbf{e_i}$$

where $\mathbf{e_i}$ and $\mathbf{e_j}$ are unit vectors in $x_i$ and $x_j$ directions, respectively, in the $x_1x_2x_3$ coordinate system. Also note that **b** and **c** are two vectors, while $[a]$ is a matrix.

One can prove that the following relation exists between these two symbols:

$$\varepsilon_{ijk}\varepsilon_{imn} = \delta_{jm}\delta_{kn} - \delta_{jn}\delta_{km}$$

**Example 1.3**
Starting from the stress transformation law, prove that $\sigma_{m'n'}\sigma_{m'n'} = \sigma_{ij}\sigma_{ij}$ where $\sigma_{m'n'}$ and $\sigma_{ij}$ are stress tensors in two different Cartesian coordinate systems.

*Solution*

$$\sigma_{m'n'}\sigma_{m'n'} = (\ell_{m'i}\ell_{n'j}\sigma_{ij})(\ell_{m'p}\ell_{n'q}\sigma_{pq}) = (\ell_{m'i}\ell_{n'j})(\ell_{m'p}\ell_{n'q})\sigma_{ij}\sigma_{pq}$$

$$= (\ell_{m'i}\ell_{m'p})(\ell_{n'j}\ell_{n'q})\sigma_{ij}\sigma_{pq} = \delta_{ip}\delta_{jq}\sigma_{ij}\sigma_{pq} = \sigma_{ij}\sigma_{ij}$$

### 1.2.6 Definition of Tensor

A Cartesian tensor of order (or rank) $r$ in $n$ dimensional space is a set of $n^r$ numbers (called the elements or components of tensor) that obey the following transformation law between two coordinate systems:

$$t_{m'n'p'q'.....} = (\ell_{m'i}\ell_{n'j}\ell_{p'k}\ell_{q'\ell}\ldots)(t_{ijk\ell\ldots}) \tag{1.32}$$

where $t_{m'n'p'q'\ldots}$ and $t_{ijk....}$ each has $r$ number of subscripts; $r$ number of direction cosines $(\ell_{m'i}\ell_{n'j}\ell_{p'k}\ell_{q'\ell}\ldots)$ are multiplied on the right-hand side. Comparing equation (1.31) with the definition of tensor transformation equation (1.32), one can conclude that the stress is a second-rank tensor.

### 1.2.7 Principal Stresses and Principal Planes

Planes on which the traction vectors are normal are called principal planes. Shear stress components on the principal planes are equal to zero. Normal stresses on the principal planes are called principal stresses.

In Figure 1.10, let **n** be the unit normal vector on the principal plane ABC and $\lambda$ the principal stress value on this plane. Therefore, the traction vector on plane ABC can be written as

$$T_i = \lambda n_i$$

Again, from equation (1.24),

$$T_i = \sigma_{ij} n_j$$

From the preceding two equations, one can write

$$\sigma_{ij} n_j - \lambda n_i = 0 \tag{1.33}$$

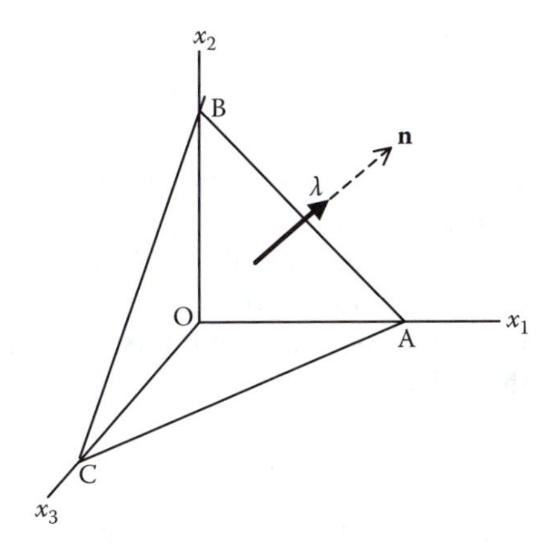


**FIGURE 1.10**
Principal stress $\lambda$ on the principal plane ABC.

The preceding equation is an eigenvalue problem that can be rewritten as

$$(\sigma_{ij} - \lambda\delta_{ij})n_j = 0 \tag{1.34}$$

The system of homogeneous equations (1.33) and (1.34) gives a nontrivial solution for $n_j$ when the determinant of the coefficient matrix is zero. Thus, for a nontrivial solution,

$$Det\begin{bmatrix} (\sigma_{11} - \lambda) & \sigma_{12} & \sigma_{13} \\ \sigma_{12} & (\sigma_{22} - \lambda) & \sigma_{23} \\ \sigma_{13} & \sigma_{23} & (\sigma_{33} - \lambda) \end{bmatrix} = 0$$

or

$$\begin{aligned} &\lambda^3 - (\sigma_{11} + \sigma_{22} + \sigma_{33})\lambda^2 + (\sigma_{11}\sigma_{22} + \sigma_{22}\sigma_{33} + \sigma_{33}\sigma_{11} - \sigma_{12}^2 - \sigma_{23}^2 - \sigma_{31}^2)\lambda \\ &\quad - (\sigma_{11}\sigma_{22}\sigma_{33} + 2\sigma_{12}\sigma_{23}\sigma_{31} - \sigma_{11}\sigma_{23}^2 - \sigma_{22}\sigma_{31}^2 - \sigma_{33}\sigma_{12}^2) = 0 \end{aligned} \tag{1.35}$$

In index notation, the preceding equation can be written as

$$\lambda^3 - \sigma_{ii}\lambda^2 + \frac{1}{2}(\sigma_{ii}\sigma_{jj} - \sigma_{ij}\sigma_{ji})\lambda - \varepsilon_{ijk}\sigma_{1i}\sigma_{2j}\sigma_{3k} = 0 \tag{1.36}$$

In equation (1.36), $\varepsilon_{ijk}$ is the permutation symbol that takes values 1, –1, or 0. If the subscripts *i, j,* and *k* have three distinct values 1, 2, and 3 (or 2, 3, and 1; or 3, 1, and 2), respectively, then its value is 1. If the values of the subscripts are in the opposite order 3, 2, and 1 (or 2, 1, and 3; or 1, 3, and 2), then $\varepsilon_{ijk}$ is –1, and if *i, j,* and *k* do not have three distinct values, then $\varepsilon_{ijk} = 0$.

Cubic equation (1.36) should have three roots of $\lambda$. Three roots correspond to the three principal stress values. After getting $\lambda$, the unit vector components $n_j$ can be obtained from equation (1.34) and, satisfying the constraint condition,

$$n_1^2 + n_2^2 + n_3^2 = 1 \tag{1.37}$$

Note that for three distinct values of $\lambda$, there are three **n** values corresponding to the three principal directions.

Since the principal stress values should be independent of the starting coordinate system, the coefficients of the cubic equation (1.36) should not change irrespective of whether we start from the $x_1x_2x_3$ coordinate system or $x_1'x_2'x_3'$ coordinate system. Thus,

$$\begin{aligned} \sigma_{ii} &= \sigma_{i'i'} \\ \sigma_{ii}\sigma_{jj} - \sigma_{ij}\sigma_{ji} &= \sigma_{i'i'}\sigma_{j'j'} - \sigma_{i'j'}\sigma_{j'i'} \\ \varepsilon_{ijk}\sigma_{1i}\sigma_{2j}\sigma_{3k} &= \varepsilon_{i'j'k'}\sigma_{1'i'}\sigma_{2'j'}\sigma_{3'k'} \end{aligned} \tag{1.38}$$

The three equations of (1.38) are known as the three stress invariants. After some algebraic manipulations, the second and third stress invariants can be further simplified and the three stress invariants can be written as

$$\sigma_{ii} = \sigma_{i'i'}$$

$$\sigma_{ij}\sigma_{ji} = \sigma_{i'j'}\sigma_{j'i'} \quad \text{or} \quad \frac{1}{2}\sigma_{ij}\sigma_{ji} = \frac{1}{2}\sigma_{i'j'}\sigma_{j'i'} \tag{1.39}$$

$$\sigma_{ij}\sigma_{jk}\sigma_{ki} = \sigma_{i'j'}\sigma_{j'k'}\sigma_{k'i'} \quad \text{or} \quad \frac{1}{3}\sigma_{ij}\sigma_{jk}\sigma_{ki} = \frac{1}{3}\sigma_{i'j'}\sigma_{j'k'}\sigma_{k'i'}$$

**Example 1.4**

(a) Obtain the principal values and principal directions for the following stress tensor:

$$[\sigma] = \begin{bmatrix} 2 & -4 & -6 \\ -4 & 4 & 2 \\ -6 & 2 & -2 \end{bmatrix} \text{MPa}$$

One given value of the principal stress is 9.739 MPa.

(b) Compute the stress state in $x_1'x_2'x_3'$ coordinate system. Direction cosines of $x_1'x_2'x_3'$ axes are:

| | $x_1'$ | $x_2'$ | $x_3'$ |
|---|---|---|---|
| $\ell_1$ | 0.7285 | 0.6601 | 0.1831 |
| $\ell_2$ | 0.4827 | −0.6843 | 0.5466 |
| $\ell_3$ | 0.4861 | −0.3098 | −0.8171 |

*Solution*

(a) A characteristic equation is obtained from equation (1.35):

$$\lambda^3 - (\sigma_{11} + \sigma_{22} + \sigma_{33})\lambda^2 + \left(\sigma_{11}\sigma_{22} + \sigma_{22}\sigma_{33} + \sigma_{33}\sigma_{11} - \sigma_{12}^2 - \sigma_{23}^2 - \sigma_{31}^2\right)\lambda$$
$$- \left(\sigma_{11}\sigma_{22}\sigma_{33} + 2\sigma_{12}\sigma_{23}\sigma_{31} - \sigma_{11}\sigma_{23}^2 - \sigma_{22}\sigma_{31}^2 - \sigma_{33}\sigma_{12}^2\right) = 0$$

For the given stress tensor it becomes

$$\lambda^3 - 4\lambda^2 - 60\lambda + 40 = 0$$

The preceding equation can be written as

$$\lambda^3 - 9.739\lambda^2 + 5.739\lambda^2 - 55.892\lambda - 4.108\lambda + 40 = 0$$

$$\Rightarrow (\lambda - 9.739)(\lambda^2 + 5.739\lambda - 4.108) = 0$$

$$\Rightarrow (\lambda - 9.739)(\lambda + 6.3825)(\lambda - 0.6435) = 0$$

whose three roots are

$$\lambda_1 = -6.3825$$

$$\lambda_2 = 9.739$$

$$\lambda_3 = 0.6435$$

These are the three principal stress values.

Principal directions are obtained from equation (1.34)

$$\begin{bmatrix} (\sigma_{11}-\lambda_1) & \sigma_{12} & \sigma_{13} \\ \sigma_{12} & (\sigma_{22}-\lambda_1) & \sigma_{23} \\ \sigma_{13} & \sigma_{23} & (\sigma_{33}-\lambda_1) \end{bmatrix} \begin{Bmatrix} \ell_{1'1} \\ \ell_{1'2} \\ \ell_{1'3} \end{Bmatrix} = 0$$

where $\ell_{1'1}$, $\ell_{1'2}$, $\ell_{1'3}$ are the direction cosines of the principal direction associated with the principal stress $\lambda_1$.

From the preceding equation one can write

$$\begin{bmatrix} (2+6.3825) & -4 & -6 \\ -4 & (4+6.3825) & 2 \\ -6 & 2 & (-2+6.3825) \end{bmatrix} \begin{Bmatrix} \ell_{1'1} \\ \ell_{1'2} \\ \ell_{1'3} \end{Bmatrix} = 0$$

The second and third equations of the preceding system of three homogeneous equations can be solved to obtain two direction cosines in terms of the third one, as given here:

$$\ell_{1'2} = 0.1333\ell_{1'1}$$

$$\ell_{1'3} = 1.3082\ell_{1'1}$$

Normalizing the direction cosines, as shown in equation (1.37), we get

$$1 = \ell_{1'1}^2 + \ell_{1'2}^2 + \ell_{1'3}^2 = \ell_{1'1}^2(1 + 0.1333^2 + 1.3082^2)$$

$$\Rightarrow \ell_{1'1} = \pm 0.605$$

$$\Rightarrow \ell_{1'2} = 0.1333\ell_{1'1} = \pm 0.081$$

$$\Rightarrow \ell_{1'3} = 1.3082\ell_{1'1} = \pm 0.791$$

Similarly, for the second principal stress $\lambda_2 = 9.739$, the direction cosines are

$$\begin{Bmatrix} \ell_{2'1} = \pm 0.657 \\ \ell_{2'2} = \mp 0.612 \\ \ell_{2'3} = \mp 0.440 \end{Bmatrix}$$

For the third principal stress $\lambda_3 = 0.6435$, the direction cosines are

$$\begin{Bmatrix} \ell_{3'1} = \pm 0.449 \\ \ell_{3'2} = \pm 0.787 \\ \ell_{3'3} = \mp 0.423 \end{Bmatrix}$$

(b) From equation (1.31), $\sigma_{m'n'} = \ell_{m'i}\ell_{n'j}\sigma_{ij}$ .
In matrix notation $[\sigma'] = [\ell][\sigma][\ell]^T$,

$$\text{where} \quad [\ell]^T = \begin{bmatrix} \ell_{1'1} & \ell_{2'1} & \ell_{3'1} \\ \ell_{1'2} & \ell_{2'2} & \ell_{3'2} \\ \ell_{1'3} & \ell_{2'3} & \ell_{3'3} \end{bmatrix} = \begin{bmatrix} 0.7285 & 0.6601 & 0.1831 \\ 0.4827 & -0.6843 & 0.5466 \\ 0.4861 & -0.3098 & -0.8171 \end{bmatrix}$$

$$\text{Thus,} \quad [\sigma'] = [\ell][\sigma][\ell]^T = \begin{bmatrix} -4.6033 & -0.8742 & 2.9503 \\ -0.8742 & 9.4682 & 1.6534 \\ 2.9503 & 1.6534 & -0.8650 \end{bmatrix} \text{MPa}$$

### 1.2.8 Transformation of Displacement and Other Vectors

The vector **V** can be expressed in two coordinate systems in the following manner (see Figure 1.11):

$$V_1e_1 + V_2e_2 + V_3e_3 = V_{1'}e_{1'} + V_{2'}e_{2'} + V_{3'}e_{3'} \tag{1.40}$$

If one adds the projections of $V_1$, $V_2$, and $V_3$ of equation (1.40) along the $x_{j'}$ direction, then the sum should be equal to the component $V_{j'}$. Thus,

$$V_{j'} = \ell_{j'1}V_1 + \ell_{j'2}V_2 + \ell_{j'3}V_3 = \ell_{j'k}V_k \tag{1.41}$$

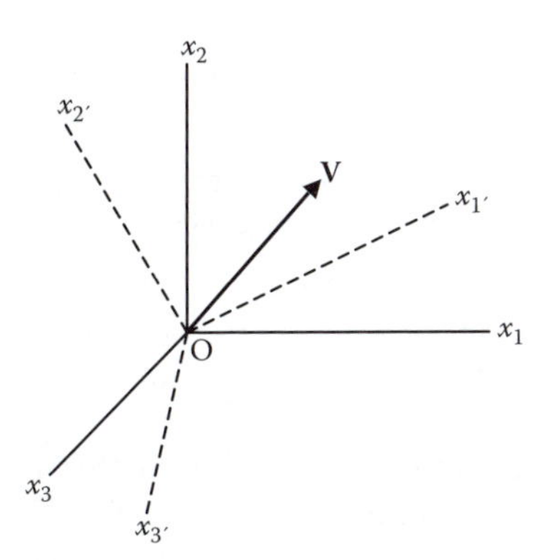


**FIGURE 1.11**
A vector **V** and two Cartesian coordinate systems.

Comparing equations (1.41) and (1.32), one can conclude that vectors are first-order tensors, or tensors of rank 1.

### 1.2.9 Strain Transformation

Equation (1.7a) gives the strain expression in the $x_1x_2x_3$ coordinate system. In the $x_{1'}x_{2'}x_{3'}$ coordinate system, the strain expression is given by $\varepsilon_{i'j'} = \frac{1}{2}(u_{i',j'} + u_{j',i'})$. Now,

$$u_{i',j'} = \frac{\partial u_{i'}}{\partial x_{j'}} = \frac{\partial(\ell_{i'm}u_m)}{\partial x_{j'}} = \ell_{i'm}\frac{\partial(u_m)}{\partial x_{j'}} = \ell_{i'm}\frac{\partial u_m}{\partial x_n}\frac{\partial x_n}{\partial x_{j'}} = \ell_{i'm}\frac{\partial u_m}{\partial x_n}\ell_{nj'} = \ell_{i'm}\ell_{j'n}\frac{\partial u_m}{\partial x_n} \tag{1.42}$$

Similarly,

$$u_{j',i'} = \ell_{j'n}\ell_{i'm}\frac{\partial u_n}{\partial x_m} \tag{1.43}$$

Therefore,

$$\begin{aligned}\varepsilon_{i'j'} &= \frac{1}{2}(u_{i',j'} + u_{j',i'}) = \frac{1}{2}(\ell_{j'm}\ell_{j'n}u_{m,n} + \ell_{j'n}\ell_{i'm}u_{n,m}) \\ &= \frac{1}{2}\ell_{i'm}\ell_{j'n}(u_{m,n} + u_{n,m}) = \ell_{i'm}\ell_{j'n}\varepsilon_{mn}\end{aligned} \tag{1.44}$$

It should be noted here that the strain transformation law (equation 1.44) is identical to the stress transformation law (equation 1.31). Therefore, strain is also a second-rank tensor.

### 1.2.10 Definition of Elastic Material and Stress–Strain Relation

Elastic (also known as conservative) material can be defined in many ways:

- The material that has one-to-one correspondence between stress and strain is called elastic material.
- The material that follows the same stress–strain path during loading and unloading is called elastic material.
- For elastic materials, the strain energy density function ($U_0$) exists and it can be expressed in terms of the state of current strain only ($U_0 = U_0(\varepsilon_{ij})$) and independent of the strain history or strain path.

If the stress–strain relation is linear, then material is called linear elastic material; otherwise, it is nonlinear elastic material. Note that elastic material does not necessarily mean that the stress–strain relation is linear, and the linear stress–strain relation does not automatically imply that the material is

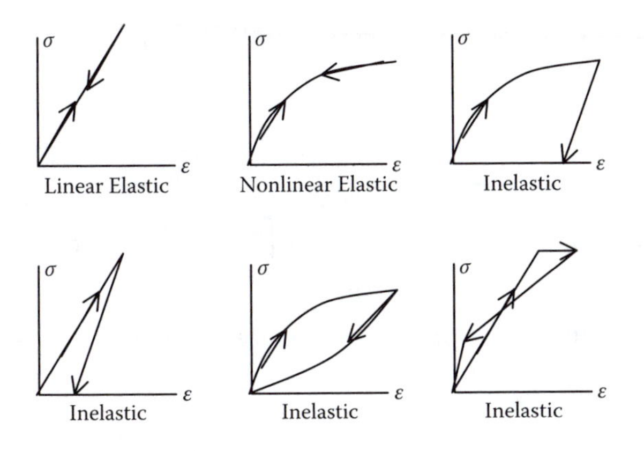


**FIGURE 1.12**
Stress–strain relations for elastic and inelastic materials.

elastic. If the stress–strain path is different during loading and unloading, then the material is no longer elastic even if the path is linear during loading and unloading. Figure 1.12 shows different stress–strain relations and indicates for each plot if the material is elastic or inelastic.

For conservative or elastic material the external work done on the material must be equal to the total increase in the strain energy of the material. If the variation of the external work done on the body is denoted by $\delta W$ and the variation of the internal strain energy stored in the body is $\delta U$, then $\delta U = \delta W$. Note that $\delta U$ can be expressed in terms of the strain energy density variation ($\delta U_0$), and $\delta W$ can be expressed in terms of the applied body force ($f_i$), the surface traction ($T_i$), and the variation of displacement ($\delta u_i$) in the following manner:

$$\begin{aligned} \delta U &= \int_V \delta U_0 dV \\ \delta W &= \int_V f_i \delta u_i dV + \int_S T_i \delta u_i dS \end{aligned} \tag{1.45}$$

In equation (1.45) integrals over $V$ and $S$ indicate volume and surface integrals, respectively. From this equation, one can write

$$\begin{aligned} \int_V \delta U_0 dV &= \int_V f_i \delta u_i dV + \int_S T_i \delta u_i dS = \int_V f_i \delta u_i dV + \int_S \sigma_{ij} n_j \delta u_i dS \\ &= \int_V f_i \delta u_i dV + \int_S (\sigma_{ij} \delta u_i) n_j dS \end{aligned}$$

Applying Gauss divergence theorem on the second integral of the right-hand side, one obtains

$$\int_V \delta U_0 dV = \int_V f_i \delta u_i dV + \int_V (\sigma_{ij}\delta u_i)_{,j}\, dV = \int_V f_i \delta u_i dV + \int_V \left(\sigma_{ij,j}\delta u_i + \sigma_{ij}\delta u_{i,j}\right) dV$$

$$= \int_V (f_i \delta u_i + \sigma_{ij,j}\delta u_i + \sigma_{ij}\delta u_{i,j}) dV = \int_V ((f_i + \sigma_{ij,j})\delta u_i + \sigma_{ij}\delta u_{i,j}) dV \tag{1.46}$$

After substituting the equilibrium equation (see equation 1.24), the preceding equation is simplified to

$$\int_V \delta U_0 dV = \int_V (\sigma_{ij}\delta u_{i,j}) dV = \int_V \frac{1}{2}(\sigma_{ij}\delta u_{i,j} + \sigma_{ij}\delta u_{i,j}) dV = \int_V \frac{1}{2}(\sigma_{ij}\delta u_{i,j} + \sigma_{ji}\delta u_{j,i}) dV$$

$$= \int_V \sigma_{ij}\frac{1}{2}(\delta u_{i,j} + \delta u_{j,i}) dV = \int_V \sigma_{ij}\delta\varepsilon_{ij} dV \tag{1.47}$$

Since equation (1.47) is valid for any arbitrary volume $V$, the integrands of the left- and right-hand sides must be equal to each other. Hence,

$$\delta U_0 = \sigma_{ij}\delta\varepsilon_{ij} \tag{1.48}$$

However, from the definition of elastic materials,

$$U_0 = U_0(\varepsilon_{ij})$$

$$\therefore \delta U_0 = \frac{\partial U_0}{\partial \varepsilon_{ij}}\delta\varepsilon_{ij} \tag{1.49}$$

For arbitrary variation of $\delta\varepsilon_{ij}$ from equations (1.48) and (1.49), one can write

$$\sigma_{ij}\delta\varepsilon_{ij} = \frac{\partial U_0}{\partial \varepsilon_{ij}}\delta\varepsilon_{ij}$$

$$\therefore \sigma_{ij} = \frac{\partial U_0}{\partial \varepsilon_{ij}} \tag{1.50}$$

From equation (1.50), the stress–strain relation can be obtained by assuming some expression of $U_0$ in terms of the strain components (Green's approach). For example, if one assumes that the strain energy density function is a quadratic function (complete second-degree polynomial) of the strain components, as shown here,

$$U_0 = D_0 + D_{kl}\varepsilon_{kl} + D_{klmn}\varepsilon_{kl}\varepsilon_{mn} \tag{1.51}$$

then,

$$\sigma_{ij} = \frac{\partial U_0}{\partial \varepsilon_{ij}} = D_{kl}\delta_{ik}$$

or

$$\sigma_{ij} = \frac{\partial U_0}{\partial \varepsilon_{ij}} = D_{kl}\delta_{ik}\delta_{jl} + D_{klmn}(\delta_{ik}\delta_{jl}\varepsilon_{mn} + \varepsilon_{kl}\delta_{im}\delta_{jn}) = D_{ij} + D_{ijmn}\varepsilon_{mn} + D_{klij}\varepsilon_{kl}$$

$$= D_{ij} + (D_{ijkl} + D_{klij})\varepsilon_{kl}$$

Substituting $(D_{ijkl} + D_{klij}) = C_{ijkl}$ and $D_{ij} = 0$ (it implies that strain is zero for zero stress, then this assumption is valid), one gets the linear stress–strain relation (or *constitutive* relation) in the following form:

$$\sigma_{ij} = C_{ijkl}\varepsilon_{kl} \tag{1.52}$$

In Cauchy's approach, equation (1.52) is obtained by relating stress tensor with strain tensor. Note that equation (1.52) is a general linear relation between two second-order tensors.

In the same manner, for a nonlinear (quadratic) material, the stress–strain relation is

$$\sigma_{ij} = C_{ij} + C_{ijkl}\varepsilon_{kl} + C_{ijklmn}\varepsilon_{kl}\varepsilon_{mn} \tag{1.53}$$

In equation (1.53) the first term on the right-hand side is the residual stress (stress for zero strain), the second term is the linear term, and the third term is the quadratic term. If one follows Green's approach, then this nonlinear stress–strain relation can be obtained from a cubic expression of the strain energy density function:

$$U_0 = D_{kl}\varepsilon_{kl} + D_{klmn}\varepsilon_{kl}\varepsilon_{mn} + D_{klmnpq}\varepsilon_{kl}\varepsilon_{mn}\varepsilon_{pq} \tag{1.54}$$

In this chapter we limit our analysis to linear materials only. Therefore, our stress–strain relation is the one given in equation (1.52).

**Example 1.5**

In the $x_1x_2x_3$ coordinate system the stress–strain relation for a general anisotropic material is given by $\sigma_{ij} = C_{ijkm}\,\varepsilon_{km}$, and in the $x_{1'}x_{2'}x_{3'}$ coordinate system the stress–strain relation for the same material is given by $\sigma_{i'j'} = C_{i'j'k'm'}\,\varepsilon_{k'm'}$.

(a) Starting from the stress and strain transformation laws, obtain a relation between $C_{ijkm}$ and $C_{i'j'k'm'}$.

(b) Is $C_{ijkm}$ a tensor? If yes, what is its rank?

*Solution*

(a) Using equations (1.52) and (1.31), one can write

$$\sigma_{i'j'} = C_{i'j'k'm'}\varepsilon_{k'm'}$$

$$\Rightarrow \ell_{i'r}\ell_{j's}\sigma_{rs} = C_{i'j'k'm'}\ell_{k'p}\ell_{m'q}\varepsilon_{pq}$$

$$\Rightarrow (\ell_{i't}\ell_{j'u})\ell_{i'r}\ell_{j's}\sigma_{rs} = (\ell_{i't}\ell_{j'u})C_{i'j'k'm'}\ell_{k'p}\ell_{m'q}\varepsilon_{pq}$$

$$\Rightarrow \delta_{tr}\delta_{us}\sigma_{rs} = (\ell_{i't}\ell_{j'u})C_{i'j'k'm'}\ell_{k'p}\ell_{m'q}\varepsilon_{pq} = \ell_{i't}\ell_{j'u}\ell_{i'p}\ell_{m'q}C_{i'j'k'm'}\varepsilon_{p}$$

$$\Rightarrow \sigma_{tu} = (\ell_{i't}\ell_{j'u}\ell_{k'p}\ell_{m'q}C_{i'j'k'm'})\varepsilon_{pq}$$

But, $\sigma_{tu} = C_{tupq}\varepsilon_{pq}$. Therefore,

$$C_{tupq} = \ell_{i't}\ell_{j'u}\ell_{k'p}\ell_{m'q}C_{i'j'k'm'} = \ell_{ti'}\ell_{uj'}\ell_{pk'}\ell_{qm'}C_{i'j'k'm'}$$

Similarly, starting with the equation $\sigma_{pq} = C_{pqrs}\varepsilon_{rs}$ and applying stress and strain transformation laws, one can show that $C_{i'j'k'm'} = \ell_{i'p}\ell_{j'q}\ell_{k'r}\ell_{m's}C_{pqrs}$.

(b) Clearly, $C_{ijkm}$ satisfies the transformation law for a fourth-order tensor. Therefore, it is a tensor of order or rank equal to 4.

### 1.2.11 Number of Independent Material Constants

In equation (1.52) the coefficient values $C_{ijkl}$ depend on the material type and are called material constants or elastic constants. Note that *i*, *j*, *k*, and *l* can each take three values: 1, 2, or 3. Thus, there are a total of 81 combinations possible. However, not all 81 material constants are independent. Since stress and strain tensors are symmetric, we can write

$$C_{ijkl} = C_{jikl} = C_{jilk} \tag{1.55}$$

The relation in equation (1.55) reduces the number of independent material constants from 81 to 36, and the stress–strain relation of equation (1.52) can be written in the following form:

$$\begin{Bmatrix} \sigma_{11} \\ \sigma_{22} \\ \sigma_{33} \\ \sigma_{23} \\ \sigma_{31} \\ \sigma_{12} \end{Bmatrix} = \begin{bmatrix} C_{1111} & C_{1122} & C_{1133} & C_{1123} & C_{1131} & C_{1112} \\ C_{2211} & C_{2222} & C_{2233} & C_{2223} & C_{2231} & C_{2212} \\ C_{3311} & C_{3322} & C_{3333} & C_{3323} & C_{3331} & C_{3312} \\ C_{2311} & C_{2322} & C_{2333} & C_{2323} & C_{2331} & C_{2312} \\ C_{3111} & C_{3122} & C_{3133} & C_{3123} & C_{3131} & C_{3112} \\ C_{1211} & C_{1222} & C_{1233} & C_{1223} & C_{1231} & C_{1212} \end{bmatrix} \begin{Bmatrix} \varepsilon_{11} \\ \varepsilon_{22} \\ \varepsilon_{33} \\ 2\varepsilon_{23} \\ 2\varepsilon_{31} \\ 2\varepsilon_{12} \end{Bmatrix} \tag{1.56}$$

In the preceding expression only six stress and strain components are shown. The other three components are not independent because of the symmetry of stress and strain tensors. The six by six C-matrix is known as the constitutive matrix. For elastic materials, the strain energy density function can be expressed as a function of only strain; then its double derivative will have the form

$$\frac{\partial^2 U_0}{\partial \varepsilon_{ij} \partial \varepsilon_{kl}} = \frac{\partial}{\partial \varepsilon_{ij}}\left(\frac{\partial U_0}{\partial \varepsilon_{kl}}\right) = \frac{\partial}{\partial \varepsilon_{ij}}(\sigma_{kl}) = \frac{\partial}{\partial \varepsilon_{ij}}(C_{klmn}\varepsilon_{mn}) = C_{klmn}\delta_{im}\delta_{jn} = C_{klij} \quad (1.57)$$

Similarly,

$$\frac{\partial^2 U_0}{\partial \varepsilon_{kl} \partial \varepsilon_{ij}} = \frac{\partial}{\partial \varepsilon_{kl}}\left(\frac{\partial U_0}{\partial \varepsilon_{ij}}\right) = \frac{\partial}{\partial \varepsilon_{kl}}(\sigma_{ij}) = \frac{\partial}{\partial \varepsilon_{kl}}(C_{ijmn}\varepsilon_{mn}) = C_{ijmn}\delta_{km}\delta_{ln} = C_{ijkl} \quad (1.58)$$

In equations (1.57) and (1.58) the order or sequence of derivative has been changed. However, since the sequence of derivative should not change the final results, one can conclude that $C_{ijkl} = C_{klij}$. In other words, the $C$-matrix of equation (1.56) must be symmetric. Then, the number of independent elastic constants is reduced from 36 to 21 and equation (1.56) is simplified to

$$\begin{Bmatrix} \sigma_1 \\ \sigma_2 \\ \sigma_3 \\ \sigma_4 \\ \sigma_5 \\ \sigma_6 \end{Bmatrix} = \begin{bmatrix} C_{11} & C_{12} & C_{13} & C_{14} & C_{15} & C_{16} \\ & C_{22} & C_{23} & C_{24} & C_{25} & C_{26} \\ & & C_{33} & C_{34} & C_{35} & C_{36} \\ & & & C_{44} & C_{45} & C_{46} \\ & symm & & & C_{55} & C_{56} \\ & & & & & C_{66} \end{bmatrix} \begin{Bmatrix} \varepsilon_1 \\ \varepsilon_2 \\ \varepsilon_3 \\ 2\varepsilon_4 \\ 2\varepsilon_5 \\ 2\varepsilon_6 \end{Bmatrix} \quad (1.59)$$

In equation (1.59), for simplicity we have denoted the six stress and strain components with only one subscript ($\sigma_i$ and $\varepsilon_i$, where $i$ varies from one to six) instead of the traditional notation of two subscripts, and the material constants have been written with two subscripts instead of four.

### 1.2.12 Material Planes of Symmetry

Equation (1.59) has 21 independent elastic constants in absence of any plane of symmetry. Such material is called general anisotropic material or *triclinic* material. However, if the material response is symmetric about a plane or an axis, then the number of independent material constants is reduced.

#### *1.2.12.1 One Plane of Symmetry*

Let the material have only one plane of symmetry: the $x_1$ plane (also denoted as the $x_2x_3$ plane); therefore, the $x_2x_3$ plane whose normal is in the $x_1$ direction is the plane of symmetry. For this material, if the stress states $\sigma_{ij}^{(1)}$ and $\sigma_{ij}^{(2)}$ are mirror images of each other with respect to the $x_1$ plane, then the corresponding strain states $\varepsilon_{ij}^{(1)}$ and $\varepsilon_{ij}^{(2)}$ should be the mirror images of each other with respect to the same plane. Following the notations of equation (1.59), we can say that the stress states $\sigma_{ij}^{(1)} = (\sigma_1, \sigma_2, \sigma_3, \sigma_4, \sigma_5, \sigma_6)$ and $\sigma_{ij}^{(2)} = (\sigma_1, \sigma_2, \sigma_3, \sigma_4, -\sigma_5, -\sigma_6)$ have mirror symmetry with respect to the $x_1$ plane. Similarly, the strain states $\varepsilon_{ij}^{(1)} = (\varepsilon_1, \varepsilon_2, \varepsilon_3, \varepsilon_4, \varepsilon_5, \varepsilon_6)$ and $\varepsilon_{ij}^{(2)} = (\varepsilon_1, \varepsilon_2, \varepsilon_3, \varepsilon_4, -\varepsilon_5, -\varepsilon_6)$ also have mirror symmetry with respect to the same plane. One can easily show by substitution that both states $(\sigma_{ij}^{(1)}, \varepsilon_{ij}^{(1)})$ and $(\sigma_{ij}^{(2)}, \varepsilon_{ij}^{(2)})$ can satisfy equation (1.59) only when a number of elastic constants of the C-matrix become zero, as shown:

$$\begin{Bmatrix} \sigma_1 \\ \sigma_2 \\ \sigma_3 \\ \sigma_4 \\ \sigma_5 \\ \sigma_6 \end{Bmatrix} = \begin{bmatrix} C_{11} & C_{12} & C_{13} & C_{14} & 0 & 0 \\ & C_{22} & C_{23} & C_{24} & 0 & 0 \\ & & C_{33} & C_{34} & 0 & 0 \\ & & & C_{44} & 0 & 0 \\ & symm & & & C_{55} & C_{56} \\ & & & & & C_{66} \end{bmatrix} \begin{Bmatrix} \varepsilon_1 \\ \varepsilon_2 \\ \varepsilon_3 \\ 2\varepsilon_4 \\ 2\varepsilon_5 \\ 2\varepsilon_6 \end{Bmatrix} \tag{1.60}$$

Material with one plane of symmetry is called *monoclinic material*. From the stress–strain relation (equation 1.60) of monoclinic materials one can see that the number of independent elastic constants is 13 for such materials.

### 1.2.12.2 Two and Three Planes of Symmetry

In addition to the $x_1$ plane, if the $x_2$ plane is also a plane of symmetry, then two stress and strain states that are symmetric with respect to the $x_2$ plane must also satisfy equation (1.59). Note that the stress states $\sigma_{ij}^{(1)} = (\sigma_1, \sigma_2, \sigma_3, \sigma_4, \sigma_5, \sigma_6)$ and $\sigma_{ij}^{(2)} = (\sigma_1, \sigma_2, \varepsilon\sigma_3, -\sigma_4, \sigma_5, -\sigma_6)$ are states of mirror symmetry with respect to the $x_2$ plane, and the strain states $\varepsilon_{ij}^{(1)} = (\varepsilon_1, \varepsilon_2, \varepsilon_3, \varepsilon_4, \varepsilon_5, \varepsilon_6)$ and $\varepsilon_{ij}^{(2)} = (\varepsilon_1, \varepsilon_2, \varepsilon_3, -\varepsilon_4, \varepsilon_5, -\varepsilon_6)$ are states of mirror symmetry with respect to the same plane. As seen before, one can easily show by substitution that both states $(\sigma_{ij}^{(1)}, \varepsilon_{ij}^{(1)})$ and $(\sigma_{ij}^{(2)}, \varepsilon_{ij}^{(2)})$ can satisfy equation (1.59) only when a number of elastic constants of the C-matrix become zero, as shown:

$$\begin{Bmatrix} \sigma_1 \\ \sigma_2 \\ \sigma_3 \\ \sigma_4 \\ \sigma_5 \\ \sigma_6 \end{Bmatrix} = \begin{bmatrix} C_{11} & C_{12} & C_{13} & 0 & C_{15} & 0 \\ & C_{22} & C_{23} & 0 & C_{25} & 0 \\ & & C_{33} & 0 & C_{35} & 0 \\ & & & C_{44} & C_{45} & 0 \\ & symm & & & C_{55} & 0 \\ & & & & & C_{66} \end{bmatrix} \begin{Bmatrix} \varepsilon_1 \\ \varepsilon_2 \\ \varepsilon_3 \\ 2\varepsilon_4 \\ 2\varepsilon_5 \\ 2\varepsilon_6 \end{Bmatrix} \tag{1.61}$$

Equation (1.60) is the constitutive relation when the $x_1$ plane is the plane of symmetry and equation (1.61) is the constitutive relation for the $x_2$ plane as the plane of symmetry. Therefore, when both $x_1$ and $x_2$ planes are planes of symmetry, the C-matrix has only nine independent material constants:

$$\begin{Bmatrix}\sigma_1\\ \sigma_2\\ \sigma_3\\ \sigma_4\\ \sigma_5\\ \sigma_6\end{Bmatrix}=\begin{bmatrix}C_{11} & C_{12} & C_{13} & 0 & 0 & 0\\ & C_{22} & C_{23} & 0 & 0 & 0\\ & & C_{33} & 0 & 0 & 0\\ & & & C_{44} & 0 & 0\\ & symm & & & C_{55} & 0\\ & & & & & C_{66}\end{bmatrix}\begin{Bmatrix}\varepsilon_1\\ \varepsilon_2\\ \varepsilon_3\\ 2\varepsilon_4\\ 2\varepsilon_5\\ 2\varepsilon_6\end{Bmatrix} \tag{1.62}$$

Note that equation (1.62) includes the case when all three planes, $x_1$, $x_2$, and $x_3$, are planes of symmetry. Thus, when two mutually perpendicular planes are planes of symmetry, the third plane automatically becomes a plane of symmetry. Materials having three planes of symmetry are called *orthotropic* (or *orthogonally anisotropic* or *orthorhombic*) materials.

#### 1.2.12.3 Three Planes of Symmetry and One Axis of Symmetry

If the material has one axis of symmetry in addition to the three planes of symmetry, then it is called *transversely isotropic* (*hexagonal*) material. If the $x_3$ axis is the axis of symmetry, then the material response in $x_1$ and $x_2$ directions must be identical. In equation (1.62), if we substitute $\varepsilon_1 = \varepsilon_0$, and all other strain components = 0, then we get the three nonzero stress components $\sigma_1 = C_{11}\varepsilon_0$, $\sigma_2 = C_{12}\varepsilon_0$, and $\sigma_3 = C_{13}\varepsilon_0$. Similarly, if the strain state has only one nonzero component, $\varepsilon_2 = \varepsilon_0$, while all other strain components are zero, then the three normal stress components are $\sigma_1 = C_{12}\varepsilon_0$, $\sigma_2 = C_{22}\varepsilon_0$, and $\sigma_3 = C_{23}\varepsilon_0$.

Since the $x_3$ axis is an axis of symmetry, $\sigma_3$ should be same for both cases and $\sigma_1$ for the first case should be equal to $\sigma_2$ for the second case, and vice versa. Thus, $C_{13} = C_{23}$ and $C_{11} = C_{22}$. Then consider two more cases: (1) $\varepsilon_{23}$ (or $\varepsilon_4$ in equation 1.62) = $\varepsilon_0$, while all other strain components are zero; and (2) $\varepsilon_{31}$ (or $\varepsilon_5$ in equation 1.62) = $\varepsilon_0$, while all other strain components are zero. From equation (1.62) one gets $\sigma_4 = C_{44}\varepsilon_0$ for case 1 and $\sigma_5 = C_{55}\varepsilon_0$. Since the $x_3$ axis is the axis of symmetry, $\sigma_4$ and $\sigma_5$ should have equal values; therefore, $C_{44} = C_{55}$. Substituting these constraint conditions in equation (1.62), one obtains

$$\begin{Bmatrix}\sigma_1\\ \sigma_2\\ \sigma_3\\ \sigma_4\\ \sigma_5\\ \sigma_6\end{Bmatrix}=\begin{bmatrix}C_{11} & C_{12} & C_{13} & 0 & 0 & 0\\ & C_{11} & C_{13} & 0 & 0 & 0\\ & & C_{33} & 0 & 0 & 0\\ & & & C_{44} & 0 & 0\\ & symm & & & C_{44} & 0\\ & & & & & C_{66}\end{bmatrix}\begin{Bmatrix}\varepsilon_1\\ \varepsilon_2\\ \varepsilon_3\\ 2\varepsilon_4\\ 2\varepsilon_5\\ 2\varepsilon_6\end{Bmatrix} \tag{1.63}$$

In equation (1.63) although there are six different material constants, only five are independent. Considering the isotropic deformation in the $x_1x_2$ plane,

$C_{66}$ can be expressed in terms of $C_{11}$ and $C_{12}$ in the following manner:

$$C_{66} = \frac{C_{11} - C_{12}}{2} \tag{1.64}$$

#### 1.2.12.4 *Three Planes of Symmetry and Two or Three Axes of Symmetry*

If we now add $x_1$ as an axis of symmetry, then, following the same arguments as before, one can show that in equation (1.63) the following three additional constraint conditions must be satisfied: $C_{12} = C_{13}$, $C_{11} = C_{33}$, and $C_{44} = C_{66}$. Thus, the constitutive matrix is simplified to

$$\begin{Bmatrix} \sigma_1 \\ \sigma_2 \\ \sigma_3 \\ \sigma_4 \\ \sigma_5 \\ \sigma_6 \end{Bmatrix} = \begin{bmatrix} C_{11} & C_{12} & C_{12} & 0 & 0 & 0 \\ & C_{11} & C_{12} & 0 & 0 & 0 \\ & & C_{11} & 0 & 0 & 0 \\ & & & C_{66} & 0 & 0 \\ & symm & & & C_{66} & 0 \\ & & & & & C_{66} \end{bmatrix} \begin{Bmatrix} \varepsilon_1 \\ \varepsilon_2 \\ \varepsilon_3 \\ 2\varepsilon_4 \\ 2\varepsilon_5 \\ 2\varepsilon_6 \end{Bmatrix} \tag{1.65}$$

Addition of the third axis of symmetry does not modify the constitutive matrix anymore. Therefore, if two mutually perpendicular axes are axes of symmetry, then the third axis must be an axis of symmetry. These materials have the same material properties in all directions and are known as *isotropic* material. From equations (1.65) and (1.64) one can see that isotropic materials have only two independent material constants. This chapter will concentrate on the analysis of the linear, elastic, isotropic materials.

**Example 1.6**

Consider an elastic orthotropic material for which the stress–strain relations are given by:

$$\varepsilon_{11} = \frac{\sigma_{11}}{E_1} - \nu_{21}\frac{\sigma_{22}}{E_2} - \nu_{31}\frac{\sigma_{33}}{E_3}$$

$$\varepsilon_{22} = \frac{\sigma_{22}}{E_2} - \nu_{12}\frac{\sigma_{11}}{E_1} - \nu_{32}\frac{\sigma_{33}}{E_3}$$

$$\varepsilon_{33} = \frac{\sigma_{33}}{E_3} - \nu_{13}\frac{\sigma_{11}}{E_1} - \nu_{23}\frac{\sigma_{22}}{E_2}$$

$$2\varepsilon_{12} = \frac{\sigma_{12}}{G_{12}} \quad 2\varepsilon_{21} = \frac{\sigma_{21}}{G_{21}}$$

$$2\varepsilon_{13} = \frac{\sigma_{13}}{G_{13}} \quad 2\varepsilon_{31} = \frac{\sigma_{31}}{G_{31}}$$

$$2\varepsilon_{23} = \frac{\sigma_{23}}{G_{23}} \quad 2\varepsilon_{32} = \frac{\sigma_{32}}{G_{32}}$$

where $E_i$ is the Young's modulus in the $x_i$ direction, and $\nu_{ij}$ and $G_{ij}$ represent Poisson's ratio and shear modulus, respectively, in different directions for different values of $i$ and $j$.

(a) How many different elastic constants do you see in the preceding relations?
(b) How many of these do you expect to be independent?
(c) How many equations or constraint relations must exist among the preceding material constants?
(d) Do you expect $G_{ij}$ to be equal to $G_{ji}$ for $i \neq j$? Justify your answer.
(e) Do you expect $\nu_{ij}$ to be equal to $\nu_{ji}$ for $i \neq j$? Justify your answer.
(f) Write down all equations (relating the material constants) that must be satisfied.
(g) If the preceding relations are proposed for an isotropic material, then how many independent relations among the material constants must exist? Do not give these equations.
(h) If the material is transversely isotropic, then how many independent relations among the material constants must exist? Do not give these equations.

*Solution*

(a) 15
(b) 9
(c) 6
(d) Yes, because $\varepsilon_{ij}$ and $\nu_{ij}$ are symmetric
(e) No, symmetry of the constitutive matrix does not require that $\nu_{ij}$ to be equal to $\nu_{ji}$.

(f)
$$\begin{Bmatrix} \varepsilon_{11} \\ \varepsilon_{22} \\ \varepsilon_{33} \\ \varepsilon_{23} \\ \varepsilon_{31} \\ \varepsilon_{12} \end{Bmatrix} = \begin{bmatrix} \frac{1}{E_1} & -\frac{\nu_{21}}{E_2} & -\frac{\nu_{31}}{E_3} & 0 & 0 & 0 \\ -\frac{\nu_{12}}{E_1} & \frac{1}{E_2} & -\frac{\nu_{32}}{E_3} & 0 & 0 & 0 \\ -\frac{\nu_{13}}{E_1} & -\frac{\nu_{23}}{E_2} & \frac{1}{E_3} & 0 & 0 & 0 \\ 0 & 0 & 0 & \frac{1}{G_{23}} & 0 & 0 \\ 0 & 0 & 0 & 0 & \frac{1}{G_{31}} & 0 \\ 0 & 0 & 0 & 0 & 0 & \frac{1}{G_{12}} \end{bmatrix} \begin{Bmatrix} \sigma_{11} \\ \sigma_{22} \\ \sigma_{33} \\ \sigma_{23} \\ \sigma_{31} \\ \sigma_{12} \end{Bmatrix}$$

From symmetry of the preceding matrix (also known as the compliance matrix),

$$\frac{\nu_{12}}{E_1} = \frac{\nu_{21}}{E_2}$$

$$\frac{\nu_{13}}{E_1} = \frac{\nu_{31}}{E_3}$$

$$\frac{\nu_{23}}{E_2} = \frac{\nu_{32}}{E_3}$$

The other three constraint conditions are $G_{12} = G_{21}$, $G_{13} = G_{31}$, and $G_{32} = G_{23}$.

(g) Thirteen constraint relations must exist because isotropic material has only two independent material constants.

(h) Ten relations should exist since the transversely isotropic solid has five independent material constants.

### 1.2.13 Stress–Strain Relation for Isotropic Materials— Green's Approach

Consider an isotropic material subjected to two states of strain as shown in Figure 1.13. The state of strain for the first case is $\varepsilon_{ij}$ in the $x_1x_2x_3$ coordinate system, as shown in the left-hand Figure; the strain state for the second case is $\varepsilon_{i'j'}$ in the $x_1x_2x_3$ coordinate system, as shown in the right-hand Figure. Note that $\varepsilon_{i'j'}$ and $\varepsilon_{ij}$ are numerically different. The numerical values for $\varepsilon_{i'j'}$ can be obtained from $\varepsilon_{ij}$ by transforming the strain components $\varepsilon_{ij}$ from the $x_1x_2x_3$ coordinate system to the $x_{1'}x_{2'}x_{3'}$ coordinate system as shown on the left-hand side of Figure 1.13. If the strain energy density function in the $x_1x_2x_3$ coordinate system is given by $U_0(\varepsilon_{ij})$, then the strain energy densities for these two cases are $U_0(\varepsilon_{ij})$ and $U_0(\varepsilon_{i'j'})$. If the material is anisotropic, then these two values can be different since the strain states are different. However, if the material is isotropic, then these two values must be the same since, in the two illustrations of Figure 1.13, identical numerical values of strain components $(\varepsilon_{i'j'})$ are applied in two different directions. For isotropic material, equal strain values applied in two different directions should not make any difference in computing the strain energy density. For $U_0(\varepsilon_{ij})$ and $U_0(\varepsilon_{i'j'})$ to be identical, $U_0$ must be a function of strain invariants because strain invariants do not change when the numerical values of the strain components are changed from $\varepsilon_{ij}$ to $\varepsilon_{i'j'}$.

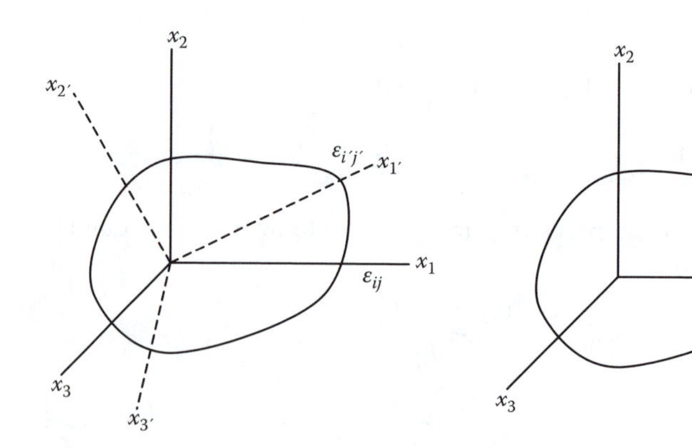


**FIGURE 1.13**
Isotropic material subjected to two states of strain.

Three stress invariants have been defined in equation (1.39). In the same manner, three strain invariants can be defined:

$$I_1 = \varepsilon_{ii}$$

$$I_2 = \frac{1}{2}\varepsilon_{ij}\varepsilon_{ji} \tag{1.66}$$

$$I_3 = \frac{1}{3}\varepsilon_{ij}\varepsilon_{jk}\varepsilon_{ki}$$

Note that $I_1$, $I_2$, and $I_3$ are linear, quadratic, and cubic functions of strain components, respectively. To obtain a linear stress–strain relation from equation (1.50), it is clear that the strain energy density function must be a quadratic function of strain as shown:

$$U_0 = C_1 I_1^2 + C_2 I_2$$

$$\therefore \sigma_{ij} = \frac{\partial U_0}{\partial \varepsilon_{ij}} = 2C_1 I_1 \frac{\partial I_1}{\partial \varepsilon_{ij}} + C_2 \frac{\partial I_2}{\partial \varepsilon_{ij}} = 2C_1 I_1 \delta_{ik}\delta_{kj} + C_2 \frac{1}{2}(\delta_{im}\delta_{jn}\varepsilon_{nm} + \varepsilon_{mn}\delta_{in}\delta_{jm})$$

$$\therefore \sigma_{ij} = 2C_1 \varepsilon_{kk}\delta_{ij} + C_2 \varepsilon_{ij} \tag{1.67}$$

In equation (1.67), if we substitute $2C_1 = \lambda$, and $C_2 = 2\mu$, then the stress–strain relation takes the following form:

$$\sigma_{ij} = \lambda \delta_{ij}\varepsilon_{kk} + 2\mu\varepsilon_{ij} \tag{1.68}$$

In equation (1.68) coefficients $\lambda$ and $\mu$ are known as Lame's first and second constants, respectively. This equation can be expressed in matrix form as in equation (1.65) to obtain

$$\begin{Bmatrix} \sigma_1 = \sigma_{11} \\ \sigma_2 = \sigma_{22} \\ \sigma_3 = \sigma_{33} \\ \sigma_4 = \sigma_{23} \\ \sigma_5 = \sigma_{31} \\ \sigma_6 = \sigma_{12} \end{Bmatrix} = \begin{bmatrix} \lambda + 2\mu & \lambda & \lambda & 0 & 0 & 0 \\ & \lambda + 2\mu & \lambda & 0 & 0 & 0 \\ & & \lambda + 2\mu & 0 & 0 & 0 \\ & & & \mu & 0 & 0 \\ & symm & & & \mu & 0 \\ & & & & & \mu \end{bmatrix} \begin{Bmatrix} \varepsilon_1 = \varepsilon_{11} \\ \varepsilon_2 = \varepsilon_{22} \\ \varepsilon_3 = \varepsilon_{33} \\ 2\varepsilon_4 = 2\varepsilon_{23} = \gamma_{23} \\ 2\varepsilon_5 = 2\varepsilon_{31} = \gamma_{31} \\ 2\varepsilon_6 = 2\varepsilon_{12} = \gamma_{12} \end{Bmatrix} \tag{1.69}$$

Note that the shear stress component ($\sigma_{ij}$) is simply equal to the engineering shear strain component ($\gamma_{ij}$) multiplied by Lame's second constant ($\mu$). Therefore, Lame's second constant is the shear modulus.

Equations (1.68) and (1.69) are also known as generalized Hooke's law in three dimensions, named after Robert Hooke, who first proposed the linear stress–strain model. Equation (1.68) can be inverted to obtain strain components in terms of the stress components as follows.

In equation (1.68), substituting the subscript $j$ by $i$, one can write:

$$\sigma_{ii} = \lambda\delta_{ii}\varepsilon_{kk} + 2\mu\varepsilon_{ii} = (3\lambda + 2\mu)\varepsilon_{ii}$$
$$\therefore \varepsilon_{ii} = \frac{\sigma_{ii}}{(3\lambda + 2\mu)} \tag{1.70}$$

Substitution of equation (1.70) back into equation (1.68) gives:

$$\sigma_{ij} = \lambda\delta_{ij}\frac{\sigma_{kk}}{(3\lambda + 2\mu)} + 2\mu\varepsilon_{ij}$$

or

$$\varepsilon_{ij} = \frac{\sigma_{ij}}{2\mu} + \delta_{ij}\frac{\lambda\sigma_{kk}}{2\mu(3\lambda + 2\mu)} \tag{1.71}$$

#### 1.2.13.1 Hooke's Law in Terms of Young's Modulus and Poisson's Ratio

In undergraduate mechanics courses, strains are expressed in terms of the stress components, Young's modulus ($E$), Poisson's ratio ($v$), and shear modulus ($\mu$) in the following form:

$$\begin{aligned}
\varepsilon_{11} &= \frac{\sigma_{11}}{E} - \frac{v\sigma_{22}}{E} - \frac{v\sigma_{33}}{E} \\
\varepsilon_{22} &= \frac{\sigma_{22}}{E} - \frac{v\sigma_{11}}{E} - \frac{v\sigma_{33}}{E} \\
\varepsilon_{33} &= \frac{\sigma_{33}}{E} - \frac{v\sigma_{22}}{E} - \frac{v\sigma_{11}}{E} \\
2\varepsilon_{12} &= \gamma_{12} = \frac{\sigma_{12}}{\mu} = \frac{2(1+v)\sigma_{12}}{E} \\
2\varepsilon_{23} &= \gamma_{23} = \frac{\sigma_{23}}{\mu} = \frac{2(1+v)\sigma_{23}}{E} \\
2\varepsilon_{31} &= \gamma_{31} = \frac{\sigma_{31}}{\mu} = \frac{2(1+v)\sigma_{31}}{E}
\end{aligned} \tag{1.72}$$

In this equation, the relations among Young's modulus ($E$), Poisson's ratio ($v$), and shear modulus ($\mu$) have been incorporated. Equation (1.72) can be expressed in index notation:

$$\varepsilon_{ij} = \frac{1+v}{E}\sigma_{ij} - \frac{v}{E}\delta_{ij}\sigma_{kk} \tag{1.73}$$

Equating the right-hand sides of equations (1.71) and (1.73), Lame's constants can be expressed in terms of Young's modulus and Poisson's ratio. Similarly, the bulk modulus $K = \frac{\sigma_{ii}}{3\varepsilon_{ii}}$ can be expressed in terms of Lame's constants from equation (1.70).

**Example 1.7**

For an isotropic material, obtain the bulk modulus $K$ in terms of (a) $E$ and $\nu$; (b) $\lambda$ and $\mu$.

*Solution*

(a) From equation 1.70, $\sigma_{ii} = (3\lambda + 2\mu)\varepsilon_{ii}$; therefore, $K = \frac{\sigma_{ii}}{3\varepsilon_{ii}} = \frac{3\lambda+2\mu}{3}$

(b) From equation 1.73, $\varepsilon_{ii} = \frac{1+\nu}{E}\sigma_{ii} - \frac{\nu}{E}\delta_{ii}\sigma_{kk} = \frac{1+\nu}{E}\sigma_{ii} - \frac{3\nu}{E}\sigma_{kk} = \frac{1-2\nu}{E}\sigma$

Thus, $K = \frac{\sigma_{ii}}{3\varepsilon_{ii}} = \frac{E}{3(1-2\nu)}$

Since the isotropic material has only two independent elastic constants, any of the five commonly used elastic constants ($\lambda$, $\mu$, E, $\nu$, and $K$) can be expressed in terms of any other two elastic constants, as shown in Table 1.1.

### 1.2.14 Navier's Equation of Equilibrium

Substituting the stress–strain relation (equation 1.68) into the equilibrium equation (equation 1.24), one obtains

$$\begin{aligned}
&(\lambda\delta_{ij}\varepsilon_{kk} + 2\mu\varepsilon_{ij})_{,j} + f_i = 0 \\
&\Rightarrow \left(\lambda\delta_{ij}u_{k,k} + 2\mu\frac{1}{2}[u_{i,j} + u_{j,i}]\right)_{,j} + f_i = 0 \\
&\Rightarrow \lambda\delta_{ij}u_{k,kj} + \mu[u_{i,jj} + u_{j,ij}] + f_i = 0 \\
&\Rightarrow \lambda u_{k,ki} + \mu[u_{i,jj} + u_{j,ji}] + f_i = 0 \\
&\Rightarrow (\lambda + \mu)u_{j,ji} + \mu u_{i,jj} + f_i = 0
\end{aligned} \tag{1.74}$$

In the vector form, this equation can be written as

$$(\lambda + \mu)\underline{\nabla}(\underline{\nabla} \bullet \mathbf{u}) + \mu\nabla^2\mathbf{u} + \mathbf{f} = 0 \tag{1.75}$$

Because of the vector identity $\nabla^2\mathbf{u} = \underline{\nabla}(\underline{\nabla} \bullet \mathbf{u}) - \underline{\nabla} \times \underline{\nabla} \times \mathbf{u}$, equation (1.75) can be also written as

$$(\lambda + 2\mu)\underline{\nabla}(\underline{\nabla} \bullet \mathbf{u}) - \mu\underline{\nabla} \times \underline{\nabla} \times \mathbf{u} + \mathbf{f} = 0 \tag{1.76}$$

In equations (1.75) and (1.76), the dot (•) is used to indicate scalar or dot product and the cross (×) is used to indicate the vector or cross-product.

**TABLE 1.1**
Relations between Different Elastic Constants for Isotropic Materials

| | $\lambda$ | $\mu$ | $\varepsilon$ | $\nu$ | $K$ |
|---|---|---|---|---|---|
| $\lambda,\mu$ | — | — | $\frac{\mu(3\lambda+2\mu)}{\lambda+\mu}$ | $\frac{\lambda}{2(\lambda+\mu)}$ | $\frac{3\lambda+2\mu}{3}$ |
| $\lambda$,E | — | $\frac{(E-3\lambda)+\sqrt{(E-3\lambda)^2+8\lambda E}}{4}$ | — | $\frac{-(E+\lambda)+\sqrt{(E+\lambda)^2+8\lambda^2}}{4\lambda}$ | $\frac{(E+3\lambda)+\sqrt{(E+3\lambda)^2-4\lambda E}}{6}$ |
| $\lambda,\nu$ | — | $\frac{\lambda(1-2\nu)}{2\nu}$ | $\frac{\lambda(1+\nu)(1-2\nu)}{\nu}$ | — | $\frac{\lambda(1+\nu)}{3\nu}$ |
| $\lambda,K$ | — | $\frac{3(K-\lambda)}{2}$ | $\frac{9K(K-\lambda)}{3K-\lambda}$ | $\frac{\lambda}{3K-\lambda}$ | — |
| $\mu,E$ | $\frac{(2\mu-E)\mu}{E-3\mu}$ | — | — | $\frac{E-2\mu}{2\mu}$ | $\frac{\mu E}{3(3\mu-E)}$ |
| $\mu,\nu$ | $\frac{2\mu\nu}{1-2\nu}$ | — | $2\mu(1+\nu)$ | — | $\frac{2\mu(1+\nu)}{3(1-2\nu)}$ |
| $\mu,K$ | $\frac{3K-2\mu}{3}$ | — | $\frac{9K\mu}{3K+\mu}$ | $\frac{3K-2\mu}{2(3K+\mu)}$ | — |
| $E,\nu$ | $\frac{\nu E}{(1+\nu)(1-2\nu)}$ | $\frac{E}{2(1+\nu)}$ | — | — | $\frac{E}{3(1-2\nu)}$ |
| $E,K$ | $\frac{3KE}{9K-E}$ | $\frac{3KE}{9K-E}$ | — | $\frac{3KE}{9K-E}$ | — |
| $\nu,K$ | $\frac{3K\nu}{1+\nu}$ | $\frac{3K(1-2\nu)}{2(1+\nu)}$ | $3K(1-2\nu)$ | — | — |

In index notations, the preceding two equations can be written as

$$(\lambda+\mu)u_{j,ji}+\mu u_{i,jj}+f_i=0$$

or

$$(\lambda+2\mu)u_{j,ji}-\mu\varepsilon_{ijk}\varepsilon_{kmn}u_{n,mj}+f_i=0 \tag{1.77}$$

where $\varepsilon_{ijk}$ and $\varepsilon_{kmn}$ are permutation symbols, defined in equation (1.36). The equilibrium equations, expressed in terms of the displacement components (equations 1.75–1.77) are known as Navier's equation.

**Example 1.8**

If a linear elastic isotropic body does not have any body force, then prove that (a) the volumetric strain is harmonic ($\varepsilon_{ii,jj}=0$), and (b) the displacement field is biharmonic ($u_{i,jjkk}=0$).

*Solution*

(a) From equation (1.77) for zero body force, one can write

$$(\lambda+\mu)u_{j,ji}+\mu u_{i,jj}=0$$
$$\Rightarrow(\lambda+\mu)u_{j,jii}+\mu u_{i,jji}=0$$

Note that

$$u_{i,jji}=u_{i,ijj}=\varepsilon_{ii,jj}$$

and

$$u_{j,jii}=\varepsilon_{jj,ii}=\varepsilon_{ii,jj}$$

Substituting $u$ by $\varepsilon$, the above Navier's equation is simplified to $(\lambda+2\mu)\varepsilon_{ii,jj}=0$.

Since $(\lambda+2\mu)\neq 0$, $\varepsilon_{ii}$ must be harmonic.

(b) Again, from equation (1.77) for zero body force, one can write

$$(\lambda+\mu)u_{j,ji}+\mu u_{i,jj}=0$$
$$\Rightarrow((\lambda+\mu)u_{j,ji}+\mu u_{i,jj})_{,kk}=(\lambda+\mu)u_{j,jikk}+\mu u_{i,jjkk}=0$$

From part (a), $\varepsilon_{jj,kk}=u_{j,jkk}=0$.

Therefore, $u_{j,jkk}=u_{j,jikk}=0$. Substituting it into the preceding equation, one obtains

$$(\lambda+\mu)u_{j,jikk}+\mu u_{i,jjkk}=\mu u_{i,jjkk}=0$$
$$\Rightarrow u_{i,jjkk}=0$$

**Example 1.9**

Obtain the governing equation of equilibrium in terms of displacement for a material whose stress–strain relation is given by $\sigma_{ij}=\alpha_{ijkl}\varepsilon_{km}\varepsilon_{ml}+\delta_{ij}\gamma$, where $\alpha_{ijkl}$ are material properties that are constants over the entire region, and $\gamma$ is the residual state of stress that varies from point to point.

*Solution*

Governing equation:

$$\sigma_{ij,j}+f_i=0$$
$$\Rightarrow\alpha_{ijkl}(\varepsilon_{km}\varepsilon_{ml})_{,j}+\delta_{ij}\gamma_{,j}+f_i=0$$
$$\Rightarrow\alpha_{ijkl}(\varepsilon_{km,j}\varepsilon_{ml}+\varepsilon_{km}\varepsilon_{ml,j})+\gamma_i+f_i=0$$
$$\Rightarrow\frac{1}{4}\alpha_{ijkl}\{(u_{k,mj}+u_{m,kj})(u_{m,l}+u_{l,m})+(u_{k,m}+u_{m,k})(u_{m,lj}+u_{l,mj})\}+\gamma_i+f_i=0$$

### 1.2.15 Fundamental Equations of Elasticity in Other Coordinate Systems

All equations derived so far have been expressed in the Cartesian coordinate system. Although the majority of elasticity problems can be solved in the Cartesian coordinate system for some problem geometries, such as axisymmetric problems, cylindrical and spherical coordinate systems are better suited for defining the problem and/or solving it. If the equation is given in the vector form (equations 1.75 and 1.76), then it can be used in any coordinate system with appropriate definitions of the vector operators in that coordinate system; however, when it is expressed in index notation in the Cartesian coordinate system (equation 1.77), then that expression cannot be used in a cylindrical or spherical coordinate system. In Table 1.2 different vector operations, strain displacement relations, and equilibrium equations are given in the three coordinate systems shown in Figure 1.14.

### 1.2.16 Time-Dependent Problems or Dynamic Problems

In all equations derived previously, it is assumed that the body is in static equilibrium. Therefore, the resultant force acting on the body is equal to zero. If the body is subjected to a nonzero resultant force, then it will have an acceleration $\ddot{u}_i$ (time derivatives are denoted by dots over the variable) and the equilibrium equation (equation 1.24) will be replaced by the following governing equation of motion:

$$\sigma_{ij,j} + f_i = \rho \ddot{u}_i \tag{1.78}$$

In the preceding equation, $\rho$ is the mass density. Therefore, Navier's equation for the dynamic case takes the following form:

$$(\lambda + 2\mu)\underline{\nabla}(\underline{\nabla} \bullet \mathbf{u}) - \mu\underline{\nabla} \times \underline{\nabla} \times \mathbf{u} + \mathbf{f} = \rho \ddot{\mathbf{u}} \tag{1.79}$$

Dynamic problems of elasticity have been solved in Kundu (2004) and are not presented here.

## 1.3 Some Classical Problems in Elasticity

Fundamental equations derived in section 1.2 are used in this section to solve a few classical two-dimensional and three-dimensional problems of elasticity. For solutions of additional problems readers are referred to Timoshenko and Goodier (1970).

**TABLE 1.2**

Important Equations in Different Coordinate Systems

| Equations | Cartesian Coordinate System | Cylindrical Coordinate System | Spherical Coordinate System |
|---|---|---|---|
| Grad $\phi$ $= \underline{\nabla}\phi$ | $\phi_{,i}$ $= \phi_{,1}e_1 + \phi_{,2}e_2 + \phi_{,3}e_3$ | $\frac{\partial \phi}{\partial r} e_r + \frac{\partial \phi}{\partial z} e_z + \frac{1}{r}\frac{\partial \phi}{\partial \theta} e_\theta$ | $\frac{\partial \phi}{\partial r} e_r + \frac{1}{r}\frac{\partial \phi}{\partial \beta} e_\beta + \frac{1}{r \sin\beta}\frac{\partial \phi}{\partial \theta} e_\theta$ |
| Div $\psi = \underline{\nabla} \bullet \psi$ | $\psi_{i,i} = \psi_{1,1} + \psi_{2,2} + \psi_{3,3}$ | $\frac{\partial \psi_r}{\partial r} + \frac{\partial \psi_z}{\partial z} + \frac{1}{r}\frac{\partial \psi_\theta}{\partial \theta}$ | $\frac{\partial \psi_r}{\partial r} + \frac{2}{r}\psi_r + \frac{1}{r}\frac{\partial \psi_\beta}{\partial \beta}$<br>$+\frac{\cot\beta}{r}\psi_\beta + \frac{1}{r\sin\beta}\frac{\partial \psi_\theta}{\partial \theta}$ |
| Curl $\psi = \underline{\nabla} \times \psi$ | $\varepsilon_{ijk}\psi_{k,j} =$<br>$\begin{vmatrix} e_1 & e_2 & e_3 \\ \frac{\partial}{\partial x_1} & \frac{\partial}{\partial x_2} & \frac{\partial}{\partial x_3} \\ \psi_1 & \psi_2 & \psi_3 \end{vmatrix}$ | $\frac{1}{r}\begin{vmatrix} e_r & re_\theta & e_z \\ \frac{\partial}{\partial r} & \frac{\partial}{\partial \theta} & \frac{\partial}{\partial z} \\ \psi_r & r\psi_\theta & \psi_z \end{vmatrix}$ | $\frac{1}{r^2 \sin\beta}\begin{vmatrix} e_r & re_\beta & r\sin\beta e_\theta \\ \frac{\partial}{\partial r} & \frac{\partial}{\partial \beta} & \frac{\partial}{\partial \theta} \\ \psi_r & r\psi_\beta & r\sin\beta\psi_\theta \end{vmatrix}$ |
| Strain–displacement relation | Equation (1.7a)<br>$\varepsilon_{ij} = \frac{1}{2}(u_{i,j} + u_{j,i})$ | $\varepsilon_{rr} = \frac{\partial u_r}{\partial r}$<br>$\varepsilon_{\theta\theta} = \frac{1}{r}\frac{\partial u_\theta}{\partial \theta} + \frac{u_r}{r}$<br>$\varepsilon_{zz} = \frac{\partial u_z}{\partial z}$<br>$\varepsilon_{rz} = \frac{1}{2}\left(\frac{\partial u_r}{\partial z} + \frac{\partial u_z}{\partial r}\right)$<br>$\varepsilon_{r\theta} = \frac{1}{2}\left(\frac{1}{r}\frac{\partial u_r}{\partial \theta} - \frac{u_\theta}{r} + \frac{\partial u_\theta}{\partial r}\right)$<br>$\varepsilon_{z\theta} = \frac{1}{2}\left(\frac{1}{r}\frac{\partial u_z}{\partial \theta} + \frac{\partial u_\theta}{\partial z}\right)$ | $\varepsilon_{rr} = \frac{\partial u_r}{\partial r}$<br>$\varepsilon_{\beta\beta} = \frac{1}{r}\frac{\partial u_\beta}{\partial \beta} + \frac{u_r}{r}$<br>$\varepsilon_{\theta\theta} = \frac{1}{r\sin\beta}\frac{\partial u_\theta}{\partial \theta} + \frac{u_r}{r} + \frac{u_\beta}{r}\cot\beta$<br>$2\varepsilon_{r\beta} = \frac{1}{r}\frac{\partial u_r}{\partial \beta} + \frac{\partial u_\beta}{\partial r} - \frac{u_\beta}{r}$<br>$2\varepsilon_{r\theta} = \frac{1}{r\sin\beta}\frac{\partial u_r}{\partial \theta} - \frac{u_\theta}{r} + \frac{\partial u_\theta}{\partial r}$<br>$2\varepsilon_{\beta\theta} = \frac{1}{r\sin\beta}\frac{\partial u_\beta}{\partial \theta} + \frac{1}{r}\frac{\partial u_\theta}{\partial \beta} - \frac{u_\theta}{r}\cot\beta$ |
| Equilibrium equations | Equations (1.24 and 1.22)<br>$\sigma_{ij,j} + f_i = 0$ | $\frac{\partial \sigma_{rr}}{\partial r} + \frac{\partial \sigma_{rz}}{\partial z} + \frac{1}{r}\frac{\partial \sigma_{r\theta}}{\partial \theta}$<br>$+\frac{1}{r}(\sigma_{rr} - \sigma_{\theta\theta}) + f_r = 0$<br>$\frac{\partial \sigma_{rz}}{\partial r} + \frac{\partial \sigma_{zz}}{\partial z} + \frac{1}{r}\frac{\partial \sigma_{z\theta}}{\partial \theta}$<br>$+\frac{1}{r}\sigma_{rz} + f_z = 0$<br>$\frac{\partial \sigma_{r\theta}}{\partial r} + \frac{\partial \sigma_{z\theta}}{\partial z} + \frac{1}{r}\frac{\partial \sigma_{\theta\theta}}{\partial \theta}$<br>$+\frac{2}{r}\sigma_{r\theta} + f_\theta = 0$ | $\frac{\partial \sigma_{rr}}{\partial r} + \frac{1}{r}\frac{\partial \sigma_{r\beta}}{\partial \beta} + \frac{1}{r\sin\beta}\frac{\partial \sigma_{r\theta}}{\partial \theta}$<br>$+\frac{1}{r}[2\sigma_{rr} + (\cot\beta)\sigma_{r\beta} - \sigma_{\beta\beta} - \sigma_{\theta\theta}] + f_r = 0$<br>$\frac{\partial \sigma_{r\beta}}{\partial r} + \frac{1}{r}\frac{\partial \sigma_{\beta\beta}}{\partial \beta} + \frac{1}{r\sin\beta}\frac{\partial \sigma_{\beta\theta}}{\partial \theta}$<br>$+\frac{1}{r}[3\sigma_{r\beta} + (\cot\theta)(\sigma_{\beta\beta} - \sigma_{\theta\theta})] + f_\beta = 0$<br>$\frac{\partial \sigma_{r\theta}}{\partial r} + \frac{1}{r}\frac{\partial \sigma_{\beta\theta}}{\partial \beta} + \frac{1}{r\sin\beta}\frac{\partial \sigma_{\theta\theta}}{\partial \theta}$<br>$+\frac{2\sigma_{\beta\theta}\cot\beta + 3\sigma_{r\theta}}{r} + f_\theta = 0$ |

*Source:* Moon, P. and Spencer, D. E. *Vectors*. Princeton, NJ: Van Nostrand Company, Inc., 1965.
*Note:* $\phi$ and $\psi$ are scalar and vector functions, respectively.

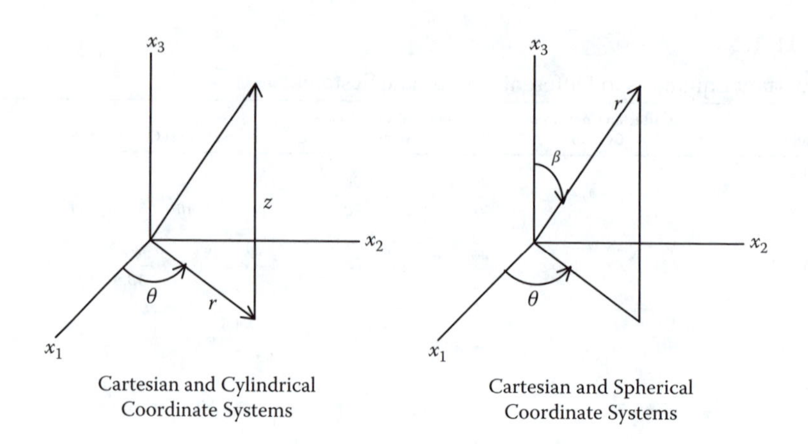


**FIGURE 1.14**
Cartesian ($x_1x_2x_3$), cylindrical ($r\theta z$), and spherical ($r\beta\theta$) coordinate systems.

### 1.3.1 In-Plane and Out-of-Plane Problems

When the elastic fields (stress, strain, and displacement) are independent of one coordinate (in other words, dependent on two coordinates only), these problems are called two-dimensional (or 2-D) problems. Two-dimensional problems can be further classified under two groups: antiplane or out-of-plane problems and in-plane problems.

If the elastic field is a function of coordinates $x_1$ and $x_2$ while the only non-zero displacement component is in the direction perpendicular to the plane containing $x_1$ and $x_2$ axes, then that problem is called the antiplane problem. Therefore, for antiplane problems,

$$\begin{aligned} u_1 &= 0 \\ u_2 &= 0 \\ u_3(x_1, x_2) &\neq 0 \end{aligned} \tag{1.80}$$

For in-plane problems, on the other hand, displacement components in the plane containing $x_1$ and $x_2$ axes are functions of $x_1$ and $x_2$ and the perpendicular displacement component ($u_3$) is not a function of $x_1$ and $x_2$; it is either zero or a constant. Thus, for in-plane problems,

$$\begin{aligned} &u_1(x_1, x_2) \neq 0 \\ &u_2(x_1, x_2) \neq 0 \\ &u_3 = 0 \text{ or a constant} \end{aligned} \tag{1.81}$$

### 1.3.2 Plane Stress and Plane Strain Problems

In-plane problems can be classified into two groups: plane stress problems and plane strain problems. If all nonzero stresses act in one plane, then the problem is called a plane stress problem; when all nonzero strains act in one plane, the problem is known as a plane strain problem. Therefore, for plane stress problems,

$$\begin{aligned}\sigma_{11}(x_1,x_2) &\neq 0\\ \sigma_{22}(x_1,x_2) &\neq 0\\ \sigma_{12}(x_1,x_2) &\neq 0\\ \sigma_{33} &= 0\\ \sigma_{13} &= 0\\ \sigma_{23} &= 0\end{aligned} \tag{1.82}$$

and for plane strain problems,

$$\begin{aligned}\varepsilon_{11}(x_1,x_2) &\neq 0\\ \varepsilon_{22}(x_1,x_2) &\neq 0\\ \varepsilon_{12}(x_1,x_2) &\neq 0\\ \varepsilon_{33} &= 0\\ \varepsilon_{13} &= 0\\ \varepsilon_{23} &= 0\end{aligned} \tag{1.83}$$

From the stress–strain relation for isotropic materials (equation 1.73), the strain components for plane stress problems can be obtained:

$$\begin{aligned}\varepsilon_{11} &= \frac{\sigma_{11}}{E} - v\frac{\sigma_{22}}{E}\\ \varepsilon_{22} &= \frac{\sigma_{22}}{E} - v\frac{\sigma_{11}}{E}\\ \varepsilon_{12} &= \frac{(1+v)\sigma_{11}}{E}\\ \varepsilon_{33} &= -\frac{v}{E}(\sigma_{11}+\sigma_{22})\\ \varepsilon_{13} &= \varepsilon_{23} = 0\end{aligned} \tag{1.84}$$

Note that the normal strain $\varepsilon_{33}$ in the $x_3$ direction is not equal to zero for plane stress problems.

The first three equations of equation (1.84) can be written in matrix form:

$$\begin{Bmatrix} \varepsilon_{11} \\ \varepsilon_{22} \\ \varepsilon_{12} \end{Bmatrix} = \frac{1}{E} \begin{bmatrix} 1 & -\nu & 0 \\ -\nu & 1 & 0 \\ 0 & 0 & (1+\nu) \end{bmatrix} \begin{Bmatrix} \sigma_{11} \\ \sigma_{22} \\ \sigma_{12} \end{Bmatrix} \tag{1.85}$$

In index notation, equation (1.85) can be written as

$$\varepsilon_{\alpha\beta} = \frac{1+\nu}{E}\sigma_{\alpha\beta} - \frac{\nu}{E}\delta_{\alpha\beta}\sigma_{\gamma\gamma} \tag{1.86}$$

Greek subscripts $\alpha$, $\beta$, and $\gamma$ in equation (1.86) and in all subsequent equations take values 1 and 2 only.

Inverting equation (1.85),

$$\begin{Bmatrix} \sigma_{11} \\ \sigma_{22} \\ \sigma_{12} \end{Bmatrix} = \frac{E}{1-\nu^2} \begin{bmatrix} 1 & \nu & 0 \\ \nu & 1 & 0 \\ 0 & 0 & (1-\nu) \end{bmatrix} \begin{Bmatrix} \varepsilon_{11} \\ \varepsilon_{22} \\ \varepsilon_{12} \end{Bmatrix} \tag{1.87}$$

Similarly, the stress–strain relation for the plane strain problems can be written as

$$\begin{aligned} \varepsilon_{11} &= \frac{\sigma_{11}}{E} - \nu\frac{\sigma_{22}}{E} - \nu\frac{\sigma_{33}}{E} \\ \varepsilon_{22} &= \frac{\sigma_{22}}{E} - \nu\frac{\sigma_{11}}{E} - \nu\frac{\sigma_{33}}{E} \\ \varepsilon_{12} &= \frac{(1+\nu)\sigma_{11}}{E} \\ \varepsilon_{33} &= \frac{\sigma_{33}}{E} - \frac{\nu}{E}(\sigma_{11}+\sigma_{22}) = 0 \\ \varepsilon_{13} &= \varepsilon_{23} = 0 \end{aligned} \tag{1.88}$$

From the fourth equation of equation set (1.88), one gets

$$\sigma_{33} = \nu(\sigma_{11}+\sigma_{22}) \tag{1.89}$$

After substituting equation (1.89) into equation (1.88), the first three equations of (1.88) can be rewritten in the following form:

$$\begin{Bmatrix} \varepsilon_{11} \\ \varepsilon_{22} \\ \varepsilon_{12} \end{Bmatrix} = \frac{1+\nu}{E} \begin{bmatrix} (1-\nu) & -\nu & 0 \\ -\nu & (1-\nu) & 0 \\ 0 & 0 & 1 \end{bmatrix} \begin{Bmatrix} \sigma_{11} \\ \sigma_{22} \\ \sigma_{12} \end{Bmatrix} \tag{1.90}$$

Inverting equation (1.90),

$$\begin{Bmatrix} \sigma_{11} \\ \sigma_{22} \\ \sigma_{12} \end{Bmatrix} = \frac{E}{(1+\nu)(1-2\nu)} \begin{bmatrix} (1-\nu) & \nu & 0 \\ \nu & (1-\nu) & 0 \\ 0 & 0 & (1-2\nu) \end{bmatrix} \begin{Bmatrix} \varepsilon_{11} \\ \varepsilon_{22} \\ \varepsilon_{12} \end{Bmatrix} \tag{1.91}$$

For plane stress and plane strain problems there are only two equilibrium equations: force equilibrium in the $x_1$ direction and force equilibrium in the $x_2$ direction. These two equations are obtained from the complete set of equilibrium equations given in equation (1.24):

$$\sigma_{\alpha\beta,\beta} + f_\alpha = 0 \tag{1.92}$$

Six compatibility equations (see example 1.1) for three-dimensional problems are reduced to only one compatibility condition in the following form for plane stress and plane strain problems:

$$\varepsilon_{11,22} + \varepsilon_{22,11} - 2\varepsilon_{12,12} = 0 \tag{1.93}$$

The preceding compatibility equation can be written in terms of the stress components using the strain–stress relations: equation (1.85) for plane stress problems and equation (1.90) for plane strain problems.

#### *1.3.2.1 Compatibility Equations for Plane Stress Problems*

Substituting equation (1.85) into equation (1.93), one gets

$$\begin{aligned} &\frac{1}{E}\left\{\sigma_{11,22} - \nu\sigma_{22,22} + \sigma_{22,11} - \nu\sigma_{11,11} - 2(1+\nu)\sigma_{12,12}\right\} = 0 \\ &\Rightarrow (\sigma_{11,22} + \sigma_{22,11}) - \nu(\sigma_{11,11} + \sigma_{22,22}) - 2(1+\nu)\sigma_{12,12} = 0 \end{aligned} \tag{1.94}$$

From equation (1.92), it is possible to obtain the following two equations after taking derivatives of the two equilibrium equations with respect to $x_1$ and $x_2$, respectively:

$$\begin{aligned} &\sigma_{11,11} + \sigma_{12,12} + f_{1,1} = 0 \\ &\sigma_{12,12} + \sigma_{22,22} + f_{2,2} = 0 \end{aligned} \tag{1.95}$$

Combining the two equations of equation (1.95),

$$\begin{aligned} &(\sigma_{11,11} + \sigma_{22,22}) + 2\sigma_{12,12} + (f_{1,1} + f_{2,2}) = 0 \\ &\Rightarrow (\sigma_{11,11} + \sigma_{22,22}) + (f_{1,1} + f_{2,2}) = -2\sigma_{12,12} \end{aligned} \tag{1.96}$$

Substituting equation (1.96) into equation (1.94),

$$(\sigma_{11,22}+\sigma_{22,11})-\nu(\sigma_{11,11}+\sigma_{22,22})-2(1+\nu)\sigma_{12,12}=0$$

$$\Rightarrow(\sigma_{11,22}+\sigma_{22,11})-\nu(\sigma_{11,11}+\sigma_{22,22})+(1+\nu)(\sigma_{11,11}+\sigma_{22,22})+(1+\nu)(f_{1,1}+f_{2,2})=0$$

$$\Rightarrow(\sigma_{11,22}+\sigma_{22,11}+\sigma_{11,11}+\sigma_{22,22})=-(1+\nu)(f_{1,1}+f_{2,2})$$

$$\Rightarrow\sigma_{\alpha\alpha,\beta\beta}=-(1+\nu)f_{\gamma,\gamma} \tag{1.97}$$

Note that in the absence of the body force, equation (1.97) is reduced to

$$\sigma_{\alpha\alpha,\beta\beta}=0 \tag{1.98}$$

#### 1.3.2.2 *Compatibility Equations for Plane Strain Problems*

Substituting equation (1.90) into equation (1.93), one obtains

$$\left(\frac{(1-\nu^2)\sigma_{11}}{E}-\nu(1+\nu)\frac{\sigma_{22}}{E}\right)_{,22}+\left(\frac{(1-\nu^2)\sigma_{22}}{E}-\nu(1+\nu)\frac{\sigma_{11}}{E}\right)_{,11}=2\left(\frac{1+\nu}{E}\sigma_{12}\right)_{,12}$$

$$\Rightarrow(1-\nu)\sigma_{11,22}-\nu\sigma_{22,22}+(1-\nu)\sigma_{22,11}-\nu\sigma_{11,11}=2\sigma_{12,12} \tag{1.99}$$

Equations (1.95) and (1.96) are the same for plane stress and plane strain problems.

Substituting equation (1.96) into equation (1.99),

$$(1-\nu)\sigma_{11,22}-\nu\sigma_{22,22}+(1-\nu)\sigma_{22,11}-\nu\sigma_{11,11}=2\sigma_{12,12}=-(\sigma_{11,11}+\sigma_{22,22})-(f_{1,1}+f_{2,2})$$

$$\Rightarrow(1-\nu)(\sigma_{11,11}+\sigma_{11,22}+\sigma_{22,11}+\sigma_{22,22})=-(f_{1,1}+f_{2,2})$$

$$\Rightarrow\sigma_{\alpha\alpha,\beta\beta}=-\frac{f_{\gamma,\gamma}}{(1-\nu)} \tag{1.100}$$

It should be noted here that in the absence of the body force, equation (1.100) is reduced to equation (1.98). Therefore, in the absence of the body force there is no difference in the compatibility equations for plane stress and plane strain problems; for both cases it is given by equation (1.98). However, in the presence of the body force there is some difference, as one can see in equations (1.97) and (1.100).

### 1.3.3 Airy Stress Function

If an elasticity problem is formulated in terms of stresses, then solution of that problem requires solving the governing equilibrium equations and compatibility conditions, subjected to the boundary conditions and regularity conditions. In problems having semi-infinite or infinite geometry, the conditions at infinity are known as regularity conditions. Note that if the problem is formulated in terms of displacements, then the compatibility conditions are automatically satisfied. Airy (1862) introduced a stress function $\phi(x_1, x_2)$ that is related to the stress field in the following manner in absence of the body force:

$$\begin{aligned} \sigma_{11} &= \phi_{22} \\ \sigma_{22} &= \phi_{11} \\ \sigma_{12} &= -\phi_{12} \end{aligned} \tag{1.101}$$

The advantage of this representation is that when these expressions are substituted into the equilibrium equation (1.92) in the absence of the body force ($f_\alpha = 0$), the simplified equilibrium equation $\sigma_{\alpha\beta,\beta} = 0$ is automatically satisfied. Substituting equation (1.101) into the compatibility equations (equation 1.98), one obtains

$$\begin{aligned} &\sigma_{\alpha\alpha,\beta\beta} = 0 \\ &\Rightarrow (\sigma_{11,22} + \sigma_{22,11} + \sigma_{11,11} + \sigma_{22,22}) = 0 \\ &\Rightarrow \phi_{22,22} + \phi_{11,11} + \phi_{22,11} + \phi_{11,22} = 0 \\ &\Rightarrow \phi_{11,11} + 2\phi_{11,22} + \phi_{22,22} = 0 \\ &\Rightarrow \nabla^4\phi = 0 \end{aligned} \tag{1.102}$$

Therefore, any biharmonic function $\phi$ satisfies the compatibility equation in absence of the body force and is a valid stress function. As a result, the elasticity problem is significantly simplified in terms of Airy stress function because in the stress formulation one needs to find three unknowns—$\sigma_{11}$, $\sigma_{22}$, and $\sigma_{12}$—that satisfy two equilibrium equations and one compatibility equation. However, in Airy stress function formulation only one unknown biharmonic function $\phi$ is to be evaluated. In both formulations the unknown functions are obtained by satisfying boundary and regularity conditions.

In the presence of the body force, the stress components and Airy stress functions are related in a slightly different manner as follows:

$$\begin{aligned} \sigma_{11} &= \Phi_{22} + V \\ \sigma_{22} &= \Phi_{11} + V \\ \sigma_{12} &= -\Phi_{12} \end{aligned} \tag{1.103}$$

where $V$ is the potential function for the body force defined as

$$f_\alpha = -V_{,\alpha} \tag{1.104}$$

When these expressions are substituted into the equilibrium equations (equation 1.92), those equations are automatically satisfied.

Substituting equations (1.103) and (1.104) into the compatibility equation (equation 1.97) for plane stress problems, one obtains

$$\begin{aligned}
&\sigma_{\alpha\alpha,\beta\beta} + (1+\nu) f_{\gamma,\gamma} = 0 \\
&\Rightarrow \sigma_{11,11} + \sigma_{11,22} + \sigma_{22,11} + \sigma_{22,22} + (1+\nu)(f_{1,1} + f_{2,2}) = 0 \\
&\Rightarrow \Phi_{22,11} + V_{11} + \Phi_{22,22} + V_{22} + \Phi_{11,11} + V_{11} + \Phi_{11,22} + V_{22} - (1+\nu)(V_{11} + V_{22}) = 0 \\
&\Rightarrow \Phi_{\alpha\alpha,\beta\beta} + (1-\nu)(V_{11} + V_{22}) = 0 \\
&\Rightarrow \nabla^4\Phi + (1-\nu)\nabla^2 V = 0 \\
&\Rightarrow \nabla^4\Phi = -(1-\nu)\nabla^2 V
\end{aligned} \tag{1.105}$$

For plane strain problems, equations (1.103) and (1.104) will have to be substituted into the appropriate compatibility equation (equation 1.100) to obtain

$$\begin{aligned}
&\sigma_{\alpha\alpha,\beta\beta} + \frac{f_{\gamma,\gamma}}{(1-\nu)} = 0 \\
&\Rightarrow \sigma_{11,11} + \sigma_{11,22} + \sigma_{22,11} + \sigma_{22,22} + \frac{1}{(1-\nu)}(f_{1,1} + f_{2,2}) = 0 \\
&\Rightarrow \Phi_{22,11} + V_{11} + \Phi_{22,22} + V_{22} + \Phi_{11,11} + V_{11} + \Phi_{11,22} + V_{22} - \frac{1}{(1-\nu)}(V_{11} + V_{22}) = 0 \\
&\Rightarrow \Phi_{\alpha\alpha,\beta\beta} = \frac{(V_{11} + V_{22})}{(1-\nu)} - 2(V_{11} + V_{22}) = \frac{1-2+2\nu}{(1-\nu)}(V_{11} + V_{22}) = \frac{2\nu-1}{(1-\nu)} V_{aa} \\
&\Rightarrow \nabla^4\Phi = \frac{2\nu-1}{(1-\nu)}\nabla^2 V
\end{aligned} \tag{1.106}$$

Clearly, in absence of the body force, the potential function $V = 0$; then equations (1.105) and (1.106) are reduced to equation (1.102). If the body force has a constant value, then $\nabla^2 V = 0$ also, and equations (1.105) and (1.106) are reduced to equation (1.102).

### 1.3.4 Some Classical Elasticity Problems in Two Dimensions

A number of two-dimensional elasticity problems are solved in this section.

#### *1.3.4.1 Plate and Beam Problems*

Solutions of some plate and beam problems are given in the following examples.

**Example 1.10**

Obtain the stress field for a cantilever plate subjected to a concentrated force at the free end as shown in Figure 1.15.

*Solution*

For this problem one can assume that the stress field is independent of the $x_3$ coordinate. For thin plate geometry, it can be also assumed that $\sigma_{33} = \sigma_{13} = \sigma_{23} = 0$, since these stress components are zero on the front and back surfaces of the plate.

Stress boundary conditions for this plate geometry are

$\sigma_{22} = \sigma_{12} = 0$ at $x_2 = \pm c$

and, at $x_1 = 0$, $\sigma_{11} = 0$ and $t\int_{-c}^{c} \sigma_{12} dx_2 = -P$, where $t$ is the plate thickness.

If we want to solve this problem in terms of the Airy stress function, then we will have to find an Airy stress function that will satisfy the following boundary conditions:

At $x_2 = \pm c$, $\phi_{,11} = \phi_{,12} = 0$.

At $x_1 = 0$, $\phi_{,22} = 0$ and $-t\int\limits_{-c}^{c} \phi_{,12}\, dx_2 = -P$.

One can easily show that if the Airy stress function $\phi$ is assumed to have a polynomial form, it requires at least up to fourth-order polynomials to satisfy all boundary conditions. The polynomial expression postulated for the Airy stress function for this problem is

$$\phi = \frac{a_2}{2}x_1^2 + b_1x_1x_2 + \frac{b_2}{2}x_2^2 + \frac{a_3}{6}x_1^3 + \frac{b_3}{2}x_1^2x_2 + \frac{c_3}{2}x_1x_2^2 + \frac{d_3}{6}x_2^3 + \frac{a_4}{12}x_1^4 + \frac{b_4}{6}x_1^3x_2 + \frac{c_4}{2}x_1^2x_2^2 + \frac{d_4}{6}x_1x_2^3 + \frac{e_4}{12}x_2^4 \tag{1.107}$$

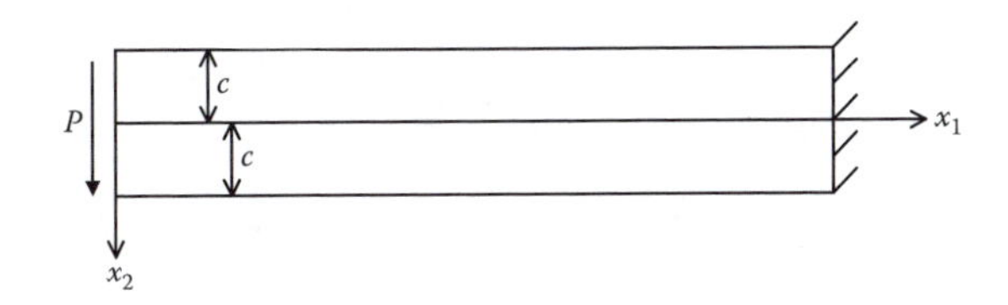


**FIGURE 1.15**
Cantilever plate.

Substituting this expression in the compatibility equation $\nabla^4\phi = 0$, one obtains

$$e_4 = -(2c_4 + a_4) \tag{1.108}$$

Note that the expression of $\phi$ starts with the quadratic polynomial. The reason for omitting the constant and linear terms is that they do not contribute to the stress fields. Numbers 2, 6, and 12 in the denominator are carefully chosen to have a minimum number of terms with fractions in the stress expressions derived from equation (1.107):

$$\begin{aligned}
\sigma_{11} &= \phi_{22} = c_2 + c_3x_1 + d_3x_2 + c_4x_1^2 + d_4x_1x_2 + e_4x_2^2 \\
\sigma_{22} &= \phi_{11} = a_2 + a_3x_1 + b_3x_2 + a_4x_1^2 + b_4x_1x_2 + c_4x_2^2 \\
-\sigma_{12} &= \phi_{12} = b_2 + b_3x_1 + c_3x_2 + \frac{b_4}{2}x_1^2 + 2c_4x_1x_2 + \frac{d_4}{2}x_2^2
\end{aligned} \tag{1.109}$$

Applying the boundary condition, at $x_2 = c$, $\sigma_{22} = 0$, we get

$$\sigma_{22}\big|_{x_2=c} = a_2 + a_3x_1 + b_3c + a_4x_1^2 + b_4x_1c + c_4c^2 = 0, \quad \text{for all } x_1$$

Therefore,

$$\begin{aligned}
a_4 &= 0 \\
a_3 + b_4c &= 0 \\
a_2 + b_3c + c_4c^2 &= 0
\end{aligned} \tag{1.110}$$

Applying the second boundary condition, at $x_2 = c$, $\sigma_{12} = 0$, one obtains

$$\sigma_{12}\big|_{x_2=c} = b_2 + b_3x_1 + c_3c + \frac{b_4}{2}x_1^2 + 2c_4x_1c + \frac{d_4}{2}c^2 = 0, \quad \text{for all } x_1$$

Therefore,

$$\begin{aligned}
b_4 &= 0 \\
b_3 + 2c_4c &= 0 \\
b_2 + c_3c + \frac{d_4}{2}c^2 &= 0
\end{aligned} \tag{1.111}$$

Applying the third boundary condition, at $x_2 = -c$, $\sigma_{22} = 0$,

$$\sigma_{22}\big|_{x_2=-c} = a_2 + a_3x_1 - b_3c + a_4x_1^2 - b_4x_1c + c_4c^2 = 0, \quad \text{for all } x_1$$

Therefore,

$$\begin{aligned} a_4 &= 0 \\ a_3 - b_4 c &= 0 \\ a_2 - b_3 c + c_4 c^2 &= 0 \end{aligned} \tag{1.112}$$

From equations (1.110) and (1.112) it is easy to see that

$$\begin{aligned} a_3 = b_4 &= 0 \\ b_3 &= 0 \\ a_2 + c_4 c^2 &= 0 \end{aligned} \tag{1.113}$$

Applying the fourth boundary condition, at $x_2 = -c$, $\sigma_{12} = 0$,

$$\sigma_{12}\big|_{x_2=-c} = b_2 + b_3 x_1 - c_3 c + \frac{b_4}{2} x_1^2 - 2c_4 x_1 c + \frac{d_4}{2} c^2 = 0, \quad \text{for all } x_1$$

Therefore,

$$\begin{aligned} b_4 &= 0 \\ b_3 - 2c_4 c &= 0 \\ b_2 - c_3 c + \frac{d_4}{2} c^2 &= 0 \end{aligned} \tag{1.114}$$

From equations (1.111) and (1.114),

$$\begin{aligned} b_3 = c_4 &= 0 \\ c_3 &= 0 \\ b_2 + \frac{d_4}{2} c^2 &= 0 \end{aligned} \tag{1.115}$$

Since $c_4 = 0$ (equation 1.115) and $a_4 = 0$ (equation 1.110), one can write from equation (1.108):

$$e_4 = 0 \tag{1.116}$$

From the fifth boundary condition, at $x_1 = 0$, $\sigma_{11} = 0$,

$$\begin{aligned} &c_2 + d_3 x_2 = 0 \text{ for all } x_2 \text{ between } -c \text{ and } +c \\ &\Rightarrow c_2 = 0 \text{ and } d_3 = 0 \end{aligned} \tag{1.117}$$

After applying the preceding five boundary conditions, the stress fields take the following form when we drop all coefficients whose values are zero and substitute $d_4 = -\frac{2b_2}{c^2}$ from equation (1.115):

$$\begin{aligned}
\sigma_{11} &= d_4 x_1 x_2 = -\frac{2b_2}{c^2} x_1 x_2 \\
\sigma_{22} &= 0 \\
\sigma_{12} &= -b_2 - \frac{d_4}{2} x_2^2 = -b_2\left(1 - \frac{x_2^2}{c^2}\right)
\end{aligned} \tag{1.118}$$

The final boundary condition at $x_1 = 0$ is $t\int_{-c}^{c} \sigma_{12} dx_2 = -P$. Therefore,

$$\begin{aligned}
P &= -t\int_{-c}^{c} \sigma_{12} dx_2 = tb_2 \int_{-c}^{c}\left(1 - \frac{x_2^2}{c^2}\right) dx_2 = tb_2\left[x_2 - \frac{x_2^3}{3c^2}\right]_{-c}^{c} \\
&= tb_2\left[2c - \frac{2c^3}{3c^2}\right] = tb_2 \frac{4c}{3} \Rightarrow b_2 = \frac{3P}{4ct}
\end{aligned} \tag{1.119}$$

Then,

$$d_4 = -\frac{2b_2}{c^2} = -\frac{2}{c^2}\frac{3P}{4ct} = -\frac{3P}{2c^3 t} \tag{1.120}$$

Substituting $b_2$ and $d_4$ in equation (1.118), one obtains

$$\begin{aligned}
\sigma_{11} &= -\frac{3P}{2c^3 t} x_1 x_2 \\
\sigma_{22} &= 0 \\
\sigma_{12} &= -\frac{3P}{4ct}\left(1 - \frac{x_2^2}{c^2}\right)
\end{aligned} \tag{1.121}$$

If the shear stress ($\sigma_{12}$) at the free end is applied in a parabolic manner as shown in equation (1.121) to produce the resultant force P, then the computed stress field (equation 1.121) is valid very close to the free end also. However, if the applied shear stress field at the free end does not follow a parabolic distribution but produces a resultant force P, then the computed stress field of equation (1.121) is valid in the plate in the region away from the free end. This is because, according to St. Venant's principle, in an elastic body statically equivalent loads give rise to the same stress field at a point far away from the region of application of the loads.

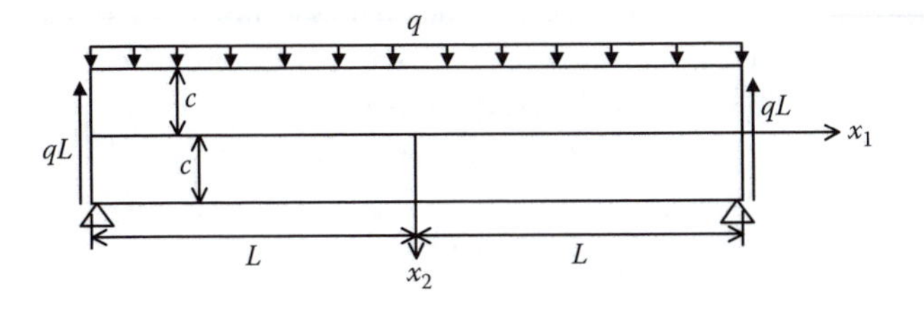


**FIGURE 1.16**
Simply supported plate.

**Example 1.11**
Obtain the stress field for a simply supported plate subjected to uniformly distributed load $q$ per unit surface area at the top surface, as shown in Figure 1.16.

*Solution*
Similar to the problem in example 1.10, here also one can assume that the stress field is independent of $x_3$ direction and $\sigma_{33} = \sigma_{13} = \sigma_{23} = 0$.

Stress boundary conditions for this plate geometry are

$$\begin{aligned}
&\sigma_{22} = \sigma_{12} = 0, \text{ at } x_2 = c \\
&\text{and, at } x_2 = -c,\ \sigma_{22} = -q,\ \sigma_{12} = 0 \\
&\text{At } x_1 = L,\ \sigma_{11} = 0 \text{ and } \int_{-c}^{c} \sigma_{12} dx_2 = -qL \\
&\text{At } x_1 = -L,\ \sigma_{11} = 0 \text{ and } \int_{-c}^{c} \sigma_{12} dx_2 = qL
\end{aligned} \tag{1.122}$$

If the Airy stress function is assumed to be composed of second-, third-, and fourth-order complete polynomials, as given in equation (1.107), then one can easily see (after following the steps of example 1.10) that all boundary conditions cannot be satisfied with that Airy stress function. However, if the fifth-order polynomial is also included in the Airy stress function expression, then these boundary conditions can be satisfied. For this reason we start with the following expression of the Airy stress function:

$$\begin{aligned}
\phi = {} & \frac{a_2}{2}x_1^2 + b_1x_1x_2 + \frac{b_2}{2}x_2^2 + \frac{a_3}{6}x_1^3 + \frac{b_3}{2}x_1^2x_2 + \frac{c_3}{2}x_1x_2^2 + \frac{d_3}{6}x_2^3 \\
& + \frac{a_4}{12}x_1^4 + \frac{b_4}{6}x_1^3x_2 + \frac{c_4}{2}x_1^2x_2^2 + \frac{d_4}{6}x_1x_2^3 + \frac{e_4}{12}x_2^4 \\
& + \frac{a_5}{20}x_1^5 + \frac{b_5}{12}x_1^4x_2 + \frac{c_5}{6}x_1^3x_2^2 + \frac{d_5}{6}x_1^2x_2^3 + \frac{e_5}{12}x_1x_2^4 + \frac{f_5}{20}x_2^5
\end{aligned} \tag{1.123}$$

From the symmetric and antisymmetric conditions of the stress field one can conclude whether $\sigma_{11}$, $\sigma_{22}$, and $\sigma_{12}$ should be even or odd functions of $x_1$. Taking into account this consideration and the boundary conditions at $x_2 = +c$ and $-c$, one can evaluate a number of unknown constants (many of these constants become equal to zero as we have seen earlier). After substituting these constants in the stress expressions, one obtains

$$\sigma_{11} = c_2 + d_3 x_2 + e_4 x_2^2 - \frac{3q}{4c^3} x_1^2 x_2 + f_5 x_2^3$$
$$\sigma_{22} = -\frac{q}{2}\left(1 - \frac{3x_2}{2c} + \frac{x_2^3}{2c^3}\right) \quad (1.124)$$
$$\sigma_{12} = -\frac{3q}{4c} x_1 \left(1 - \frac{x_2^2}{c^2}\right)$$

Note that at $x_1 = L$, $\int_{-c}^{c} \sigma_{12} dx_2 = -qL$ and, at $x_1 = -L$, $\int_{-c}^{c} \sigma_{12} dx_2 = qL$ are automatically satisfied by equation (1.124). However, $\sigma_{11} = 0$ at $x_1 = L$ and $x_1 = -L$ will give additional conditions:

$$c_2 + d_3 x_2 + e_4 x_2^2 - \frac{3q}{4c^3} L^2 x_2 + f_5 x_2^3 = 0 \quad (1.125)$$

Therefore,

$$c_2 = 0$$
$$e_4 = 0$$
$$f_5 = 0$$
$$d_3 = \frac{3qL^2}{4c^3}$$

Thus, the final solution becomes

$$\sigma_{11} = \frac{3q}{4c^3} x_2 \left(L^2 - x_1^2\right)$$
$$\sigma_{22} = -\frac{q}{2}\left(1 - \frac{3x_2}{2c} + \frac{x_2^3}{2c^3}\right) \quad (1.126)$$
$$\sigma_{12} = -\frac{3q}{4c} x_1 \left(1 - \frac{x_2^2}{c^2}\right)$$

If, at the two end surfaces $x_1 = L$ and $x_1 = -L$, the $\sigma_{11} = 0$ condition cannot be explicitly satisfied for all $x_2$ values, but the following two conditions can be satisfied:

$$\int_{-c}^{c} \sigma_{11} dx_2 = 0$$
$$\int_{-c}^{c} \sigma_{11} x_2 dx_2 = 0 \qquad (1.127)$$

then this is also acceptable because the preceding two equations imply that the axial force and the bending moment at the two ends are equal to zero even if $\sigma_{11} \neq 0$ for all values of $x_2$. Since two equations given in equation (1.127) are statically equivalent to the $\sigma_{11} = 0$ case, from St. Venant's principle we can state that at a distance away from the two ends, equation (1.127) and the $\sigma_{11} = 0$ case will give identical solutions.

#### 1.3.4.2 Half-Plane Problems

The full space is extended to infinity in all directions and does not have any boundary. The half space is half of the full space. Therefore, the half space has a boundary at $x_j = c$, and is extended to infinity either in the $x_j > c$ or $x_j < c$ direction. Note that $c$ is a constant, and $j$ is 1, 2, or 3. For two-dimensional problems, the half space is known as the half plane. In other words, two-dimensional half spaces are called half planes. In this section, the solution steps are given for solving half-plane problems that are subjected to external loads applied at the boundary, as shown in Figure 1.17. The problem geometry and applied loads are independent of the out-of-plane or $x_3$ direction. Therefore, the solution should be also independent of $x_3$. This problem can be formulated in terms of three stress components: $\sigma_{11}(x_1, x_2)$, $\sigma_{22}(x_1, x_2)$, and $\sigma_{12}(x_1, x_2)$.

Boundary conditions for this problem are at $x_2 = 0$, $\sigma_{22} = -p(x_1)$, and $\sigma_{12} = 0$. In such problems where the problem geometry extends to infinity, one also needs to satisfy the conditions at infinity that are called *regularity conditions*. For this specific problem, the regularity condition is that all stress

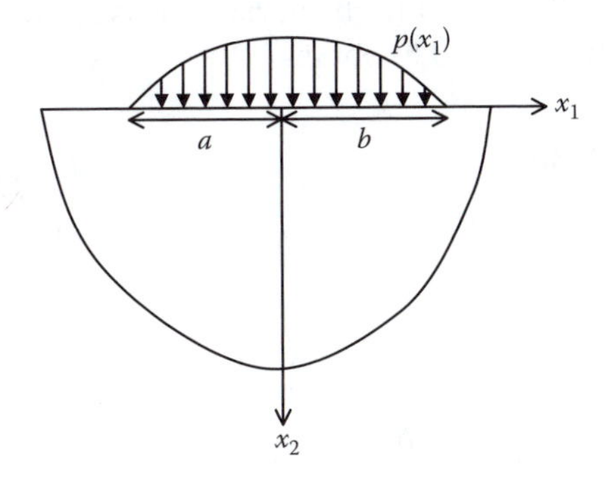


**FIGURE 1.17**
Half plane subjected to surface traction.

components must vanish [$\sigma_{11}(x_1, x_2) = \sigma_{22}(x_1, x_2) = \sigma_{12}(x_1, x_2) = 0$] as $x_1$ and $x_2$ go to infinity. The equilibrium equations ($\sigma_{\alpha\beta,\beta} = 0$; see equation 1.92) and the compatibility condition ($\sigma_{\alpha\alpha,\beta\beta} = 0$; see equation 1.98) constitute the governing equations of this problem.

The half-plane problem is solved by introducing the Airy stress function (see equation 1.101) that must satisfy the biharmonic equation (see equation 1.102). This partial differential equation can be solved by integral transform techniques such as the Fourier transform method.

#### 1.3.4.2.1 *Fundamentals of Fourier Transform*

Fourier transform of a function $g(x_1)$ is denoted by $G(k)$, where

$$G(k) = \Im(g(x_1)) = \int_{-\infty}^{\infty} g(x_1)e^{ikx_1}dx_1 \tag{1.128}$$

$g(x_1)$ is obtained from $G(k)$ by applying the inverse Fourier transform operation:

$$g(x_1) = \Im^{-1}(G(k)) = \frac{1}{2\pi}\int_{-\infty}^{\infty} G(k)e^{-ikx_1}dk \tag{1.129}$$

Fourier transforms of the derivatives of a function $g(x_1)$ are expressed in terms of its Fourier transform $G(k)$ in the following manner:

$$\begin{aligned} \Im\left(\frac{dg}{dx_1}\right) &= -ikG(k) \\ \Im\left(\frac{d^n g}{dx_1^n}\right) &= (-ik)^n G(k) \end{aligned} \tag{1.130}$$

#### 1.3.4.2.2 *Solution of Half-Plane Problems by Fourier Transform Technique*

Fourier transform with respect to the variable $x_1$ is applied to the governing compatibility equation $\nabla^4\phi = 0$ to obtain

$$\Im(\nabla^2\nabla^2\phi) = \Im\left(\left(\frac{\partial^2}{\partial x_1^2} + \frac{\partial^2}{\partial x_2^2}\right)\left(\frac{\partial^2}{\partial x_1^2} + \frac{\partial^2}{\partial x_2^2}\right)\phi\right) = \left(-k^2 + \frac{\partial^2}{\partial x_2^2}\right)\left(-k^2 + \frac{\partial^2}{\partial x_2^2}\right)\Phi = 0$$

$$\Rightarrow \frac{\partial^4\Phi}{\partial x_2^4} - 2k^2\frac{\partial^2\Phi}{\partial x_2^2} + k^4\Phi = 0 \tag{1.131}$$

In equation (1.131) $\Phi(k, x_2)$ is the Fourier transform of $\phi(x_1, x_2)$. Solution of equation (1.131) is given by

$$\Phi(k, x_2) = ae^{kx_2} + be^{-kx_2} + cx_2e^{kx_2} + dx_2e^{-kx_2} \tag{1.132}$$

Alternately, the solution can be expressed as

$$\Phi(k,x_2) = Ae^{|k|x_2} + Be^{-|k|x_2} + Cx_2e^{|k|x_2} + Dx_2e^{-|k|x_2} \tag{1.133}$$

Comparing equations (1.132) and (1.133), it can be concluded that $A = a$ for positive $k$ and $A = b$ for negative $k$. Similarly, $B$, $C$, and $D$ can be defined.

In the Fourier transformed domain the three stress components can be expressed as

$$\begin{aligned}
&\Im(\sigma_{11} = \phi_{22}) \Rightarrow \quad \Sigma_{11} = \frac{\partial^2\Phi}{\partial x_2^2} \\
&\Im(\sigma_{22} = \phi_{11}) \Rightarrow \quad \Sigma_{22} = -k^2\Phi \\
&\Im(\sigma_{12} = -\phi_{12}) \Rightarrow \quad \Sigma_{12} = ik\frac{\partial\Phi}{\partial x_2}
\end{aligned} \tag{1.134}$$

From equation (1.134) and the regularity condition ($\Sigma_{11}$, $\Sigma_{22}$, and $\Sigma_{12}$ should approach zero as $x_2 \to \infty$), one can conclude that $\Phi$ and its first and second derivatives with respect to $x_2$ must vanish as $x_2$ approaches infinity. Therefore, constants $A$ and $C$ of equation (1.133) must be equal to zero. Thus, after applying the regularity condition, equation (1.133) is simplified to

$$\Phi(k,x_2) = Be^{-|k|x_2} + Dx_2e^{-|k|x_2} \tag{1.135}$$

Applying Fourier transform to the boundary conditions at $x_2 = 0$, one obtains

$$\begin{aligned}
&\Im(\sigma_{22} = \phi_{11} = -p(x_1)) \Rightarrow \quad \Sigma_{22} = -k^2\Phi(k,0) = -P(k) \Rightarrow \quad \Phi(k,0) = \frac{P(k)}{k^2} \\
&\Im(\sigma_{12} = -\phi_{12} = 0) \Rightarrow \quad \Sigma_{12} = ik\frac{\partial\Phi}{\partial x_2} = 0 \Rightarrow \quad \left.\frac{\partial\Phi(k,x_2)}{\partial x_2}\right|_{x_2=0} = 0
\end{aligned} \tag{1.136}$$

where

$$P(k) = \Im(p(x_1)) = \int_{-\infty}^{\infty} p(x_1)e^{ikx_1}dx_1 = \int_{-a}^{b} p(x_1)e^{ikx_1}dx_1 \tag{1.137}$$

From equations (1.135) and (1.136), one gets

$$\begin{aligned}
B &= \frac{P(k)}{k^2} \\
D &= \frac{P(k)}{|k|}
\end{aligned} \tag{1.138}$$

Substituting equation (1.138) into equation (1.135), $\Phi(k, x_2)$ is obtained:

$$\Phi(k,x_2)=\frac{P(k)}{k^2}e^{-|k|x_2}+\frac{P(k)}{|k|}x_2e^{-|k|x_2}=P(k)\left(\frac{1}{k^2}+\frac{x_2}{|k|}\right)e^{-|k|x_2} \tag{1.139}$$

Substituting equation (1.139) into equation (1.134), the three stress components in the Fourier transformed domain are obtained:

$$\begin{aligned}
\Sigma_{11}&=\frac{\partial^2\Phi}{\partial x_2^2}=-P(k)[1-|k|\,x_2]e^{-|k|x_2}\\
\Sigma_{22}&=-k^2\Phi=-P(k)[1+|k|\,x_2]e^{-|k|x_2}\\
\Sigma_{12}&=ik\frac{\partial\Phi}{\partial x_2}=-iP(k)kx_2e^{-|k|x_2}
\end{aligned} \tag{1.140}$$

To obtain the stress field in the $x_1$–$x_2$ space, one must take the inverse transform of the preceding equations:

$$\begin{aligned}
\sigma_{11}(x_1,x_2)&=\Im^{-1}(\Sigma_{11})=-\frac{1}{2\pi}\int_{-\infty}^{\infty}P(k)\big[1-|k|\,x_2\big]e^{-|k|x_2-ikx_1}dk\\
\sigma_{22}(x_1,x_2)&=\Im^{-1}(\Sigma_{22})=-\frac{1}{2\pi}\int_{-\infty}^{\infty}P(k)\big[1+|k|\,x_2\big]e^{-|k|x_2-ikx_1}dk\\
\sigma_{12}(x_1,x_2)&=\Im^{-1}(\Sigma_{12})=-\frac{i}{2\pi}\int_{-\infty}^{\infty}P(k)kx_2e^{-|k|x_2-ikx_1}dk
\end{aligned} \tag{1.141}$$

#### 1.3.4.2.3 *Solution of the Half-Plane Problem for a Uniform Load over a Finite Region*

If the applied load in Figure 1.17 has a constant value of $p_0$ and is applied in the region extending from $x_1 = -a$ to $x_1 = a$ (see Figure 1.18), then the Fourier transform of the applied load is given by

$$P(k)=\Im(p(x_1))=\int_{-\infty}^{\infty}p(x_1)e^{ikx_1}dx_1=\int_{-a}^{a}p_0e^{ikx_1}dx_1=\frac{p_0}{ik}[e^{ikx_1}]_{-a}^{a}=2p_0\frac{\sin(ka)}{k} \tag{1.142}$$

Note that here $p(x_1)$ is an even function of $x_1$ and $P(k)$ is an even function of $k$. When $P(k)$ is an even function of $k$, then the inverse transform can be expressed in the form of a semi-infinite integral instead of an infinite integral as follows:

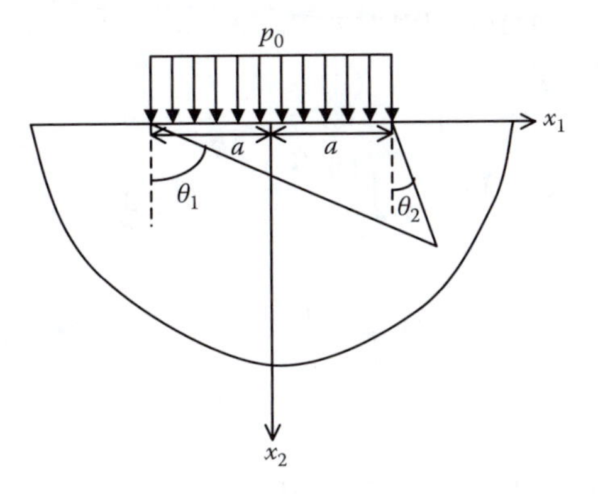


**FIGURE 1.18**
Half plane subjected to uniform traction over a finite region.

$$p(x_1) = \Im^{-1}(P(k)) = \frac{1}{2\pi}\int_{-\infty}^{\infty} P(k)e^{-ikx_1}dk = \frac{1}{2\pi}\left(\int_{-\infty}^{0} P(k)e^{-ikx_1}dk + \int_{0}^{\infty} P(k)e^{-ikx_1}dk\right)$$

$$= \frac{1}{2\pi}\left(-\int_{\infty}^{0} P(-k)e^{ikx_1}dk + \int_{0}^{\infty} P(k)e^{-ikx_1}dk\right) = \frac{1}{2\pi}\left(\int_{0}^{\infty} P(-k)e^{ikx_1}dk + \int_{0}^{\infty} P(k)e^{-ikx_1}dk\right)$$

$$= \frac{1}{2\pi}\left(\int_{0}^{\infty} P(k)(e^{ikx_1} + e^{-ikx_1})dk\right) = \frac{1}{\pi}\int_{0}^{\infty} P(k)\cos(kx_1)dk \tag{1.143}$$

Similarly, the direct transform can be expressed in the form of a semi-infinite integral when $p(x_1)$ is an even function of $x_1$:

$$P(k) = \Im(p(x_1)) = \int_{-\infty}^{\infty} p(x_1)e^{ikx_1}dx_1 = 2\int_{0}^{\infty} p(x_1)\cos(kx_1)dx_1 \tag{1.144}$$

When $p(x_1)$ is an odd function of $x_1$, then the direct and inverse transforms take the following forms:

$$P(k) = \Im(p(x_1)) = \int_{-\infty}^{\infty} p(x_1)e^{ikx_1}dx_1 = 2i\int_{0}^{\infty} p(x_1)\sin(kx_1)dx_1$$

$$p(x_1) = \Im^{-1}(P(k)) = -\frac{i}{\pi}\int_{0}^{\infty} P(k)\sin(kx_1)dk \tag{1.145}$$

From equations (1.141) and (1.142), the stress fields can be obtained in the form of infinite integrals

$$\sigma_{11}(x_1,x_2) = -\frac{p_0}{\pi}\int_{-\infty}^{\infty}\frac{\sin(ka)}{k}\left[1-|k|\,x_2\right]e^{-|k|x_2-ikx_1}dk$$

$$\sigma_{22}(x_1,x_2) = -\frac{p_0}{\pi}\int_{-\infty}^{\infty}\frac{\sin(ka)}{k}\left[1+|k|\,x_2\right]e^{-|k|x_2-ikx_1}dk \tag{1.146}$$

$$\sigma_{12}(x_1,x_2) = -\frac{ip_0}{\pi}\int_{-\infty}^{\infty}x_2\sin(ka)e^{-|k|x_2-ikx_1}dk$$

Alternately, recognizing that the problem geometry and the applied load are symmetric with respect to the $x_2$ axis, it is concluded that the generated stress field should be symmetric also. Therefore, the normal stresses $\sigma_{11}$, $\sigma_{22}$ must be an even function of $x_1$ while the shear stress $\sigma_{12}$ should be an odd function of $x_1$. Using the transformation laws (equations 1.143–1.145) for even and odd functions, the stress fields can be expressed in terms of the semi-infinite integrals:

$$\sigma_{11}(x_1,x_2) = -\frac{p_0}{\pi}\int_0^{\infty}\frac{1}{k}[1+kx_2]\{\sin[k(x_1+a)]-\sin[k(x_1-a)]\}e^{-kx_2}dk$$

$$\sigma_{22}(x_1,x_2) = -\frac{p_0}{\pi}\int_0^{\infty}\frac{1}{k}[1+kx_2]\{\sin[k(x_1+a)]-\sin[k(x_1-a)]\}e^{-kx_2}dk \tag{1.147}$$

$$\sigma_{12}(x_1,x_2) = \frac{p_0}{\pi}\int_0^{\infty}x_2\{\cos[k(x_1+a)]-\cos[k(x_1-a)]\}e^{-kx_2}dk$$

After integration, the stress field takes the following form:

$$\sigma_{11} = -\frac{p_0}{\pi}\left\{(\theta_1-\theta_2)-\frac{1}{2}[\sin(2\theta_1)-\sin(2\theta_2)]\right\}$$

$$\sigma_{22} = -\frac{p_0}{\pi}\left\{(\theta_1-\theta_2)+\frac{1}{2}[\sin(2\theta_1)-\sin(2\theta_2)]\right\} \tag{1.148}$$

$$\sigma_{12} = -\frac{p_0}{2\pi}[\sin(2\theta_1)-\sin(2\theta_2)]$$

where angles $\theta_1$ and $\theta_2$ are shown in Figure 1.18.

For the stress state given in equation (1.148), it is possible to obtain the maximum shear stress at a rotated coordinate system $x_{1'}x_{2'}$ using the stress transformation law (or from the Mohr's circle for this two-dimensional stress state):

$$\begin{aligned}
\sigma_{1'2'}\big|_{\max} &= \tau_{\max} = \left\{\left(\frac{\sigma_{11}-\sigma_{22}}{2}\right)^2 + \sigma_{12}^2\right\}^{\frac{1}{2}} \\
&= \frac{p_0}{\pi}\left\{\left(\frac{\sin 2\theta_1 - \sin 2\theta_2}{2}\right)^2 + \left(\frac{\cos 2\theta_1 - \cos 2\theta_2}{2}\right)^2\right\}^{\frac{1}{2}} \\
&= \frac{p_0}{2\pi}\left\{\begin{matrix}\sin^2 2\theta_1 + \sin^2 2\theta_2 - 2\sin 2\theta_1 \sin 2\theta_2 + \cos^2 2\theta_1 \\ + \cos^2 2\theta_2 - 2\cos 2\theta_1 \cos 2\theta_2\end{matrix}\right\}^{\frac{1}{2}} \\
&= \frac{p_0}{2\pi}\{2 - 2(\sin 2\theta_1 \sin 2\theta_2 + \cos 2\theta_1 \cos 2\theta_2)\}^{\frac{1}{2}} = \frac{p_0\sqrt{2}}{2\pi}\{1 - \cos 2(\theta_1 - \theta_2)\}^{\frac{1}{2}} \\
&= \frac{p_0\sqrt{2}}{2\pi}\{2\sin^2(\theta_1 - \theta_2)\}^{\frac{1}{2}} = \frac{p_0}{\pi}\sin(\theta_1 - \theta_2)
\end{aligned} \tag{1.149}$$

Note that $\tau_{\max} = \frac{p_0}{\pi}\sin(\theta_1 - \theta_2) = C$ (where $C$ is a constant) corresponds to all points on a circle going through the points $x_1 = a$ and $x_1 = -a$. For different values of $C$, different circles are obtained. Among all these maximum shear stress contours, the absolute maximum shear stress value is $\frac{p_0}{\pi}$ for $\sin(\theta_1 - \theta_2) = 1$.

#### 1.3.4.2.4 *Half Plane Subjected to a Concentrated Load*

When the half plane is subjected to a concentrated load of magnitude $P_0$ as shown in Figure 1.19, the solution field for this problem can be obtained from the preceding expressions.

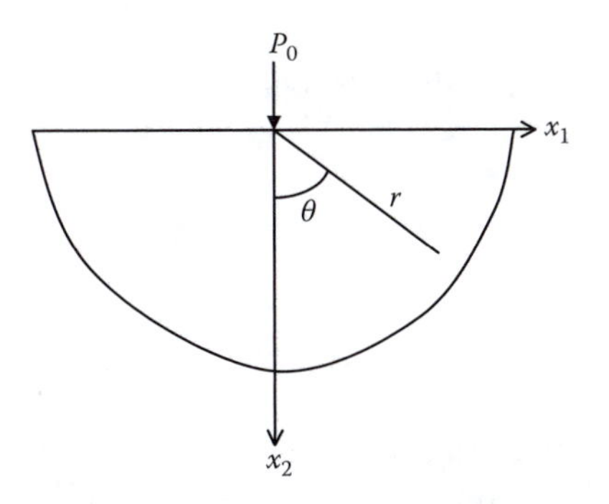


**FIGURE 1.19**
Half plane subjected to a concentrated force.

Applied load distribution at the $x_2 = 0$ boundary is given by $p(x_1) = P_0\delta(x_1)$, where $\delta(x_1)$ is the Dirac delta function. Therefore, its Fourier transform is

$$P(k) = \int_{-\infty}^{\infty} P_0\delta(x_1)e^{ikx_1}dx_1 = P_0$$

and, from equation (1.141),

$$\begin{aligned}
\sigma_{11}(x_1,x_2) &= -\frac{P_0}{2\pi}\int_{-\infty}^{\infty}[1-|k|\,x_2]e^{-|k|x_2-ikx_1}dk \\
\sigma_{22}(x_1,x_2) &= -\frac{P_0}{2\pi}\int_{-\infty}^{\infty}[1+|k|\,x_2]e^{-|k|x_2-ikx_1}dk \\
\sigma_{12}(x_1,x_2) &= -\frac{iP_0}{2\pi}\int_{-\infty}^{\infty}kx_2e^{-|k|x_2-ikx_1}dk
\end{aligned} \tag{1.150}$$

or

$$\begin{aligned}
\sigma_{11}(x_1,x_2) &= -\frac{2P_0}{\pi}\frac{x_1^2x_2}{(x_1^2+x_2^2)^2} \\
\sigma_{22}(x_1,x_2) &= -\frac{2P_0}{\pi}\frac{x_2^3}{(x_1^2+x_2^2)^2} \\
\sigma_{12}(x_1,x_2) &= -\frac{2P_0}{\pi}\frac{x_1x_2^2}{(x_1^2+x_2^2)^2}
\end{aligned} \tag{1.151}$$

In the polar coordinate system (if $\theta$ is measured from the vertical axis as shown in Figure 1.19), the stress field is given by

$$\begin{aligned}
\sigma_{rr} &= -\frac{2P\cos\theta}{\pi r} \\
\sigma_{r\theta} &= \sigma_{\theta\theta} = 0
\end{aligned} \tag{1.152}$$

It should be noted here that if we know the point force solution (equation 1.151; Figure 1.19), then the solution for a general load distribution $p(x_1)$, applied on the boundary from $x_1 = -a$ to $x_1 = b$, can be obtained simply by linear superposition of point force solutions. Thus, the stress field for this general load distribution $p(x_1)$ on the boundary is given by

$$\sigma_{11}(x_1,x_2) = -\frac{2}{\pi}\int_{-a}^{b}\frac{p(k)(x_1-k)^2 x_2}{[(x_1-k)^2+x_2^2]^2}dk$$

$$\sigma_{22}(x_1,x_2) = -\frac{2}{\pi}\int_{-a}^{b}\frac{p(k)x_2^3}{\left[(x_1-k)^2+x_2^2\right]^2}dk \tag{1.153}$$

$$\sigma_{12}(x_1,x_2) = -\frac{2}{\pi}\int_{-a}^{b}\frac{p(k)(x_1-k)x_2^2}{[(x_1-k)^2+x_2^2]^2}dk$$

#### *1.3.4.3 Circular Hole, Disk, and Cylindrical Pressure Vessel Problems*

Problem geometries such as a circular hole in an infinite sheet, disks with or without concentric holes, and cylindrical vessels subjected to surface tractions are solved in this section. These problems can be solved relatively easily by the complex variable technique shown next.

##### *1.3.4.3.1 Compatibility Equation in Terms of Complex Variables*

A complex variable $z$ and its conjugate $\bar{z}$ are defined in terms of the real part $x$ and imaginary part $y$ in the following manner:

$$z = x + iy$$
$$\bar{z} = x - iy \tag{1.154}$$

Therefore, $x$ and $y$ can be written in terms of $z$ and $\bar{z}$ as

$$x = \frac{1}{2}(z+\bar{z})$$
$$y = \frac{1}{2i}(z-\bar{z}) \tag{1.155}$$

Substituting the preceding expressions of $x$ and $y$ in the Airy function $\phi(x, y)$ (described in section 1.3.3), one obtains the stress function in terms of the complex variable $z$ and its conjugate $\bar{z}$. For notational simplicity in this section, $x_1$ and $x_2$ are substituted by $x$ and $y$, respectively. $\phi(x, y)$ and $\Phi(z,\bar{z})$ represent the same Airy function in terms of two different sets of variables.

Therefore,

$$\frac{\partial\phi(x,y)}{\partial x}=\frac{\partial}{\partial x}\Phi(z,\bar{z})=\frac{\partial\Phi}{\partial z}\frac{\partial z}{\partial x}+\frac{\partial\Phi}{\partial\bar{z}}\frac{\partial\bar{z}}{\partial x}=\frac{\partial\Phi}{\partial z}+\frac{\partial\Phi}{\partial\bar{z}}=\left(\frac{\partial}{\partial z}+\frac{\partial}{\partial\bar{z}}\right)\Phi$$

$$\frac{\partial\phi(x,y)}{\partial y}=\frac{\partial}{\partial y}\Phi(z,\bar{z})=\frac{\partial\Phi}{\partial z}\frac{\partial z}{\partial y}+\frac{\partial\Phi}{\partial\bar{z}}\frac{\partial\bar{z}}{\partial y}=i\left(\frac{\partial\Phi}{\partial z}-\frac{\partial\Phi}{\partial\bar{z}}\right)=i\left(\frac{\partial}{\partial z}-\frac{\partial}{\partial\bar{z}}\right)\Phi$$

$$\frac{\partial^2\phi(x,y)}{\partial x^2}=\left(\frac{\partial}{\partial z}+\frac{\partial}{\partial\bar{z}}\right)\left(\frac{\partial}{\partial z}+\frac{\partial}{\partial\bar{z}}\right)\Phi=\left(\frac{\partial^2\Phi}{\partial z^2}+2\frac{\partial^2\Phi}{\partial z\partial\bar{z}}+\frac{\partial^2\Phi}{\partial\bar{z}^2}\right)$$

$$\frac{\partial^2\phi(x,y)}{\partial y^2}=-\left(\frac{\partial}{\partial z}-\frac{\partial}{\partial\bar{z}}\right)\left(\frac{\partial}{\partial z}-\frac{\partial}{\partial\bar{z}}\right)\Phi=\left(-\frac{\partial^2\Phi}{\partial z^2}+2\frac{\partial^2\Phi}{\partial z\partial\bar{z}}-\frac{\partial^2\Phi}{\partial\bar{z}^2}\right) \tag{1.156}$$

$$\frac{\partial^2\phi(x,y)}{\partial x\partial y}=i\left(\frac{\partial}{\partial z}+\frac{\partial}{\partial\bar{z}}\right)\left(\frac{\partial}{\partial z}-\frac{\partial}{\partial\bar{z}}\right)\Phi=i\left(\frac{\partial^2}{\partial z^2}-\frac{\partial^2}{\partial\bar{z}^2}\right)\Phi$$

$$\therefore\nabla^2\phi(x,y)=\frac{\partial^2\phi(x,y)}{\partial x^2}+\frac{\partial^2\phi(x,y)}{\partial y^2}=4\frac{\partial^2\Phi(z,\bar{z})}{\partial z\partial\bar{z}}$$

$$\therefore\nabla^4\phi(x,y)=16\frac{\partial^4\Phi(z,\bar{z})}{\partial z^2\partial\bar{z}^2}$$

From equation (1.102) it is known that the compatibility condition implies that the Airy stress function must be biharmonic. Therefore, the compatibility condition takes the following form:

$$\nabla^4\phi(x,y)=16\frac{\partial^4\Phi(z,\bar{z})}{\partial z^2\partial\bar{z}^2}=0 \tag{1.157}$$

*1.3.4.3.2 Stress Fields in Terms of Complex Potential Functions*

Note that if $\chi_1$, $\chi_2$, $\chi_3$, and $\chi_4$ are four analytic (or differentiable) functions of $z$ or $\bar{z}$, then the following $\Phi(z,\bar{z})$ should satisfy equation (1.157):

$$\Phi(z,\bar{z})=z\chi_1(\bar{z})+\bar{z}\chi_2(z)+\chi_3(z)+\chi_4(\bar{z})=z\bar{\chi}_1(z)+\bar{z}\chi_2(z)+\chi_3(z)+\bar{\chi}_4(z) \tag{1.158}$$

Since Airy stress function must be a real function, one can write

$$\chi_1(z)=\chi_2(z)=\frac{1}{2}\phi(z)$$
$$\chi_3(z)=\chi_4(z)=\frac{1}{2}\psi(z) \tag{1.159}$$

Substituting equation (1.159) into equation (1.158), one gets

$$\Phi(z,\bar{z}) = z\chi_1(\bar{z}) + \bar{z}\chi_2(z) + \chi_3(z) + \chi_4(\bar{z}) = \frac{1}{2}[z\phi(\bar{z}) + \bar{z}\phi(z) + \psi(z) + \psi(\bar{z})] \quad (1.160)$$

Therefore,

$$\frac{\partial\Phi(z,\bar{z})}{\partial z} = \frac{1}{2}\left[\phi(\bar{z}) + \bar{z}\frac{\partial\phi(z)}{\partial z} + \frac{\partial\psi(z)}{\partial z}\right] = \frac{1}{2}[\phi(\bar{z}) + \bar{z}\phi'(z) + \psi'(z)]$$

$$\frac{\partial^2\Phi(z,\bar{z})}{\partial z^2} = \frac{1}{2}[\bar{z}\phi''(z) + \psi''(z)] \quad (1.161)$$

$$\frac{\partial^2\Phi(z,\bar{z})}{\partial z\partial\bar{z}} = \frac{1}{2}[\phi'(\bar{z}) + \phi'(z)]$$

From equations (1.101), (1.156), and (1.161), one obtains

$$\sigma_{xx} + \sigma_{yy} = \frac{\partial^2\phi}{\partial y^2} + \frac{\partial^2\phi}{\partial x^2} = 4\frac{\partial^2\Phi}{\partial z\partial\bar{z}} = 2[\phi'(z) + \phi'(\bar{z})] = 2[\phi'(z) + \bar{\phi}'(z)] \quad (1.162)$$

and

$$\sigma_{yy} - \sigma_{xx} + 2i\sigma_{xy} = \frac{\partial^2\phi}{\partial x^2} - \frac{\partial^2\phi}{\partial y^2} - 2i\frac{\partial^2\phi}{\partial x\partial y}$$

$$= \left(\frac{\partial^2\Phi}{\partial z^2} + 2\frac{\partial^2\Phi}{\partial z\partial\bar{z}} + \frac{\partial^2\Phi}{\partial\bar{z}^2}\right) - \left(-\frac{\partial^2\Phi}{\partial z^2} + 2\frac{\partial^2\Phi}{\partial z\partial\bar{z}} - \frac{\partial^2\Phi}{\partial\bar{z}^2}\right) + 2\left(\frac{\partial^2\Phi}{\partial z^2} - \frac{\partial^2\Phi}{\partial\bar{z}^2}\right) \quad (1.163)$$

$$= 4\frac{\partial^2\Phi}{\partial z^2} = 2[\bar{z}\phi''(z) + \psi''(z)]$$

From the stress transformation law it is possible to show that the stress fields in the $(r, \theta)$ polar coordinate system are given by

$$\sigma_{rr} + \sigma_{\theta\theta} = \sigma_{xx} + \sigma_{yy} = 2[\phi'(z) + \phi'(\bar{z})] = 2[\phi'(z) + \bar{\phi}'(z)] \quad (1.164)$$

$$\sigma_{\theta\theta} - \sigma_{rr} + 2i\sigma_{r\theta} = [\sigma_{yy} - \sigma_{xx} + 2i\sigma_{xy}]e^{2i\theta} = 2[\bar{z}\phi''(z) + \psi''(z)]e^{2i\theta} \quad (1.165)$$

Subtracting equation (1.165) from equation (1.164), one obtains

$$\sigma_{rr} - i\sigma_{r\theta} = \phi'(z) + \bar{\phi}'(z) - [\bar{z}\phi''(z) + \psi''(z)]e^{2i\theta} \quad (1.166)$$

Let us specialize the expressions given in equations (1.164) to (1.166) for the following two special cases: (a) $\phi'(z) = A_n z^n$, $\psi''(z) = 0$, and (b) $\psi''(z) = B_m z^m$, $\phi'(z) = 0$.

***Stress fields for $\phi'(z) = A_n z^n$, $\psi''(z) = 0$:***

$$\phi'(z) = A_n z^n = A_n r^n e^{in\theta}$$

Therefore, if $A_n$ is a real number,

$$\begin{aligned}
\sigma_{rr} + \sigma_{\theta\theta} &= 2[\phi'(z) + \bar{\phi}'(z)] = 2A_n r^n [e^{in\theta} + e^{-in\theta}] = 4A_n r^n \cos n\theta \\
\sigma_{\theta\theta} - \sigma_{rr} + 2i\sigma_{r\theta} &= 2[\bar{z}\phi''(z) + \psi''(z)]e^{2i\theta} \\
&= 2[re^{-i\theta} nA_n r^{n-1} e^{i(n-1)\theta}]e^{2i\theta} = 2nA_n r^n e^{in\theta} \\
\sigma_{rr} - i\sigma_{r\theta} &= \phi'(z) + \bar{\phi}'(z) - [\bar{z}\phi''(z) + \psi''(z)]e^{2i\theta} \\
&= A_n r^n [e^{in\theta} + e^{-in\theta}] - nA_n r^n e^{in\theta} = A_n r^n [(1-n)e^{in\theta} + e^{-in\theta}]
\end{aligned} \tag{1.167}$$

If $A_n$ is an imaginary number,

$$\begin{aligned}
\sigma_{rr} + \sigma_{\theta\theta} &= 2A_n r^n [e^{in\theta} - e^{-in\theta}] = 4iA_n r^n \sin n\theta \\
\sigma_{\theta\theta} - \sigma_{rr} + 2i\sigma_{r\theta} &= 2[re^{-i\theta} nA_n r^{n-1} e^{i(n-1)\theta}]e^{2i\theta} = 2nA_n r^n e^{in\theta} \\
\sigma_{rr} - i\sigma_{r\theta} &= \phi'(z) + \bar{\phi}'(z) - [\bar{z}\phi''(z) + \psi''(z)]e^{2i\theta} \\
&= A_n r^n [e^{in\theta} - e^{-in\theta}] - nA_n r^n e^{in\theta} = A_n r^n [(1-n)e^{in\theta} - e^{-in\theta}]
\end{aligned} \tag{1.168}$$

Equations (1.167) and (1.168) can be combined to obtain

$$\begin{aligned}
\sigma_{rr} + \sigma_{\theta\theta} &= 2A_n r^n [e^{in\theta} \pm e^{-in\theta}] \\
\sigma_{\theta\theta} - \sigma_{rr} + 2i\sigma_{r\theta} &= 2nA_n r^n e^{in\theta} \\
\sigma_{rr} - i\sigma_{r\theta} &= A_n r^n [(1-n)e^{in\theta} \pm e^{-in\theta}]
\end{aligned} \tag{1.169}$$

In this equation, the plus (+) sign is for the case when $A_n$ is real and the minus (–) sign is for imaginary $A_n$.

***Stress fields for $\psi''(z) = B_m z^m$, $\theta'(z) = 0$:***

For the case $\psi''(z) = B_m z^m = B_m r^m e^{im\theta}$ and $\phi'(z) = 0$,

$$\begin{aligned}
\sigma_{rr} + \sigma_{\theta\theta} &= 2[\phi'(z) + \bar{\phi}'(z)] = 0 \\
\sigma_{\theta\theta} - \sigma_{rr} + 2i\sigma_{r\theta} &= 2[\bar{z}\phi''(z) + \psi''(z)]e^{2i\theta} = 2[B_m r^m e^{im\theta}]e^{2i\theta} = 2B_m r^m e^{i(m+2)\theta} \\
\sigma_{rr} - i\sigma_{r\theta} &= \phi'(z) + \bar{\phi}'(z) - [\bar{z}\phi''(z) + \psi''(z)]e^{2i\theta} = -B_m r^m e^{i(m+2)\theta}
\end{aligned} \tag{1.170}$$

Note that equation (1.170) is valid for both real and imaginary values of $B_m$.

**Example 1.12**

Obtain the complex potential functions $\theta'(z)$ and $\psi''(z)$, given in equations (1.162) and (1.163), for a plate under uniaxial tension $\sigma_{yy} = \sigma_0$, $\sigma_{xx} = \sigma_{xy} = 0$.

*Solution*

From equations (1.162) and (1.163),

$$2[\phi'(z)+\bar{\phi}'(z)]=\sigma_{xx}+\sigma_{yy}=\sigma_0$$

$$2[\bar{z}\phi''(z)+\psi''(z)]=\sigma_{yy}-\sigma_{xx}+2i\sigma_{xy}=\sigma_0$$

Note that these two equations are satisfied when

$$\phi'(z)=\frac{\sigma_0}{4},\quad \psi''(z)=\frac{\sigma_0}{2} \tag{1.171}$$

**Example 1.13**

An infinite plate with a circular hole of radius $a$ is subjected to a biaxial state of stress $\sigma_{xx}=\sigma_{yy}=\sigma_0$ at the far field (far away from the hole) as shown in Figure 1.20. Obtain the stress field in the entire plate.

*Solution*

The original problem can be decomposed into two problems (I and II) as shown at the bottom of Figure 1.20.

Solution of problem I: Solution of problem I is straightforward: $\sigma_{xx}=\sigma_{yy}=\sigma_0$ and $\sigma_{xy}=0$.

From equations (1.162) and (1.163),

$$2[\phi'(z)+\bar{\phi}'(z)]=\sigma_{xx}+\sigma_{yy}=2\sigma_0$$

$$2[\bar{z}\phi''(z)+\psi''(z)]=\sigma_{yy}-\sigma_{xx}+2i\sigma_{xy}=0$$

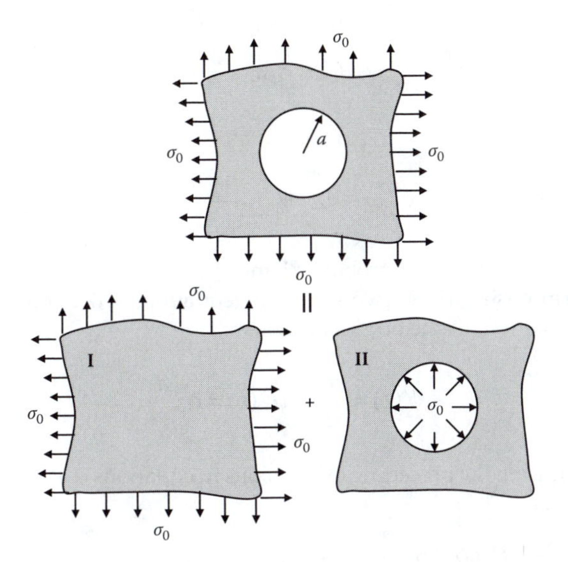


**FIGURE 1.20**
Infinite plate with a circular hole subjected to a biaxial state of stress at the far field (top figure) is decomposed into two problems (I and II) as shown at the bottom.

These two equations are satisfied when

$$\phi'(z) = \frac{\sigma_0}{2}, \quad \psi''(z) = 0 \tag{1.172}$$

Solution of problem II: For problem II the boundary condition is given by

$$\sigma_{rr} = -\sigma_0 = \sigma_{r\theta} = 0 \text{ at } r = a, \qquad \text{or} \qquad \sigma_{rr} = -i\sigma_{r\theta} = -\sigma_0 = 0 \text{ at } r = a.$$

Regularity conditions (conditions at infinity) are given by

$$\sigma_{rr}, \sigma_{\theta\theta}, \sigma_{r\theta} \to 0 \quad \text{as } r \to \infty.$$

Note that both boundary and regularity conditions are independent of $\theta$. Therefore, the solution should be independent of $\theta$ also. Keeping only the $\theta$ independent terms of equations (1.169) and (1.170), we can construct the complex potential functions $\theta'(z)$ and $\psi''(z)$ for problem II:

$$\phi'(z) = A_0, \quad \psi''(z) = B_{-2}z^{-2}$$

Since $\phi'(z) = A_0$ does not give a decaying stress field as $r$ increases, it violates the regularity condition. Therefore, $A_0$ must be equal to zero. Keeping only the $\psi''(z)$ term, the boundary condition takes the following form (see equation 1.170):

$$\sigma_{rr} - i\sigma_{r\theta} = -\sigma_0 = -B_{-2}a^{-2}$$

$$\therefore B_{-2} = \sigma_0 a^2$$

Therefore,

$$\phi'(z) = 0, \quad \psi''(z) = \sigma_0 \frac{a^2}{z^2} \tag{1.173}$$

Combined solution of problems I and II:

Superimposing these two solutions (equations 1.172 and 1.173), one gets the solution for the original problem:

$$\phi'(z) = \frac{\sigma_0}{2}, \quad \psi''(z) = \sigma_0 \frac{a^2}{z^2} \tag{1.174}$$

Substituting the preceding expressions in equations (1.167) and (1.170), one obtains

$$\sigma_{rr} + \sigma_{\theta\theta} = 4A_n r^n \cos n\theta = 4A_0 = 2\sigma_0$$

$$\sigma_{rr} - i\sigma_{r\theta} = \phi'(z) + \bar{\phi}'(z) - [\bar{z}\phi''(z) + \psi''(z)]e^{2i\theta} = \sigma_0 - \sigma_0 a^2 r^{-2} = \sigma_0\left(1 - \frac{a^2}{r^2}\right)$$

From these two equations it is easy to see that

$$\sigma_{rr} = \sigma_0\left(1 - \frac{a^2}{r^2}\right)$$

$$\sigma_{\theta\theta} = \sigma_0\left(1 + \frac{a^2}{r^2}\right)$$

$$\sigma_{r\theta} = 0$$

Note that at the circular boundary ($r = a$), $\sigma_{\theta\theta} = 2\sigma_0$. Therefore, the hole increases the normal stress value by a factor of 2. This factor is called the stress concentration factor.

**Example 1.14**

An infinite plate with a circular hole of radius $a$ is subjected to uniaxial state of stress $\sigma_{yy} = \sigma_0$ at the far field (far away from the hole) as shown in Figure 1.21. Compute the stress field in the entire plate.

*Solution*

The original problem can be decomposed into two problems (I and II) as shown at the bottom of Figure 1.21. The stress field at the circular boundary of problem II is obtained after solving problem I.

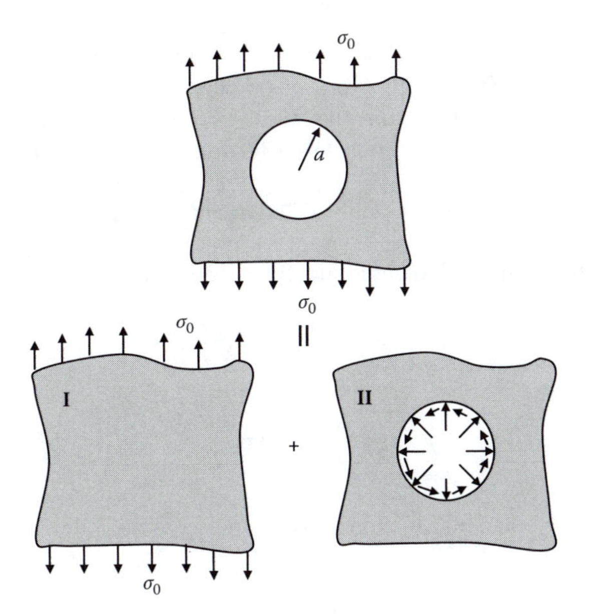


**FIGURE 1.21**
An infinite plate with a circular hole subjected to the uniaxial state of stress at the far field (top figure) is decomposed into two problems (I and II) as shown at the bottom.

Solution of problem I: Problem I has been solved in example 1.12; see equation (1.171):

$$\phi'(z) = \frac{\sigma_0}{4}, \quad \psi''(z) = \frac{\sigma_0}{2} \tag{1.175}$$

After knowing the potential functions $\phi''(z)$ and $\psi''(z)$, the stress field at the circular boundary ($r = a$) is obtained from equations (1.167) and (1.170):

$$\sigma_{rr} - i\sigma_{r\theta} = \phi'(z) + \bar{\phi}'(z) - [\bar{z}\phi''(z) + \psi''(z)]e^{2i\theta} = A_n r^n[(1-n)e^{in\theta} + e^{-in\theta}]$$

$$- B_m r^m e^{i(2+m)\theta} = 2A_0 - B_0 e^{2i\theta} = \frac{\sigma_0}{2} - \frac{\sigma_0}{2}e^{2i\theta} = \frac{\sigma_0}{2}(1 - e^{2i\theta})$$

Solution of problem II: From the preceding equation, one can clearly see that the boundary stresses at $r = a$ are not zero in problem I. To make the boundary stresses zero at $r = a$ in the original problem, the following stress field is assigned in problem II as the boundary condition:

$$\sigma_{rr} - i\sigma_{r\theta} = -\frac{\sigma_0}{2}(1 - e^{2i\theta}) = -\frac{\sigma_0}{2} + \frac{\sigma_0}{2}e^{2i\theta} \qquad \text{at } r = a$$

Regularity conditions (conditions at infinity) for the preceding applied stress field are

$$\sigma_{rr}, \sigma_{\theta\theta}, \sigma_{r\theta} \to 0 \quad \text{as } r \to \infty$$

Note that the boundary conditions have two terms; one is $\theta$ independent term and the second one has $\theta$ dependence in the form of $e^{2i\theta}$ or $\cos 2\theta$ and $\sin 2\theta$. Therefore, the solution should have some terms that are independent of $\theta$ and some terms having $\theta$ dependence of the form $\cos 2\theta$ and $\sin 2\theta$.

In example 1.13, in equation (1.173) the solution of problem II is given for the case when $\sigma_{rr} - i\sigma_{r\theta} = -\sigma_0$ at $r = a$. Utilizing that solution, one obtains the following potential functions for the boundary stress field:

$$\sigma_{rr} - i\sigma_{r\theta} = -\frac{\sigma_0}{2}$$

$$\phi'(z) = 0, \quad \psi''(z) = \frac{\sigma_0}{2}\frac{a^2}{z^2} \tag{1.176}$$

Next, the potential functions corresponding to the following boundary condition:

$$\sigma_{rr} - i\sigma_{r\theta} = \frac{\sigma_0}{2}e^{2i\theta} \qquad \text{at } r = a$$

are obtained in the following manner.

From equations (1.167) and (1.170) we see that the terms associated with $A_2$, $A_{-2}$, $B_0$, and $B_{-4}$ give the desired angular dependence required by the boundary conditions. Out of these four terms, only $A_{-2}$ and $B_{-4}$ also satisfy the regularity conditions while $A_2$ and $B_0$ do not. Therefore,

$$\phi'(z) = A_{-2}z^{-2}, \quad \psi''(z) = B_{-4}z^{-4}$$

Substituting these expressions in the boundary condition equation, one obtains

$$\sigma_{rr} - i\sigma_{r\theta} = A_n r^n[(1-n)e^{in\theta} + e^{-in\theta}] - B_m r^m e^{i(2+m)\theta} = A_{-2}a^{-2}[3e^{-2i\theta} + e^{2i\theta}] - B_{-4}a^{-4}e^{-2i\theta} = \frac{\sigma_0}{2}e^{2i\theta}$$

Equating the coefficients of $e^{2i\theta}$ and $e^{-2i\theta}$ on both sides of the preceding equation,

$$A_{-2} = \frac{\sigma_0}{2}a^2$$

$$B_{-4} = \frac{3\sigma_0}{2}a^4$$

Therefore,

$$\phi'(z) = \frac{\sigma_0 a^2}{2z^2}, \quad \psi''(z) = \frac{3\sigma_0 a^4}{2z^4} \tag{1.177}$$

Combined solution of problems I and II: Adding equations (1.175), (1.176), and (1.177), the complete potential functions for the original problem are obtained:

$$\phi'(z) = \frac{\sigma_0}{4} + \frac{\sigma_0 a^2}{2z^2}, \quad \psi''(z) = \frac{\sigma_0}{2} + \frac{\sigma_0 a^2}{2z^2} + \frac{3\sigma_0 a^4}{2z^4} \tag{1.178}$$

From equations (1.178), (1.167), and (1.170), the stress fields are obtained:

$$\begin{aligned}
\sigma_{rr} - i\sigma_{r\theta} &= A_n r^n[(1-n)e^{in\theta} + e^{-in\theta}] - B_m r^m e^{i(2+m)\theta} \\
&= 2A_0 + A_{-2}r^{-2}[3e^{-2i\theta} + e^{2i\theta}] - B_0 e^{2i\theta} - B_{-2}r^{-2} - B_{-4}r^{-4}e^{-2i\theta} \\
&= \frac{\sigma_0}{2} + \frac{\sigma_0 a^2}{2r^2}[3e^{-2i\theta} + e^{2i\theta}] - \frac{\sigma_0}{2}e^{2i\theta} - \frac{\sigma_0 a^2}{2r^2} - \frac{3\sigma_0 a^4}{2r^4}e^{-2i\theta} \\
&= \frac{\sigma_0}{2} + \frac{\sigma_0 a^2}{2r^2}\left[3e^{-2i\theta} + e^{2i\theta} - \frac{r^2}{a^2}e^{2i\theta} - 1 - \frac{3a^2}{r^2}e^{-2i\theta}\right] \\
&= \frac{\sigma_0}{2} + \frac{\sigma_0 a^2}{2r^2}\left[-1 + \left(1 - \frac{r^2}{a^2}\right)e^{2i\theta} + 3\left(1 - \frac{a^2}{r^2}\right)e^{-2i\theta}\right]
\end{aligned} \tag{1.179}$$

$$\sigma_{rr} + \sigma_{\theta\theta} = 4A_n r^n \cos n\theta = 4A_0 + 4A_{-2} r^{-2} \cos 2\theta = \sigma_0 + \frac{2\sigma_0 a^2}{r^2} \cos 2\theta$$

$$= \sigma_0 \left(1 + \frac{2a^2}{r^2} \cos 2\theta\right)$$

From the preceding equations one obtains, at $r = a$,

$$\sigma_{rr} - i\sigma_{r\theta} = \frac{\sigma_0}{2} + \frac{\sigma_0 a^2}{2a^2}[-1 + 0 + 0] = 0$$

$$\therefore \sigma_{rr} = 0, \ \sigma_{r\theta} = 0$$

and

$$\sigma_{rr} + \sigma_{\theta\theta} = \sigma_0 \left(1 + \frac{2a^2}{a^2} \cos 2\theta\right) = \sigma_0(1 + 2\cos 2\theta)$$

$$\therefore \sigma_{\theta\theta} = \sigma_0(1 + 2\cos 2\theta)$$

Clearly, at $r = a$, the maximum value of $\sigma_{\theta\theta}$ is $3\sigma_0$ at $\theta = 0°$ and 180°. Therefore, the stress concentration factor for the uniaxial state of stress is 3.

**Example 1.15**

An annular plate (or a cylindrical tube) of inner radius $a$ and outer radius $b$ is subjected to a pressure $p_0$ at the inner surface, while the outer surface is stress free, as shown in Figure 1.22. Obtain the stress field in the entire plate.

*Solution*

Boundary conditions for this problem are

$$\sigma_{rr} - i\sigma_{r\theta} = -p_0 \quad \text{at} \quad r = a$$

$$\sigma_{rr} - i\sigma_{r\theta} = 0 \quad \text{at} \ r = b$$

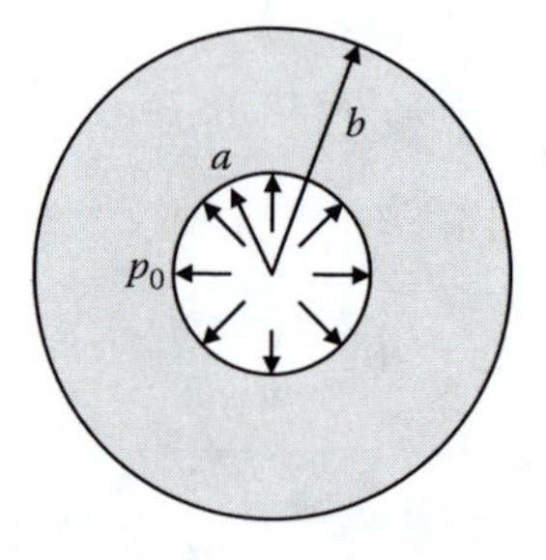


**FIGURE 1.22**
A cylindrical tube or an annular plate or a spherical pressure vessel is subjected to a compressive stress or pressure $p_0$ at the inner surface at $r = a$ while the outer surface at $r = b$ is stress free.

Since the boundary conditions are independent of the angular position, the potential functions should have the form

$$\phi'(z) = A_0, \quad \psi''(z) = \frac{B_{-2}}{z^2}$$

Therefore,

$$\sigma_{rr} - i\sigma_{r\theta} = A_n r^n[(1-n)e^{in\theta} + e^{-in\theta}] - B_m r^m e^{i(2+m)\theta} = 2A_0 - B_{-2}r^{-2}$$

For simplicity we substitute $2A_0 = A$ and $B_{-2} = B$ to obtain

$$\sigma_{rr} - i\sigma_{r\theta} = 2A_0 - B_{-2}r^{-2} = A - \frac{B}{r^2}$$

Applying the boundary conditions at $r = a$ and $b$,

$$A - \frac{B}{a^2} = -p_0$$

$$A - \frac{B}{b^2} = 0$$

From the preceding two equations, one obtains

$$A = \frac{B}{b^2}$$

$$\therefore \frac{B}{b^2} - \frac{B}{a^2} = -p_0$$

$$\Rightarrow B = \frac{p_0 a^2 b^2}{b^2 - a^2}, \quad A = \frac{p_0 a^2}{b^2 - a^2}$$

Therefore,

$$\sigma_{rr} - i\sigma_{r\theta} = A - \frac{B}{r^2} = \frac{p_0 a^2}{b^2 - a^2}\left(1 - \frac{b^2}{r^2}\right)$$

and

$$\sigma_{rr} + \sigma_{\theta\theta} = 4A_n r^n \cos n\theta = 4A_0 = 2A = \frac{2p_0 a^2}{b^2 - a^2}$$

From these two equations, the three stress components are obtained:

$$\sigma_{rr} = \frac{p_0 a^2}{b^2 - a^2}\left(1 - \frac{b^2}{r^2}\right)$$

$$\sigma_{\theta\theta} = \frac{p_0 a^2}{b^2 - a^2}\left(1 + \frac{b^2}{r^2}\right) \tag{1.180}$$

$$\sigma_{r\theta} = 0$$

**Example 1.16**

An annular plate (or a cylindrical tube) of inner radius $a$ and outer radius $b$ is subjected to a nonaxisymmetric stress field $\sigma_{rr} = p_0 \cos 2\theta$, $\sigma_{r\theta} = q_0 \sin 2\theta$ at the inner surface, while the outer surface is stress free. Obtain the stress field in the entire plate.

*Solution*

Boundary conditions for this problem are

$$\sigma_{rr} - i\sigma_{r\theta} = p_0 \cos 2\theta - iq_0 \sin 2\theta \quad \text{at } r = a$$

$$\sigma_{rr} - i\sigma_{r\theta} = 0 \quad \text{at } r = b$$

Note that this loading is symmetric about the line $\theta = 0$.

Since the angular dependence of the boundary conditions is of the form $\cos 2\theta$ and $\sin 2\theta$ in the $\sigma_{rr} - i\sigma_{r\theta}$ expression, only the terms containing $e^{\pm 2i\theta}$ are kept. Thus,

$$\sigma_{rr} - i\sigma_{r\theta} = A_n r^n[(1-n)e^{in\theta} + e^{-in\theta}] - B_m r^m e^{i(2+m)\theta}$$

$$= A_2 r^2[-e^{2i\theta} + e^{-2i\theta}] + A_{-2} r^{-2}[3e^{-2i\theta} + e^{2i\theta}] - B_0 e^{2i\theta} - B_{-4} r^{-4} e^{-2i\theta}$$

$$= [-A_2 r^2 + A_{-2} r^{-2} - B_0]e^{2i\theta} + [A_2 r^2 + 3A_{-2} r^{-2} - B_{-4} r^{-4}]e^{-2i\theta}$$

$$= [-A_2 r^2 + A_{-2} r^{-2} - B_0](\cos 2\theta + i \sin 2\theta)$$

$$+ [A_2 r^2 + 3A_{-2} r^{-2} - B_{-4} r^{-4}](\cos 2\theta - i \sin 2\theta)$$

$$= [-A_2 r^2 + A_{-2} r^{-2} - B_0 + A_2 r^2 + 3A_{-2} r^{-2} - B_{-4} r^{-4}]\cos 2\theta$$

$$i[-A_2 r^2 + A_{-2} r^{-2} - B_0 - A_2 r^2 - 3A_{-2} r^{-2} + B_{-4} r^{-4}]\sin 2\theta$$

$$= [4A_{-2} r^{-2} - B_0 - B_{-4} r^{-4}]\cos 2\theta + i[-2A_2 r^2 - 2A_{-2} r^{-2} - B_0 + B_{-4} r^{-4}]\sin 2\theta$$

Applying the boundary conditions at the two boundaries,

$$[\sigma_{rr} - i\sigma_{r\theta}]_{r=a}$$

$$= [4A_{-2}a^{-2} - B_0 - B_{-4}a^{-4}]\cos 2\theta + i[-2A_2a^2 - 2A_{-2}a^{-2} - B_0 + B_{-4}a^{-4}]$$

$$\sin 2\theta = p_0 \cos 2\theta - iq_0 \sin 2\theta$$

$$[\sigma_{rr} - i\sigma_{r\theta}]_{r=b}$$

$$= [4A_{-2}b^{-2} - B_0 - B_{-4}b^{-4}]\cos 2\theta + i[-2A_2b^2 - 2A_{-2}b^{-2} - B_0 + B_{-4}b^{-4}]\sin 2\theta = 0$$

Note that from these two complex equations the following four algebraic equations are obtained to solve for the four unknowns $A_2$, $A_{-2}$, $B_0$, and $B_{-4}$:

$$4A_{-2}a^{-2} - B_0 - B_{-4}a^{-4} = p_0$$

$$2A_2a^2 + 2A_{-2}a^{-2} + B_0 - B_{-4}a^{-4} = q_0$$

$$4A_{-2}b^{-2} - B_0 - B_{-4}b^{-4} = 0$$

$$2A_2b^2 + 2A_{-2}b^{-2} + B_0 - B_{-4}b^{-4} = 0$$

It should be noted here that if the applied stress field at the inner surface is changed to $\sigma_{rr} = p_0 \sin 2\sigma$, $\sigma_{r\theta} = q_0 \cos 2\theta$, then all four coefficients—$A_2$, $A_{-2}$, $B_0$, and $B_{-4}$—will be imaginary. In that case, equations (1.168) and (1.170) should be used to obtain

$$\sigma_{rr} - i\sigma_{r\theta} = A_n r^n \left[(1-n)e^{in\theta} - e^{-in\theta}\right] - B_m r^m e^{i(2+m)\theta}$$

$$= A_2r^2[-e^{2i\theta} - e^{-2i\theta}] + A_{-2}r^{-2}[3e^{-2i\theta} - e^{2i\theta}] - B_0e^{2i\theta} - B_{-4}r^{-4}e^{-2i\theta}$$

$$= \left[-A_2r^2 - A_{-2}r^{-2} - B_0\right]e^{2i\theta} + \left[-A_2r^2 + 3A_{-2}r^{-2} - B_{-4}r^{-4}\right]e^{-2i\theta}$$

$$= \left[-A_2r^2 - A_{-2}r^{-2} - B_0\right](\cos 2\theta + i\sin 2\theta)$$

$$+\left[-A_2r^2 + 3A_{-2}r^{-2} - B_{-4}r^{-4}\right](\cos 2\theta - i\sin 2\theta)$$

$$= \left[-A_2r^2 - A_{-2}r^{-2} - B_0 - A_2r^2 + 3A_{-2}r^{-2} - B_{-4}r^{-4}\right]\cos 2\theta$$

$$i\left[-A_2r^2 - A_{-2}r^{-2} - B_0 + A_2r^2 - 3A_{-2}r^{-2} + B_{-4}r^{-4}\right]\sin 2\theta$$

$$= \left[-2A_2r^2 + 2A_{-2}r^{-2} - B_0 - B_{-4}r^{-4}\right]\cos 2\theta + i\left[-4A_{-2}r^{-2} - B_0 + B_{-4}r^{-4}\right]\sin 2\theta$$

Applying the boundary conditions at the two boundaries,

$$[\sigma_{rr} - i\sigma_{r\theta}]_{r=a}$$

$$= [-2A_2a^2 + 2A_{-2}a^{-2} - B_0 - B_{-4}a^{-4}]\cos 2\theta + i[-4A_{-2}a^{-2} - B_0 + B_{-4}a^{-4}]$$

$$\sin 2\theta = p_0 \sin 2\theta - iq_0 \cos 2\theta$$

$$[\sigma_{rr} - i\sigma_{r\theta}]_{r=b}$$

$$= [-2A_2b^2 + 2A_{-2}b^{-2} - B_0 - B_{-4}b^{-4}]\cos 2\theta + i[-4A_{-2}b^{-2} - B_0 + B_{-4}b^{-4}]\sin 2\theta = 0$$

These two complex equations give four algebraic equations to solve for the four unknowns.

If the applied stress field at the inner surface is changed to $\sigma_{rr} = p_0 \cos 2\theta$ and $\sigma_{r\theta} = q_0 \cos 2\theta$, then all four coefficients—$A_2$, $A_{-2}$, $B_0$, and $B_{-4}$—become complex equations that have both real and imaginary components. In this case the problem should be solved in two steps, first considering $\sigma_{rr} = p_0 \cos 2\theta$, $\sigma_{r\theta} = 0$ that give real coefficients and then considering $\sigma_{rr} = 0$, $\sigma_{r\theta} = q_0 \cos 2\theta$ that should give imaginary coefficients. The final solution is then obtained by superimposing these two solutions.

### 1.3.5 Thick Wall Spherical Pressure Vessel

A stress field in a spherical pressure vessel of inner radius $a$ and outer radius $b$ subjected to an internal pressure $p_0$ as shown in Figure 1.22 is computed in this section. In the spherical coordinate system (Figure 1.14), the strain–displacement relations are given in Table 1.2.

$$
\begin{aligned}
\varepsilon_{rr} &= \frac{\partial u_r}{\partial r} \\
\varepsilon_{\beta\beta} &= \frac{1}{r}\frac{\partial u_\beta}{\partial \beta} + \frac{u_r}{r} \\
\varepsilon_{\theta\theta} &= \frac{1}{r\sin\beta}\frac{\partial u_\theta}{\partial \theta} + \frac{u_r}{r} + \frac{u_\beta}{r}\cot\beta \\
2\varepsilon_{r\beta} &= \frac{1}{r}\frac{\partial u_r}{\partial \beta} + \frac{\partial u_\beta}{\partial r} - \frac{u_\beta}{r} \\
2\varepsilon_{r\theta} &= \frac{1}{r\sin\beta}\frac{\partial u_r}{\partial \theta} - \frac{u_\theta}{r} + \frac{\partial u_\theta}{\partial r} \\
2\varepsilon_{\beta\theta} &= \frac{1}{r\sin\beta}\frac{\partial u_\beta}{\partial \theta} + \frac{1}{r}\frac{\partial u_\theta}{\partial \beta} - \frac{u_\theta}{r}\cot\beta
\end{aligned}
\tag{1.181}
$$

Since the boundary conditions are independent of angles $\beta$ and $\theta$, the displacement field inside the spherical shell should have the following form:

$$u_\theta = u_\beta = 0$$
$$u_r = u(r) \tag{1.182}$$

Substituting equation (1.182) into equation (1.181), one gets

$$\varepsilon_{rr} = \frac{\partial u_r}{\partial r} = \frac{\partial u}{\partial r}$$

$$\varepsilon_{\beta\beta} = \frac{1}{r}\frac{\partial u_\beta}{\partial \beta} + \frac{u_r}{r} = \frac{u}{r}$$

$$\varepsilon_{\theta\theta} = \frac{1}{r\sin\beta}\frac{\partial u_\theta}{\partial \theta} + \frac{u_r}{r} + \frac{u_\beta}{r}\cot\beta = \frac{u}{r}$$

$$2\varepsilon_{r\beta} = \frac{1}{r}\frac{\partial u_r}{\partial \beta} + \frac{\partial u_\beta}{\partial r} - \frac{u_\beta}{r} = 0 \tag{1.183}$$

$$2\varepsilon_{r\theta} = \frac{1}{r\sin\beta}\frac{\partial u_r}{\partial \theta} - \frac{u_\theta}{r} + \frac{\partial u_\theta}{\partial r} = 0$$

$$2\varepsilon_{\beta\theta} = \frac{1}{r\sin\beta}\frac{\partial u_\beta}{\partial \theta} + \frac{1}{r}\frac{\partial u_\theta}{\partial \beta} - \frac{u_\theta}{r}\cot\beta = 0$$

From equation (1.182) and Table 1.2, one obtains

$$\text{curl}\, u = \underline{\nabla} \times \mathbf{u} = 0$$

$$\text{div}\, u = \underline{\nabla} \bullet \mathbf{u} = \frac{\partial u}{\partial r} + \frac{2}{r}u \tag{1.184}$$

$$\text{grad}(\text{div}\, u) = \underline{\nabla}(\underline{\nabla} \bullet \mathbf{u}) = \frac{\partial^2 u}{\partial r^2} + \frac{2}{r}\frac{\partial u}{\partial r} - \frac{2}{r^2}u$$

Substituting equation (1.184) into Navier's equation of equilibrium,

$$(\lambda + 2\mu)\underline{\nabla}(\underline{\nabla} \bullet \mathbf{u}) - \mu\underline{\nabla} \times \underline{\nabla} \times \mathbf{u} = 0$$

$$\Rightarrow (\lambda + 2\mu)\left(\frac{\partial^2 u}{\partial r^2} + \frac{2}{r}\frac{\partial u}{\partial r} - \frac{2}{r^2}u\right) = 0 \tag{1.185}$$

$$\Rightarrow \frac{\partial^2 u}{\partial r^2} + \frac{2}{r}\frac{\partial u}{\partial r} - \frac{2}{r^2}u = 0$$

Equation (1.185) can be solved by assuming $u = A_n r^n$. Substituting this expression in equation (1.185),

$$[n(n-1)+2n-2]A_n r^{n-2} = 0 \tag{1.186}$$

$$\therefore (n-1)(n+2)A_n r^{n-2} = 0$$

Therefore, $n = 1$ and $-2$, and the general solution is

$$u = A_1 r + \frac{A_{-2}}{r^2} \tag{1.187}$$

Substituting equation (1.187) into equation (1.183),

$$\begin{aligned} \varepsilon_{rr} &= \frac{\partial u}{\partial r} = A_1 - \frac{2A_{-2}}{r^3} \\ \varepsilon_{\beta\beta} &= \varepsilon_{\theta\theta} = \frac{u}{r} = A_1 + \frac{A_{-2}}{r^3} \\ \varepsilon_{r\beta} &= \varepsilon_{r\theta} = \varepsilon_{\beta\theta} = 0 \end{aligned} \tag{1.188}$$

From the stress–strain relation,

$$\begin{aligned} \sigma_{rr} &= \lambda(\varepsilon_{rr} + \varepsilon_{\theta\theta} + \varepsilon_{\beta\beta}) + 2\mu\varepsilon_{rr} \\ &= \lambda\left(A_1 - \frac{2A_{-2}}{r^3} + A_1 + \frac{A_{-2}}{r^3} + A_1 + \frac{A_{-2}}{r^3}\right) + 2\mu\left(A_1 - \frac{2A_{-2}}{r^3}\right) \\ &= (3\lambda + 2\mu)A_1 - \frac{4\mu A_{-2}}{r^3} \\ \sigma_{\theta\theta} &= \lambda(\varepsilon_{rr} + \varepsilon_{\theta\theta} + \varepsilon_{\beta\beta}) + 2\mu\varepsilon_{\theta\theta} \\ &= \lambda\left(A_1 - \frac{2A_{-2}}{r^3} + A_1 + \frac{A_{-2}}{r^3} + A_1 + \frac{A_{-2}}{r^3}\right) + 2\mu\left(A_1 + \frac{A_{-2}}{r^3}\right) \\ &= (3\lambda + 2\mu)A_1 + \frac{2\mu A_{-2}}{r^3} \\ \sigma_{\beta\beta} &= \lambda(\varepsilon_{rr} + \varepsilon_{\theta\theta} + \varepsilon_{\beta\beta}) + 2\mu\varepsilon_{\beta\beta} = (3\lambda + 2\mu)A_1 + \frac{2\mu A_{-2}}{r^3} \end{aligned} \tag{1.189}$$

From the boundary conditions,

$$\begin{aligned} (3\lambda + 2\mu)A_1 - \frac{4\mu A_{-2}}{a^3} &= -p_0 \\ (3\lambda + 2\mu)A_1 - \frac{4\mu A_{-2}}{b^3} &= 0 \end{aligned} \tag{1.190}$$

Coefficients $A_1$ and $A_{-2}$ are obtained from equation (1.190) and then those values are substituted into equation (1.189) to obtain

$$\sigma_{rr} = -p_0 \frac{\left(\frac{b^3}{r^3} - 1\right)}{\left(\frac{b^3}{a^3} - 1\right)}$$
$$\sigma_{\theta\theta} = \sigma_{\beta\beta} = p_0 \frac{\left(\frac{b^3}{2r^3} + 1\right)}{\left(\frac{b^3}{a^3} - 1\right)} \tag{1.191}$$

Note that $\sigma_{rr}$ is compressive, while $\sigma_{\theta\theta}$ and $\sigma_{\beta\beta}$ are tensile and maximum at the inner radius.

## 1.4 Concluding Remarks

This chapter has given a brief review of the fundamentals of the mechanics of deformable solids. The chapter started with the derivation of basic equations of the theory of elasticity and continuum mechanics and ended after solving some classical problems of elasticity. It is an important chapter because different equations of fracture mechanics given in the following chapters are derived based on the fundamental knowledge of the theory of elasticity presented in this chapter.

## References

Airy, G. B., British Association for the Advancement of Science report, 1862.

Kundu, T., ed., *Ultrasonic nondestructive evaluation: Engineering and biological material characterization*. Boca Raton, FL: CRC Press, 2004.

Moon, P. and Spencer, D. E. *Vectors*. Princeton, NJ: D. Van Nostrand Company, Inc., 1965.

Timoshenko, S. P. and Goodier, J. N. *Theory of elasticity*. New York: McGraw–Hill, 1970.

## Exercise Problems

Problem 1.1: Simplify the following expressions (note that $\delta_{ij}$ is the Kronecker delta, repeated index means summation, and comma represents derivative):

(a) $\delta_{mm}$

(b) $\delta_{ij}\delta_{kj}$

(c) $\delta_{ij}u_{k,kj}$

(d) $\delta_{ij}\delta_{ij}$

(e) $\delta_{mm}\delta_{ij}x_j$

(f) $\delta_{km}u_{i,jk}u_{i,jm}$

(g) $\frac{\partial x_m}{\partial x_k}\frac{\partial x_m}{\partial x_k}$ .

Problem 1.2: Express the following mathematical operations in index notation:

(a) $\underline{\nabla}\bullet\mathbf{u}$

(b) $\nabla^2\phi$

(c) $\nabla^2\mathbf{u}$

(d) $\underline{\nabla}\phi$

(e) $\underline{\nabla}\times\mathbf{u}$

(f) $\underline{\nabla}\times(\underline{\nabla}\times\mathbf{u})$

(g) $[C]=[A][B]$

(h) $[A]^T[B]\neq[A][B]^T$

(i) $\{c\}=[A]^T\{b\}$

(j) $\underline{\nabla}\bullet(\underline{\nabla}\times\mathbf{u})$

(k) $\underline{\nabla}\times(\underline{\nabla}\phi)$

(l) $\underline{\nabla}(\underline{\nabla}\bullet\mathbf{u})$

(m) $\underline{\nabla}\bullet(\underline{\nabla}\phi)$

where $\mathbf{u}$ is a vector quantity and $\phi$ is a scalar quantity; $A$, $B$, and $C$ are $3\times3$ matrices; and $c$ and $b$ are $3\times1$ vectors.

Problem 1.3: Consider two surfaces passing through point $P$ (see Figure 1.23). The unit normal vectors on these two surfaces at point $P$ are $m_j$ and $n_j$, respectively. The traction vectors on the two surfaces at point $P$ are denoted by $\overset{m}{T}_i$ and $\overset{n}{T}_i$, respectively. Check whether the dot product between $\overset{m}{\underline{T}}$ and $\underline{n}$ is same as or different from the dot product between $\overset{n}{\underline{T}}$ and $\underline{m}$.

Problem 1.4:

(a) A thin triangular plate is fixed along the boundary OA and is subjected to a uniformly distributed horizontal load $p_0$ per unit

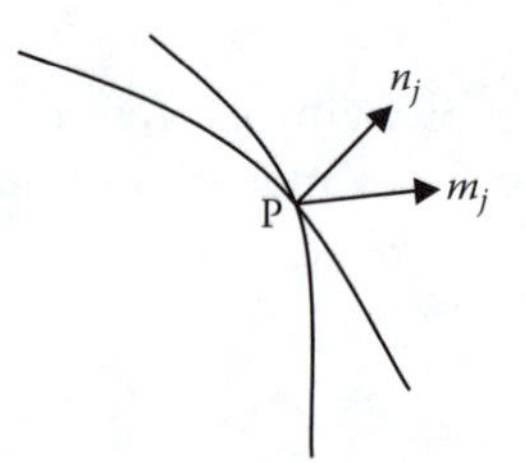


**FIGURE 1.23**

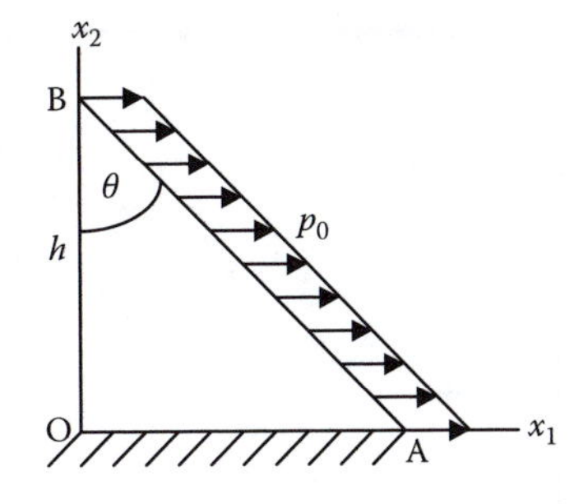


**FIGURE 1.24**

area along the boundary AB as shown in Figure 1.24. Give all boundary conditions in terms of displacement or stress components in the $x_1x_2$ coordinate system.

(b) If $p_0$ acts normal to the boundary AB, what will be the stress boundary conditions along line AB?

Problem 1.5: The quarter disk of radius $a$, shown in Figure 1.25, is subjected to a linearly varying shear stress, which varies from 0 to $T_0$ along boundaries AO and CO and a uniform pressure $p_0$ along the boundary ABC. Assume that all out-of-plane stress components are zero.

(a) Give all stress boundary conditions along the boundaries OA and OC in terms of stress components $\sigma_{11}$, $\sigma_{22}$, and $\sigma_{12}$ in the Cartesian coordinate system.

(b) Give all stress boundary conditions along the boundaries OA and OC in terms of stress components $\sigma_{rr}$, $\sigma_{\theta\theta}$, and $\sigma_{r\theta}$ in the cylindrical coordinate system.

(c) Give all stress boundary conditions at point B in terms of stress components $\sigma_{11}$, $\sigma_{22}$, and $\sigma_{12}$ in the Cartesian coordinate system.

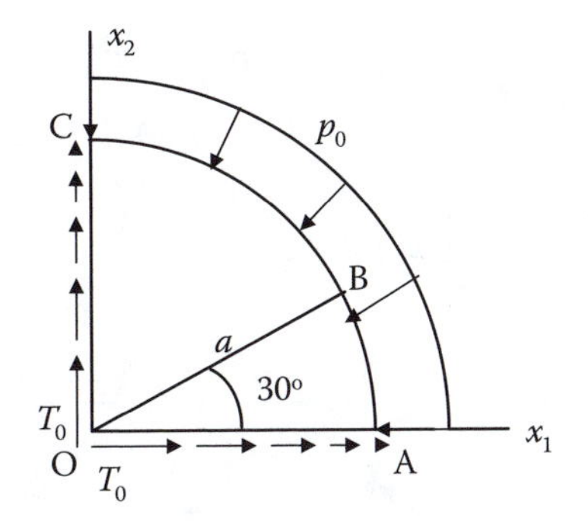


**FIGURE 1.25**

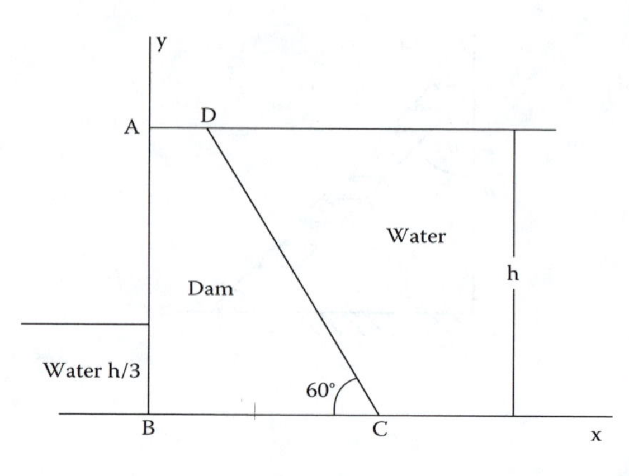


**FIGURE 1.26**

(d) Give all stress boundary conditions along the boundary ABC in terms of stress components $\theta_{rr}$, $\sigma_{\theta\theta}$, and $\sigma_{r\theta}$ in the cylindrical coordinate system.

Problem 1.6:

(a) A dam made of isotropic material has two different water heads on two sides, as shown in Figure 1.26. Define all boundary conditions along the boundaries AB and CD in terms of stress components $\sigma_{xx}$, $\sigma_{yy}$, and $\tau_{xy}$.

(b) If the dam is made of orthotropic material, what changes, if any, should be in your answer to part (a)?

Problem 1.7: Express the surface integral $\int_c x_i n_j dS$ in terms of the volume $V$ bounded by the surface $S$ (see Figure 1.27). $n_j$ is the $j$th component of the outward unit normal vector on the surface.

Problem 1.8: Obtain the principal stresses and their directions for the following stress state:

$$[\sigma] = \begin{bmatrix} 5 & -3 & 0 \\ -3 & 2 & 0 \\ 0 & 0 & 10 \end{bmatrix}$$

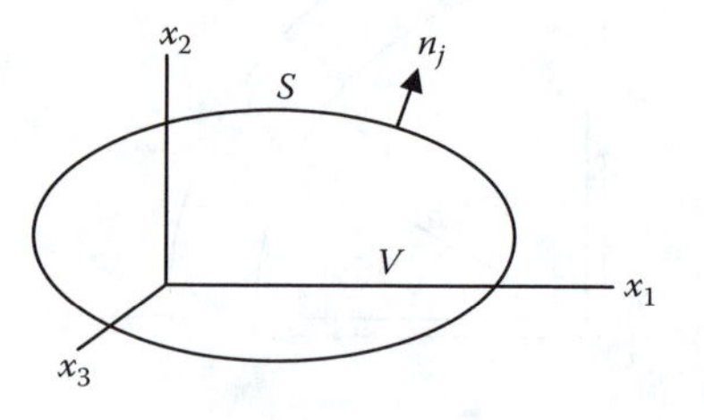


**FIGURE 1.27**

Problem 1.9: An anisotropic elastic solid is subjected to some load that gives a strain state $\varepsilon_{ij}$ in the $x_1x_2x_3$ coordinate system. In a different (rotated) coordinate system $x_{1'}x_{2'}x_{3'}$, the strain state is transformed to $\varepsilon_{m'n'}$.

(a) Do you expect the strain energy density function $U_0$ to be a function of strain invariants only?

(b) Do you expect the same or different expressions of $U_0$ when it is expressed in terms of $\varepsilon_{ij}$ or $\varepsilon_{m'n'}$?

(c) Do you expect the same or different numerical values of $U_0$ when you compute it from its expression in terms of $\varepsilon_{ij}$ and from its expression in terms of $\varepsilon_{m'n'}$?

(d) Justify your answers.

Answer parts (a), (b), and (c) if the material is isotropic.

Problem 1.10: Stress–strain relation for a linear elastic material is given by $\sigma_{ij} = C_{ijkl}\varepsilon_{kl}$. Starting from the stress–strain relation for an isotropic material, prove that $C_{ijkl}$ for the isotropic material is given by $\lambda\delta_{ij}\delta_{k\ell} + \mu(\delta_{ik}\delta_{j\ell} + \delta_{i\ell}\delta_{jk})$.

Problem 1.11: Obtain the governing equation of equilibrium in terms of displacement for a material whose stress–strain relation is given by

$$\sigma_{ij} = \alpha_{ijkl}\varepsilon_{km}\varepsilon_{ml} + \beta_{ijkl}\varepsilon_{kl} + \delta_{ij}\gamma$$

where $\alpha_{ijkl}$ and $\beta_{ijkl}$ are material properties that are constants over the entire region, and $\gamma$ is the residual hydrostatic state of stress that varies from point to point.

Problem 1.12: Starting from the three-dimensional stress transformation law $\sigma_{i'j'} = l_{i'm}l_{j'n}\sigma_{mn}$, prove that for two-dimensional stress transformation the following equations hold good (see Figure 1.28):

$$\sigma_{1'1'} = \sigma_{11}\cos^2\theta + \sigma_{22}\sin^2\theta + 2\sigma_{12}\sin\theta\cos\theta$$

$$\sigma_{2'2'} = \sigma_{11}\sin^2\theta + \sigma_{22}\cos^2\theta - 2\sigma_{12}\sin\theta\cos\theta$$

$$\sigma_{1'2'} = (-\sigma_{11} + \sigma_{22})\sin\theta\cos\theta + \sigma_{12}(\cos^2\theta - \sin^2\theta)$$

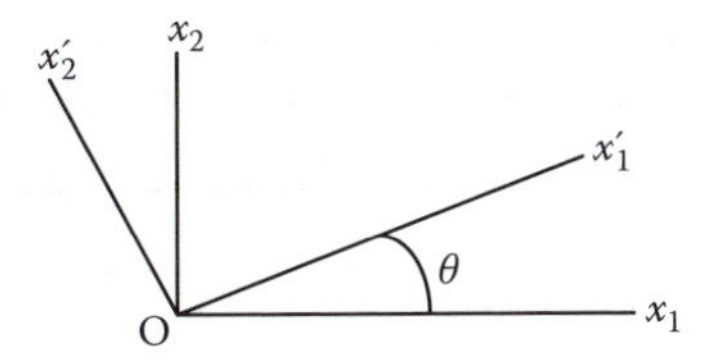


**FIGURE 1.28**

Problem 1.13: Derive the constraint condition (if any) that must be satisfied for the following displacement states to be valid solutions of linear, elastic, isotropic material of volume $V$ and boundary surface $S$, when only surface tractions (no body forces) are applied on the body. Note that $m$, $n$, and $p$ are different constants.

(a) $u_1 = mx_1^4 + nx_1^2x_2^2 + px_2^2, \quad u_2 = u_3 = 0$

(b) $u_2 = mx_1^2 + nx_1x_2^2 + px_2^2, \quad u_1 = u_3 = 0$

Problem 1.14: Prove that the compatibility equations for plane stress and plane strain problems can be written as

(a) $\sigma_{\alpha\alpha,\beta\beta} = -(1+\nu)f_{\gamma,\gamma}$ for plane stress problems and

(b) $\sigma_{\alpha\alpha,\beta\beta} = -\frac{f_{\gamma,\gamma}}{(1-\nu)}$ for plane strain problems

where $\nu$ is the Poisson's ratio, $f_\gamma$ is the body force per unit volume, and $\alpha$, $\beta$, $\gamma$ can take values 1 or 2.

Problem 1.15: Consider a cantilever beam of thickness $t$, depth $2c$, and length $L$ ($L >> c$, $L >> t$), subjected to a uniform pressure (force per unit area $\sigma_0$) on the top surface, as shown in Figure 1.29. The following two solution states are proposed for this problem:

Solution state 1:

$$\sigma_{11} = \frac{\sigma_0 x_2}{2c}\left\{\frac{x_2^2}{c^2} - \frac{3x_1^2}{2c^2} - \frac{3}{5}\right\}$$

$$\sigma_{22} = \frac{\sigma_0}{2}\left\{-1 + \frac{3x_2}{2c} - \frac{x_2^3}{2c^3}\right\}$$

$$\sigma_{12} = -\frac{3\sigma_0 x_1}{4c}\left\{1 - \frac{x_2^2}{c^2}\right\}$$

Solution state 2:

$$\sigma_{11} = \frac{\sigma_0 x_2}{2c}\left\{-\frac{3x_1^2}{2c^2}\right\}$$

$$\sigma_{22} = \frac{\sigma_0}{2}\left\{-1 + \frac{3x_2}{2c} - \frac{x_2^3}{2c^3}\right\}$$

$$\sigma_{12} = -\frac{3\sigma_0 x_1}{4c}\left\{1 - \frac{x_2^2}{c^2}\right\}$$

(a) Check if the preceding solution states satisfy the stress boundary conditions and governing equations for this problem.

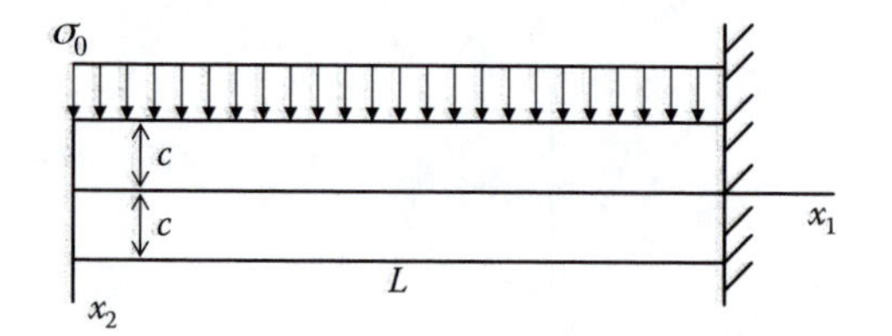


**FIGURE 1.29**

(b) In your undergraduate mechanics of materials class, you learned under beam theory that the bending stress and shear stress formulae have the following forms:

$$\sigma_{11} = \sigma = \frac{Mx_2}{I} \quad \text{and} \quad \sigma_{12} = \tau = \frac{VQ}{It},$$

where $I$ is the area moment of inertia, $Q$ is the first moment of the part of the cross-sectional area, and $M$ and $V$ are bending moment and shear force at the cross section. Check if either of the two solution states corresponds to the beam theory formulations.

(c) For a point not too close to the ends (at a distance greater than $2c$ from both ends), which solution state do you think is closer to the true solution and why?

(d) For large values of $x_1$, should both solution states give almost identical results? Explain your answer.

Problem 1.16: A half plane ($x_1 > 0$) is subjected to the shear load $P_0$ per unit area on its surface in the region $-a < x_2 < +a$, as shown in Figure 1.30. Calculate the stress field at a general point $(x_1, x_2)$ inside the half plane. Express your results in terms of some infinite integrals.

Problem 1.17: Two linear elastic half planes are subjected to force couples, as shown in Figure 1.31.

(a) Give three Cartesian components of stress ($\sigma_{11}$, $\sigma_{22}$, $\sigma_{12}$) for these two problems.

(b) Specialize these expressions along two lines: (1) $x_1 = 0$, and (2) $x_1 = x_2$.

(c) Prove St. Venant's principle by computing the stress field for the two geometries along the $x_1 = 0$ line.

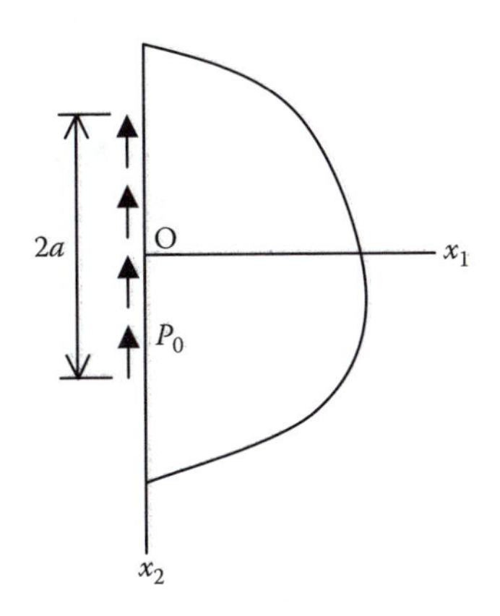


**FIGURE 1.30**

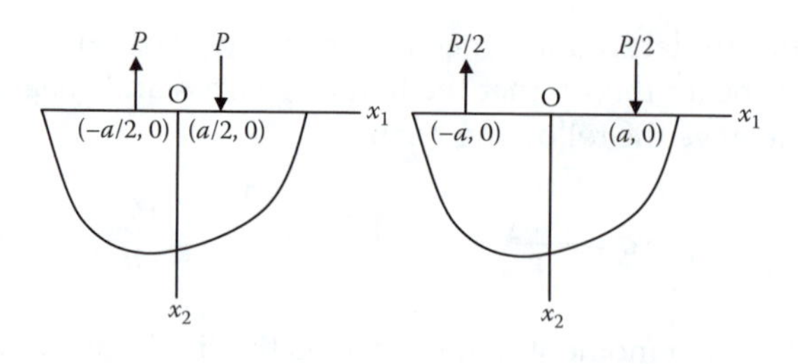


**FIGURE 1.31**

(d) At what depth along the line $x_1 = 0$ do the results for the two problems become almost identical (difference is less than 1%)?

Problem 1.18: A thin plate of infinite extent is subjected to a far field stress distribution $\sigma_{xx} = \sigma_{yy} = 0$ and $\sigma_{xy} = \tau$. The plate has a circular hole of radius $a$ at the origin which has stress free boundary. Find the stress field $\sigma_{rr}$, $\sigma_{\theta\theta}$, and $\sigma_{r\theta}$ at a general point $P(r,\theta)$ (see Figure 1.32).

Problem 1.19: A cylindrical tube is subjected to an uniform pressure $p$ at its inner boundary ($r = a$) and a $\theta$ dependent pressure $p \cos 2\theta$ at its outer boundary ($r = b$) (see Figure 1.33).

(a) Give expressions of the three stress components $\sigma_{rr}$, $\sigma_{\theta\theta}$, and $\sigma_{r\theta}$ at a general point $P(r,\theta)$ in terms of some unknown constants.

(b) Give all equations that you must satisfy to solve for the unknown constants of part (a). Note that for $n$ unknown constants you need $n$ equations. Do not try to solve these equations for the unknown constants.

Problem 1.20: Prove that if $\sigma_{yy} - \sigma_{xx} + 2i\sigma_{xy} = 2\left[\bar{z}\phi''(z) + \psi''(z)\right]$, then $\sigma_{\theta\theta} - \sigma_{rr} + 2i\sigma_{r\theta} = 2[\bar{z}\phi''(z) + \psi''(z)]e^{2i\theta}$, where $z$ is the complex variable $z = x + iy$.

Problem 1.21: Air is pumped out of a hollow sphere of outer radius 2 $m$ and inner radius 1 $m$; then it is placed at a depth of 2 km in the

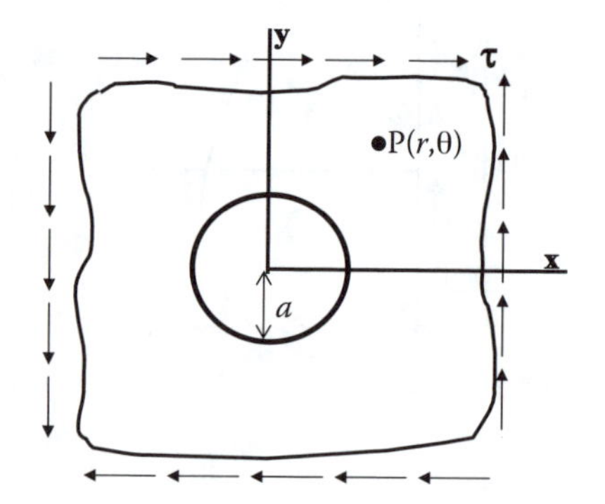


**FIGURE 1.32**

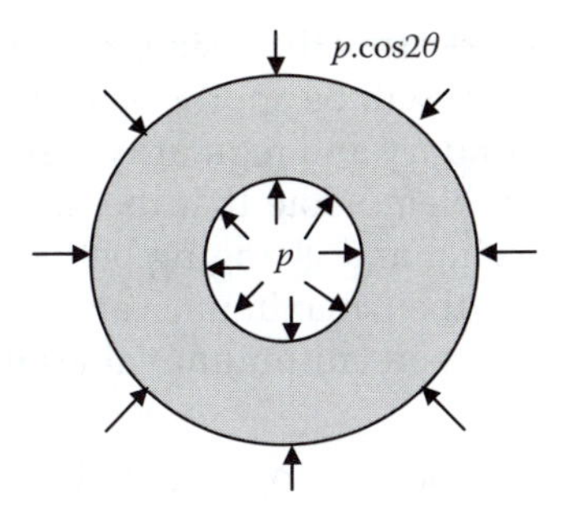


**FIGURE 1.33**

ocean (density of water is 1000 kg/m$^3$). Compute the stress field in the sphere material. If the sphere is taken deeper into the ocean, state whether the failure will start from the outer surface or inner surface of the sphere if the sphere material is

(a) weak in tension but infinitely strong in compression and shear

(b) weak in compression but infinitely strong in tension and shear

(c) weak in shear but infinitely strong in tension and compression

Problem 1.22: Starting from the stress expressions (Equation 1.191) for a spherical thick wall pressure vessel subjected to an internal pressure $p_0$, prove that, for a thin wall spherical pressure vessel of wall thickness $t$ [$t = (b - a)$], where $t \ll a$, (a) the circumferential stress $\sigma_{\theta\theta} = \sigma_{\phi\phi} = p_0 a/(2t)$ and (b) the radial stress $\sigma_{rr}$, varies linearly from $-p_0$ at the inner surface to zero at the outer surface.

Problem 1.23: An infinite plate with a circular hole of radius $a$ is subjected to a varying normal stress at $r = a$, as shown in Figure 1.34. Let the applied stress field at $r = a$ be approximately represented as one of the trigonometric functions (sine, cosine, or tangent).

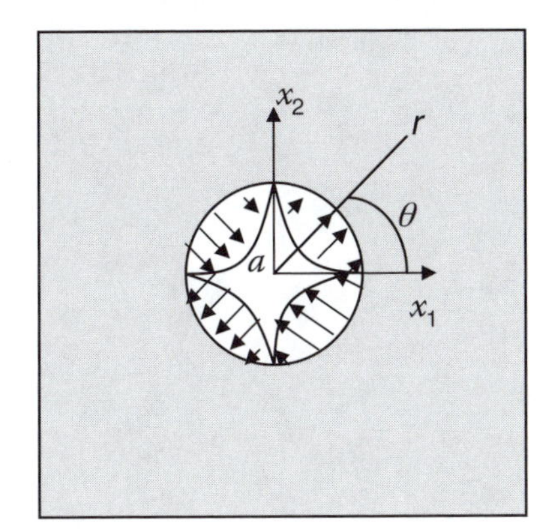


**FIGURE 1.34**

(a) Which trigonometric function (sine, cosine, or tan) and what angular dependence will be appropriate for this problem geometry? Give the boundary and regularity conditions that you must satisfy for this problem. Note that the normal stress at $r = a$ is continuously varying and changing between tension and compression, reaching their maximum values ($\pm p_0$) at $\theta = 45°$, 135°, 225°, and 315° and their minimum values (0) at $\theta = 0$, 90°, 180°, and 270°.

(b) Give the expression of $\sigma_{rr} - i\sigma_{r\theta}$ in the plate material in terms of some unknown constants $A_n$ and $B_m$.

(c) For this problem give all acceptable values of $n$ and $m$ of part (b). There is no need to solve for $A_n$ and $B_m$.

# 2

# *Elastic Crack Model*

## 2.1 Introduction

Airy stress function technique can be followed to solve many two-dimensional elasticity problems as discussed in chapter 1. Williams (1957) solved the fundamental problem of the stress field computation near a crack tip in an elastic, isotropic material using the Airy stress function as described in the following section.

## 2.2 Williams' Method to Compute the Stress Field near a Crack Tip

We are interested in computing the stress field near a crack tip, located at the origin; the problem geometry is shown in Figure 2.1. The stresses are defined in terms of the Airy stress function $\phi$:

$$\begin{aligned}\sigma_{rr} &= \frac{1}{r}\frac{\partial\phi}{\partial r} + \frac{1}{r^2}\frac{\partial^2\phi}{\partial\theta^2} \\ \sigma_{\theta\theta} &= \frac{\partial^2\phi}{\partial r^2} \\ \sigma_{r\theta} &= -\frac{\partial}{\partial r}\left(\frac{1}{r}\frac{\partial\phi}{\partial\theta}\right)\end{aligned} \tag{2.1}$$

Note that with the preceding definition of the stress field, the governing equilibrium equation is automatically satisfied in a two-dimensional polar coordinate system. However, satisfaction of the compatibility equation requires the Airy stress function $\phi$ to be biharmonic (see equation 1.102). Therefore, $\phi$ must satisfy the following equation:

$$\nabla^4\phi = \nabla^2\nabla^2\phi = \left(\frac{\partial^2}{\partial r^2} + \frac{1}{r}\frac{\partial}{\partial r} + \frac{1}{r^2}\frac{\partial^2}{\partial\theta^2}\right)\left(\frac{\partial^2}{\partial r^2} + \frac{1}{r}\frac{\partial}{\partial r} + \frac{1}{r^2}\frac{\partial^2}{\partial\theta^2}\right)\phi = 0 \tag{2.2}$$

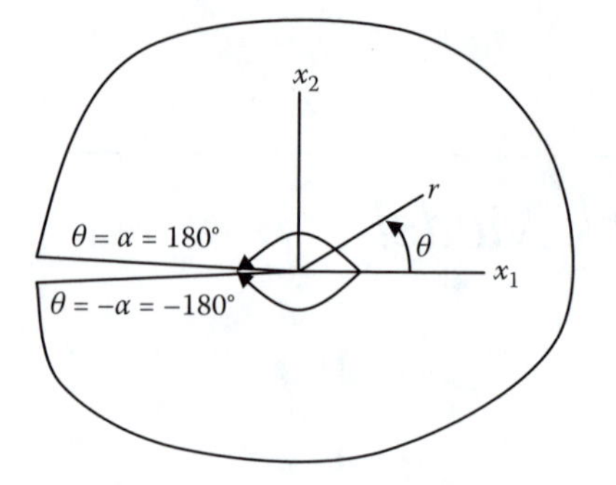


**FIGURE 2.1**
Infinitely sharp smooth crack in a solid. Crack surfaces are at $\theta = \pm\alpha = \pm 180°$.

Williams postulated the solution in the following form:

$$\phi = r^{\lambda+1}F(\theta) \tag{2.3}$$

Therefore,

$$\begin{aligned}
\nabla^2\phi = \nabla^2[r^{\lambda+1}F(\theta)] &= \left(\frac{\partial^2}{\partial r^2} + \frac{1}{r}\frac{\partial}{\partial r} + \frac{1}{r^2}\frac{\partial^2}{\partial\theta^2}\right)[r^{\lambda+1}F(\theta)] \\
&= (\lambda+1)\lambda r^{\lambda-1}F(\theta) + (\lambda+1)r^{\lambda-1}F(\theta) + r^{\lambda-1}F''(\theta) \\
&= [(\lambda+1)^2 F(\theta) + F''(\theta)]r^{\lambda-1} = \left[\frac{\partial^2}{\partial\theta^2} + (\lambda+1)^2\right]r^{\lambda-1}F(\theta)
\end{aligned}$$

Hence,

$$\begin{aligned}
\nabla^2\nabla^2\phi = \nabla^2\nabla^2[r^{\lambda+1}F(\theta)] &= \left(\frac{\partial^2}{\partial r^2} + \frac{1}{r}\frac{\partial}{\partial r} + \frac{1}{r^2}\frac{\partial^2}{\partial\theta^2}\right)\left[\frac{\partial^2}{\partial\theta^2} + (\lambda+1)^2\right]r^{\lambda-1}F(\theta) \\
&= \left[\frac{\partial^2}{\partial\theta^2} + (\lambda+1)^2\right]\left(\frac{\partial^2}{\partial r^2} + \frac{1}{r}\frac{\partial}{\partial r} + \frac{1}{r^2}\frac{\partial^2}{\partial\theta^2}\right)r^{\lambda-1}F(\theta) \\
&= \left[\frac{\partial^2}{\partial\theta^2} + (\lambda+1)^2\right][(\lambda-1)(\lambda-2)r^{\lambda-3}F(\theta) + (\lambda-1)r^{\lambda-3}F(\theta) + r^{\lambda-3}F''(\theta)] \\
&= \left[\frac{\partial^2}{\partial\theta^2} + (\lambda+1)^2\right]\left[\frac{\partial^2}{\partial\theta^2} + (\lambda-1)^2\right]r^{\lambda-3}F(\theta) = 0
\end{aligned} \tag{2.4}$$

Since $r^{\lambda-3} \neq 0$ in general, the preceding equation will be satisfied if

$$\left[\frac{d^2}{d\theta^2} + (\lambda+1)^2\right]\left[\frac{d^2}{d\theta^2} + (\lambda-1)^2\right]F(\theta) = 0 \tag{2.5}$$

Note that equation (2.5) will be satisfied if either

$$\left[\frac{d^2}{d\theta^2}+(\lambda-1)^2\right]F(\theta)=0 \tag{2.6}$$

or

$$\left[\frac{d^2}{d\theta^2}+(\lambda+1)^2\right]F(\theta)=0 \tag{2.7}$$

Solution of equation (2.6) is

$$F(\theta)=c_1\cos\{(\lambda-1)\theta\}+c_2\sin\{(\lambda-1)\theta\} \tag{2.8}$$

and that of equation (2.7) is

$$F(\theta)=c_3\cos\{(\lambda+1)\theta\}+c_4\sin\{(\lambda+1)\theta\} \tag{2.9}$$

Therefore, the general solution of equation (2.5) is given by

$$F(\theta)=c_1\cos\{(\lambda-1)\theta\}+c_2\sin\{(\lambda-1)\theta\}+c_3\cos\{(\lambda+1)\theta\}+c_4\sin\{(\lambda+1)\theta\} \tag{2.10}$$

Thus,

$$\begin{aligned}\phi(r,\theta)=r^{\lambda+1}F(\theta)&=r^{\lambda+1}[c_1\cos\{(\lambda-1)\theta\}+c_2\sin\{(\lambda-1)\theta\}\\ &+c_3\cos\{(\lambda+1)\theta\}+c_4\sin\{(\lambda+1)\theta\}]\end{aligned} \tag{2.11}$$

is an acceptable Airy stress function because it is biharmonic and therefore satisfies the compatibility condition.

Then, from equation (2.1) three components of stress can be obtained:

$$\begin{aligned}\sigma_{rr}&=\frac{1}{r^2}\frac{\partial^2\phi}{\partial\theta^2}+\frac{1}{r}\frac{\partial\phi}{\partial r}=r^{\lambda-1}F''(\theta)+(\lambda+1)r^{\lambda-1}F(\theta)=r^{\lambda-1}\{F''(\theta)+(\lambda+1)F(\theta)\}\\ \sigma_{\theta\theta}&=\frac{\partial^2\phi}{\partial r^2}=\lambda(\lambda+1)r^{\lambda-1}F(\theta)\\ \sigma_{r\theta}&=-\frac{\partial}{\partial r}\left(\frac{1}{r}\frac{\partial\phi}{\partial\theta}\right)=-\frac{\partial}{\partial r}(r^{\lambda}F'(\theta))=-\lambda r^{\lambda-1}F'(\theta)\end{aligned} \tag{2.12}$$

In the preceding expressions,

$$\begin{aligned}F'(\theta)&=-c_1(\lambda-1)\sin\{(\lambda-1)\theta\}+c_2(\lambda-1)\cos\{(\lambda-1)\theta\}\\ &\quad-c_3(\lambda+1)\sin\{(\lambda+1)\theta\}+c_4(\lambda+1)\cos\{(\lambda+1)\theta\}\\ F''(\theta)&=-c_1(\lambda-1)^2\cos\{(\lambda-1)\theta\}-c_2(\lambda-1)^2\sin\{(\lambda-1)\theta\}\\ &\quad-c_3(\lambda+1)^2\cos\{(\lambda+1)\theta\}-c_4(\lambda+1)^2\sin\{(\lambda+1)\theta\}\end{aligned} \tag{2.13}$$

### 2.2.1 Satisfaction of Boundary Conditions

For the problem geometry shown in Figure 2.1 with stress-free crack surfaces, the boundary conditions are given by:

$$\text{at } \theta = \pm\alpha, \ \sigma_{\theta\theta} = \sigma_{r\theta} = 0 \tag{2.14}$$

Substituting the stress expressions in equation (2.14), we get, at $\theta = +\alpha$,

$$\begin{aligned} \sigma_{\theta\theta} &= \lambda(\lambda+1)r^{\lambda-1}[c_1 \cos\{(\lambda-1)\alpha\} + c_2 \sin\{(\lambda-1)\alpha\} + c_3 \cos\{(\lambda+1)\alpha\} \\ &\quad + c_4 \sin\{(\lambda+1)\alpha\}] = 0 \\ \sigma_{r\theta} &= -\lambda r^{\lambda-1}[-c_1(\lambda-1)\sin\{(\lambda-1)\alpha\} + c_2(\lambda-1)\cos\{(\lambda-1)\alpha\} \\ &\quad - c_3(\lambda+1)\sin\{(\lambda+1)\alpha\} + c_4(\lambda+1)\cos\{(\lambda+1)\alpha\}] = 0 \end{aligned} \tag{2.15}$$

and, at $\theta = -\alpha$,

$$\begin{aligned} \sigma_{\theta\theta} &= \lambda(\lambda+1)r^{\lambda-1}[c_1 \cos\{(\lambda-1)\alpha\} - c_2 \sin\{(\lambda-1)\alpha\} + c_3 \cos\{(\lambda+1)\alpha\} \\ &\quad - c_4 \sin\{(\lambda+1)\alpha\}] = 0 \\ \sigma_{r\theta} &= -\lambda r^{\lambda-1}[c_1(\lambda-1)\sin\{(\lambda-1)\alpha\} + c_2(\lambda-1)\cos\{(\lambda-1)\alpha\} \\ &\quad + c_3(\lambda+1)\sin\{(\lambda+1)\alpha\} + c_4(\lambda+1)\cos\{(\lambda+1)\alpha\}] = 0 \end{aligned} \tag{2.16}$$

Carrying out addition and subtraction operations between equations (2.15) and (2.16), one gets

$$\begin{aligned} c_1 \cos\{(\lambda-1)\alpha\} + c_3 \cos\{(\lambda+1)\alpha\} &= 0 \\ c_1(\lambda-1)\sin\{(\lambda-1)\alpha\} + c_3(\lambda+1)\sin\{(\lambda+1)\alpha\} &= 0 \\ c_2 \sin\{(\lambda-1)\alpha\} + c_4 \sin\{(\lambda+1)\alpha\} &= 0 \\ c_2(\lambda-1)\cos\{(\lambda-1)\alpha\} + c_4(\lambda+1)\cos\{(\lambda+1)\alpha\} &= 0 \end{aligned} \tag{2.17}$$

The preceding four equations can be written in matrix form:

$$\begin{bmatrix} \cos\{(\lambda-1)\alpha\} & \cos\{(\lambda+1)\alpha\} \\ (\lambda-1)\sin\{(\lambda-1)\alpha\} & (\lambda+1)\sin\{(\lambda+1)\alpha\} \end{bmatrix} \begin{Bmatrix} c_1 \\ c_3 \end{Bmatrix} = \begin{Bmatrix} 0 \\ 0 \end{Bmatrix}$$

and (2.18)

$$\begin{bmatrix} \sin\{(\lambda-1)\alpha\} & \sin\{(\lambda+1)\alpha\} \\ (\lambda-1)\cos\{(\lambda-1)\alpha\} & (\lambda+1)\cos\{(\lambda+1)\alpha\} \end{bmatrix} \begin{Bmatrix} c_2 \\ c_4 \end{Bmatrix} = \begin{Bmatrix} 0 \\ 0 \end{Bmatrix}$$

For the nontrivial solution of $c_1$, $c_2$, $c_3$, and $c_4$, the determinant of the coefficient matrices must vanish. Therefore,

$$\begin{aligned}&(\lambda+1)\sin\{(\lambda+1)\alpha\}\cos\{(\lambda-1)\alpha\}-(\lambda-1)\sin\{(\lambda-1)\alpha\}\cos\{(\lambda+1)\alpha\}=0\\&(\lambda+1)\cos\{(\lambda+1)\alpha\}\sin\{(\lambda-1)\alpha\}-(\lambda-1)\cos\{(\lambda-1)\alpha\}\sin\{(\lambda+1)\alpha\}=0\end{aligned} \tag{2.19}$$

The first equation of (2.19) is simplified to yield

$$\begin{aligned}&\lambda[\sin\{(\lambda+1)\alpha\}\cos\{(\lambda-1)\alpha\}-\sin\{(\lambda-1)\alpha\}\cos\{(\lambda+1)\alpha\}]\\&\quad+[\sin\{(\lambda+1)\alpha\}\cos\{(\lambda-1)\alpha\}+\sin\{(\lambda-1)\alpha\}\cos\{(\lambda+1)\alpha\}]=0\\&\Rightarrow \lambda\sin[(\lambda+1)\alpha-(\lambda-1)\alpha]+\sin[(\lambda+1)\alpha+(\lambda-1)\alpha]=0\\&\Rightarrow \lambda\sin[2\alpha]+\sin[2\lambda\alpha]=0\end{aligned} \tag{2.20}$$

The second equation of (2.19) gives

$$\begin{aligned}&\lambda[\cos\{(\lambda+1)\alpha\}\sin\{(\lambda-1)\alpha\}-\cos\{(\lambda-1)\alpha\}\sin\{(\lambda+1)\alpha\}]\\&\quad+\lambda[\cos\{(\lambda+1)\alpha\}\sin\{(\lambda-1)\alpha\}+\cos\{(\lambda-1)\alpha\}\sin\{(\lambda+1)\alpha\}]=0\\&\Rightarrow \lambda\sin[\{(\lambda-1)\alpha\}-\{(\lambda+1)\alpha\}]+\sin[\{(\lambda-1)\alpha\}+\{(\lambda+1)\alpha\}]=0\\&\Rightarrow -\lambda\sin[2\alpha]+\sin[2\lambda\alpha]=0\end{aligned} \tag{2.21}$$

From equations (2.20) and (2.21), one obtains

$$\begin{aligned}&\sin(2\lambda\alpha)=0\\&\lambda\sin(2\alpha)=0\end{aligned} \tag{2.22}$$

For $\alpha=\pi$, the second equation of (2.22) is automatically satisfied and the first equation becomes

$$\sin(2\pi\lambda)=0 \tag{2.23}$$

Equation (2.23) is satisfied for $\lambda=\frac{n}{2}$, where $n=\pm0, \pm1, \pm2, \pm3, \ldots$.

For different values of $\lambda$, the function given in equation (2.10) has different expressions with different constants, $c_{in}$, $i = 1, 2, 3$, and 4:

$$\begin{aligned}F_n(\theta)=c_{1n}\cos\left\{\left(\frac{n}{2}-1\right)\theta\right\}+c_{2n}\sin\left\{\left(\frac{n}{2}-1\right)\theta\right\}\\+c_{3n}\cos\left\{\left(\frac{n}{2}+1\right)\theta\right\}+c_{4n}\sin\left\{\left(\frac{n}{2}+1\right)\theta\right\}\end{aligned} \tag{2.24}$$

### 2.2.2 Acceptable Values of $n$ and $\lambda$

From equation (2.12) one can see that the stress components have the $r$ dependence, $r^{\lambda-1}$. So strains should also have the same $r$ dependence, $r^{\lambda-1}$. Since displacements are obtained by integrating strains, the displacements should have the $r$ dependence, $r^{\lambda}$. Therefore, $\lambda$ cannot have any negative value since the displacement cannot be infinity at the crack tip where $r = 0$. $\lambda$ cannot be zero either because then the displacement components become a function of $\theta$, which can give rise to multiple values of displacement at the crack tip. This is because at the origin $r = 0$, $\theta$ can have any value between $+\alpha$ and $-\alpha$. To have a single value of the displacement at the crack tip, only positive values of $\lambda$ are acceptable. Thus,

$$n = 1,\ 2,\ 3,\ 4\ \ldots$$
$$\lambda = \frac{1}{2},\ 1,\ \frac{3}{2},\ 2\ \ldots \tag{2.25}$$

Substituting $\lambda = \frac{n}{2}$ and $\alpha = \pi$ in equation (2.17), one obtains

$$
\begin{aligned}
&c_{1n}\cos\left\{\left(\frac{n}{2}-1\right)\pi\right\} + c_{3n}\cos\left\{\left(\frac{n}{2}+1\right)\pi\right\} = 0\\
&c_{1n}\left(\frac{n}{2}-1\right)\sin\left\{\left(\frac{n}{2}-1\right)\pi\right\} + c_{3n}\left(\frac{n}{2}+1\right)\sin\left\{\left(\frac{n}{2}+1\right)\pi\right\} = 0\\
&c_{2n}\sin\left\{\left(\frac{n}{2}-1\right)\pi\right\} + c_{4n}\sin\left\{\left(\frac{n}{2}+1\right)\pi\right\} = 0\\
&c_{2n}\left(\frac{n}{2}-1\right)\cos\left\{\left(\frac{n}{2}-1\right)\pi\right\} + c_{4n}\left(\frac{n}{2}+1\right)\cos\left\{\left(\frac{n}{2}+1\right)\pi\right\} = 0
\end{aligned} \tag{2.26}
$$

For odd values of $n$ (= 1, 3, 5...), the first and fourth equations of (2.26) are automatically satisfied and the remaining two equations can be simplified after substituting the following relations:

$$\sin\left\{\pi + \frac{n\pi}{2}\right\} = -\sin\left(\frac{n\pi}{2}\right)$$

$$\sin\left\{\pi - \frac{n\pi}{2}\right\} = \sin\left(\frac{n\pi}{2}\right)$$

$$\Rightarrow \sin\left\{\frac{n\pi}{2} - \pi\right\} = \sin\left\{-\left(\pi - \frac{n\pi}{2}\right)\right\} = -\sin\left\{\pi - \frac{n\pi}{2}\right\} = -\sin\left(\frac{n\pi}{2}\right)$$

$$c_{1n}\left(\frac{n}{2}-1\right)+c_{3n}\left(\frac{n}{2}+1\right)=0 \quad \Rightarrow c_{3n}=-\frac{n-2}{n+2}c_{1n}$$
$$c_{2n}+c_{4n}=0 \quad \Rightarrow c_{4n}=-c_{2n} \tag{2.27}$$

For even values of $n$ (=2, 4, 6…), the second and third equations of (2.26) are automatically satisfied and the remaining two equations can be simplified after substituting the following relations:

$$\cos\left\{\pi+\frac{n\pi}{2}\right\}=-\cos\left(\frac{n\pi}{2}\right)$$

$$\cos\left\{\pi-\frac{n\pi}{2}\right\}=-\cos\left(\frac{n\pi}{2}\right)$$

$$\Rightarrow \cos\left\{\frac{n\pi}{2}-\pi\right\}=\cos\left\{-\left(\pi-\frac{n\pi}{2}\right)\right\}=\cos\left\{\pi-\frac{n\pi}{2}\right\}=-\cos\left(\frac{n\pi}{2}\right)$$

$$c_{1n}+c_{3n}=0 \quad \Rightarrow c_{3n}=-c_{1n}$$
$$c_{2n}\left(\frac{n}{2}-1\right)+c_{4n}\left(\frac{n}{2}+1\right)=0 \quad \Rightarrow c_{4n}=-\frac{n-2}{n+2}c_{2n} \tag{2.28}$$

Using equations (2.12), (2.24), (2.27), and (2.28), the stress components can be expressed as a series expression:

$$\sigma_{\theta\theta}=\sum_{n=1,2,3}\frac{n}{2}\left(\frac{n}{2}+1\right)r^{\frac{n}{2}-1}\left[c_{1n}\cos\left\{\left(\frac{n}{2}-1\right)\theta\right\}+c_{2n}\sin\left\{\left(\frac{n}{2}-1\right)\theta\right\}\right.$$
$$\left.+c_{3n}\cos\left\{\left(\frac{n}{2}+1\right)\theta\right\}+c_{4n}\sin\left\{\left(\frac{n}{2}+1\right)\theta\right\}\right]$$
$$=\sum_{n=1,3,5}\frac{n}{2}\left(\frac{n}{2}+1\right)r^{\frac{n}{2}-1}\left[c_{1n}\left(\cos\left\{\left(\frac{n}{2}-1\right)\theta\right\}-\frac{n-2}{n+2}\cos\left\{\left(\frac{n}{2}+1\right)\theta\right\}\right)\right. \tag{2.29a}$$
$$\left.+c_{2n}\left(\sin\left\{\left(\frac{n}{2}-1\right)\theta\right\}-\sin\left\{\left(\frac{n}{2}+1\right)\theta\right\}\right)\right]$$
$$+\sum_{n=2,4,6}\frac{n}{2}\left(\frac{n}{2}+1\right)r^{\frac{n}{2}-1}\left[c_{1n}\left(\cos\left\{\left(\frac{n}{2}-1\right)\theta\right\}-\cos\left\{\left(\frac{n}{2}+1\right)\theta\right\}\right)\right.$$
$$\left.+c_{2n}\left(\sin\left\{\left(\frac{n}{2}-1\right)\theta\right\}-\frac{n-2}{n+2}\sin\left\{\left(\frac{n}{2}+1\right)\theta\right\}\right)\right]$$

and

$$\sigma_{r\theta} = -\sum_{n=1,2,3} \frac{n}{2} r^{\frac{n}{2}-1}\left[-c_{1n}\left(\frac{n}{2}-1\right)\sin\left\{\left(\frac{n}{2}-1\right)\theta\right\} + c_{2n}\left(\frac{n}{2}-1\right)\cos\left\{\left(\frac{n}{2}-1\right)\theta\right\}\right.$$

$$\left. -c_{3n}\left(\frac{n}{2}+1\right)\sin\left\{\left(\frac{n}{2}+1\right)\theta\right\} + c_{4n}\left(\frac{n}{2}+1\right)\cos\left\{\left(\frac{n}{2}+1\right)\theta\right\}\right]$$

$$= \sum_{n=1,3,5} \frac{n}{2} r^{\frac{n}{2}-1}\left[c_{1n}\left(\left(\frac{n}{2}-1\right)\sin\left\{\left(\frac{n}{2}-1\right)\theta\right\} - \frac{n-2}{n+2}\left(\frac{n}{2}+1\right)\sin\left\{\left(\frac{n}{2}+1\right)\theta\right\}\right)\right.$$

$$\left. + c_{2n}\left(-\left(\frac{n}{2}-1\right)\cos\left\{\left(\frac{n}{2}-1\right)\theta\right\} + \left(\frac{n}{2}+1\right)\cos\left\{\left(\frac{n}{2}+1\right)\theta\right\}\right)\right] \tag{2.29b}$$

$$+ \sum_{n=2,4,6} \frac{n}{2} r^{\frac{n}{2}-1}\left[c_{1n}\left(\left(\frac{n}{2}-1\right)\sin\left\{\left(\frac{n}{2}-1\right)\theta\right\} - \left(\frac{n}{2}+1\right)\sin\left\{\left(\frac{n}{2}+1\right)\theta\right\}\right)\right.$$

$$\left. + c_{2n}\left(-\left(\frac{n}{2}-1\right)\cos\left\{\left(\frac{n}{2}-1\right)\theta\right\} + \frac{n-2}{n+2}\left(\frac{n}{2}+1\right)\cos\left\{\left(\frac{n}{2}+1\right)\theta\right\}\right)\right]$$

### 2.2.3 Dominant Term

Among all acceptable values of $\lambda$ the term that gives maximum stress value near the crack tip corresponds to $n = 1$ or $\lambda = ½$, because then the stress field near the crack tip is of the order of $r^{-1/2}$. Clearly, for small values of $r$ (when the point is located very close to the crack tip), the stress values become very large and, at the crack tip, become infinity. For other values of $\lambda$ and $n$ ($n$ = 2, 3, 4, 5…), one gets bounded stress field at the crack tip. Naturally, those terms (corresponding to $n$ = 2, 3, 4, 5…) are not as important as the term corresponding to $\lambda = ½$ (or $n = 1$).

Substituting $n = 1$ in equation (2.29),

$$\sigma_{\theta\theta} = \frac{1}{2}\left(\frac{1}{2}+1\right) r^{\frac{1}{2}-1}\left[c_{11}\left(\cos\left\{\left(\frac{1}{2}-1\right)\theta\right\} - \frac{1-2}{1+2}\cos\left\{\left(\frac{1}{2}+1\right)\theta\right\}\right)\right.$$

$$\left. + c_{21}\left(\sin\left\{\left(\frac{1}{2}-1\right)\theta\right\} - \sin\left\{\left(\frac{1}{2}+1\right)\theta\right\}\right)\right] \tag{2.30a}$$

$$= \frac{3}{4\sqrt{r}}\left[c_{11}\left\{\cos\left(\frac{\theta}{2}\right) + \frac{1}{3}\cos\left(\frac{3\theta}{2}\right)\right\} - c_{21}\left\{\sin\left(\frac{\theta}{2}\right) + \sin\left(\frac{3\theta}{2}\right)\right\}\right]$$

and

$$\sigma_{r\theta} = \frac{1}{2} r^{\frac{1}{2}-1} \left[ c_{11} \left( \left( \frac{1}{2} - 1 \right) \sin\left\{ \left( \frac{1}{2} - 1 \right) \theta \right\} - \frac{1-2}{1+2} \left( \frac{1}{2} + 1 \right) \sin\left\{ \left( \frac{1}{2} + 1 \right) \theta \right\} \right) \right.$$

$$\left. + c_{21} \left( -\left( \frac{1}{2} - 1 \right) \cos\left\{ \left( \frac{1}{2} - 1 \right) \theta \right\} + \left( \frac{1}{2} + 1 \right) \cos\left\{ \left( \frac{1}{2} + 1 \right) \theta \right\} \right) \right] \tag{2.30b}$$

$$= \frac{1}{4\sqrt{r}} \left[ c_{11} \left\{ \sin\left( \frac{\theta}{2} \right) + \sin\left( \frac{3\theta}{2} \right) \right\} + c_{21} \left\{ \cos\left( \frac{\theta}{2} \right) + 3\cos\left( \frac{3\theta}{2} \right) \right\} \right]$$

From equation (2.30) one can clearly see that if the problem geometry and loading are symmetric about the $x_1$ axis, then only the first term with the coefficient $c_{11}$ should be considered; for antisymmetric loadings, only the second term with the coefficient $c_{21}$ should be considered.

The even and odd terms of equation (2.30) can be further simplified as

$$\sigma_{\theta\theta}\big|_{symm} = \frac{3c_{11}}{4\sqrt{r}} \left\{ \cos\left( \frac{\theta}{2} \right) + \frac{1}{3} \cos\left( \frac{3\theta}{2} \right) \right\}$$

$$= \frac{3c_{11}}{4\sqrt{r}} \left\{ \cos\left( \frac{\theta}{2} \right) + \frac{1}{3} \left[ 4\cos^3\left( \frac{\theta}{2} \right) - 3\cos\left( \frac{\theta}{2} \right) \right] \right\} \tag{2.31}$$

$$= \frac{c_{11}}{\sqrt{r}} \cos^3\left( \frac{\theta}{2} \right) = \frac{K_I}{\sqrt{2\pi r}} \cos^3\left( \frac{\theta}{2} \right)$$

where $K_I = c_{11}\sqrt{2\pi}$. Similarly,

$$\sigma_{\theta\theta}\big|_{antisymm} = -\frac{3c_{21}}{4\sqrt{r}} \left\{ \sin\left( \frac{\theta}{2} \right) + \sin\left( \frac{3\theta}{2} \right) \right\}$$

$$= -\frac{3c_{21}}{4\sqrt{r}} \left\{ \sin\left( \frac{\theta}{2} \right) + 3\sin\left( \frac{\theta}{2} \right) - 4\sin^3\left( \frac{\theta}{2} \right) \right\}$$

$$= -\frac{3c_{21}}{\sqrt{r}} \sin\left( \frac{\theta}{2} \right) \left\{ 1 - \sin^2\left( \frac{\theta}{2} \right) \right\} \tag{2.32}$$

$$= -\frac{3c_{21}}{\sqrt{r}} \sin\left( \frac{\theta}{2} \right) \cos^2\left( \frac{\theta}{2} \right) = -\frac{K_{II}}{\sqrt{2\pi r}} 3\sin\left( \frac{\theta}{2} \right) \cos^2\left( \frac{\theta}{2} \right)$$

where $K_{II} = c_{21}\sqrt{2\pi}$.

$$\sigma_{r\theta}\big|_{symm} = \frac{c_{11}}{4\sqrt{r}}\left\{\sin\left(\frac{\theta}{2}\right)+\sin\left(\frac{3\theta}{2}\right)\right\} = \frac{c_{11}}{4\sqrt{r}}\left\{\sin\left(\frac{\theta}{2}\right)+3\sin\left(\frac{\theta}{2}\right)-4\sin^3\left(\frac{\theta}{2}\right)\right\}$$

$$= \frac{c_{11}}{\sqrt{r}}\left\{\sin\left(\frac{\theta}{2}\right)-\sin^3\left(\frac{\theta}{2}\right)\right\} \tag{2.33}$$

$$= \frac{c_{11}}{\sqrt{r}}\sin\left(\frac{\theta}{2}\right)\cos^2\left(\frac{\theta}{2}\right) = \frac{K_I}{\sqrt{2\pi r}}\sin\left(\frac{\theta}{2}\right)\cos^2\left(\frac{\theta}{2}\right)$$

$$\sigma_{r\theta}\big|_{antisymm} = \frac{c_{21}}{4\sqrt{r}}\left\{\cos\left(\frac{\theta}{2}\right)+3\cos\left(\frac{3\theta}{2}\right)\right\}$$

$$= \frac{c_{21}}{4\sqrt{r}}\left\{\cos\left(\frac{\theta}{2}\right)+3\left(4\cos^3\left(\frac{\theta}{2}\right)-3\cos\left(\frac{\theta}{2}\right)\right)\right\}$$

$$= \frac{c_{21}}{4\sqrt{r}}\left\{12\cos^3\left(\frac{\theta}{2}\right)-8\cos\left(\frac{\theta}{2}\right)\right\} = \frac{c_{21}}{\sqrt{r}}\cos\left(\frac{\theta}{2}\right)\left\{3\cos^2\left(\frac{\theta}{2}\right)-2\right\}$$

$$= \frac{c_{21}}{\sqrt{r}}\cos\left(\frac{\theta}{2}\right)\left\{1-3\sin^2\left(\frac{\theta}{2}\right)\right\} = \frac{K_{II}}{\sqrt{2\pi r}}\cos\left(\frac{\theta}{2}\right)\left\{1-3\sin^2\left(\frac{\theta}{2}\right)\right\} \tag{2.34}$$

In the same manner, the $\sigma_{rr}$ expression can be derived. From equation (2.12),

$$\sigma_{rr} = r^{\lambda-1}\{F''(\theta)+(\lambda+1)F(\theta)\}$$

$$= \sum_n r^{\lambda_n-1}\{F_n''(\theta)+(\lambda_n+1)F_n(\theta)\} \qquad \text{since } \lambda \text{ has multiple values } \lambda_n$$

$$= \sum_n r^{\left(\frac{n}{2}-1\right)}\left[\left\{\left(\frac{n}{2}+1\right)-\left(\frac{n}{2}-1\right)^2\right\}\left\{c_{1n}\cos\left(\frac{n}{2}-1\right)\theta+c_{2n}\sin\left(\frac{n}{2}-1\right)\theta\right\}\right.$$

$$\left.+\left\{\left(\frac{n}{2}+1\right)-\left(\frac{n}{2}+1\right)^2\right\}\left\{c_{3n}\cos\left(\frac{n}{2}+1\right)\theta+c_{4n}\sin\left(\frac{n}{2}+1\right)\theta\right\}\right] \tag{2.35}$$

For $n = 1$,

$$\sigma_{rr} = r^{\left(\frac{1}{2}-1\right)}\left[\left\{\left(\frac{1}{2}+1\right)-\left(\frac{1}{2}-1\right)^2\right\}\left\{c_{11}\cos\left(\frac{1}{2}-1\right)\theta - c_{21}\sin\left(\frac{1}{2}-1\right)\theta\right\}\right.$$
$$\left.+\left\{\left(\frac{1}{2}+1\right)-\left(\frac{1}{2}+1\right)^2\right\}\left\{c_{31}\cos\left(\frac{1}{2}+1\right)\theta + c_{41}\sin\left(\frac{1}{2}+1\right)\theta\right\}\right]$$
$$= r^{-\frac{1}{2}}\left[\frac{5}{4}c_{11}\cos\left(\frac{\theta}{2}\right) - \frac{5}{4}c_{21}\sin\left(\frac{\theta}{2}\right) - \frac{3}{4}c_{31}\cos\left(\frac{3\theta}{2}\right) - \frac{3}{4}c_{41}\sin\left(\frac{3\theta}{2}\right)\right]$$
$$= \frac{1}{\sqrt{r}}\left[\left\{\frac{5}{4}\cos\left(\frac{\theta}{2}\right) + \frac{3}{4}\left(\frac{1-2}{1+2}\right)\cos\left(\frac{3\theta}{2}\right)\right\}c_{11} + \left\{-\frac{5}{4}\sin\left(\frac{\theta}{2}\right) + \frac{3}{4}\sin\left(\frac{3\theta}{2}\right)\right\}c_{21}\right]$$
$$= \frac{1}{\sqrt{r}}\left[\left\{\frac{5}{4}\cos\left(\frac{\theta}{2}\right) - \frac{1}{4}\cos\left(\frac{3\theta}{2}\right)\right\}c_{11} + \left\{-\frac{5}{4}\sin\left(\frac{\theta}{2}\right) + \frac{3}{4}\sin\left(\frac{3\theta}{2}\right)\right\}c_{21}\right] \tag{2.36}$$

The symmetric component of equation (2.36) is

$$\sigma_{rr}\big|_{symm} = \frac{c_{11}}{\sqrt{r}}\left\{\frac{5}{4}\cos\left(\frac{\theta}{2}\right) - \frac{1}{4}\cos\left(\frac{3\theta}{2}\right)\right\}$$
$$= \frac{c_{11}}{\sqrt{r}}\left\{\frac{5}{4}\cos\left(\frac{\theta}{2}\right) - \frac{1}{4}\left[4\cos^3\left(\frac{\theta}{2}\right) - 3\cos\left(\frac{\theta}{2}\right)\right]\right\}$$
$$= \frac{c_{11}}{\sqrt{r}}\left\{2\cos\left(\frac{\theta}{2}\right) - \cos^3\left(\frac{\theta}{2}\right)\right\} = \frac{K_I}{\sqrt{2\pi r}}\left\{2\cos\left(\frac{\theta}{2}\right) - \cos^3\left(\frac{\theta}{2}\right)\right\} \tag{2.37}$$
$$= \frac{K_I}{\sqrt{2\pi r}}\cos\left(\frac{\theta}{2}\right)\left\{2 - \cos^2\left(\frac{\theta}{2}\right)\right\}$$

and the antisymmetric component is given by

$$\sigma_{rr}\big|_{antisymm} = \frac{c_{21}}{\sqrt{r}}\left\{-\frac{5}{4}\sin\left(\frac{\theta}{2}\right) + \frac{3}{4}\sin\left(\frac{3\theta}{2}\right)\right\}$$
$$= \frac{c_{21}}{\sqrt{r}}\left\{-\frac{5}{4}\sin\left(\frac{\theta}{2}\right) + \frac{3}{4}\left[3\sin\left(\frac{\theta}{2}\right) - 4\sin^3\left(\frac{\theta}{2}\right)\right]\right\}$$
$$= \frac{c_{21}}{\sqrt{r}}\left\{2\sin\left(\frac{\theta}{2}\right) - 3\sin^3\left(\frac{\theta}{2}\right)\right\} = \frac{K_{II}}{\sqrt{2\pi r}}\left\{2\sin\left(\frac{\theta}{2}\right) - 3\sin^3\left(\frac{\theta}{2}\right)\right\} \tag{2.38}$$
$$= \frac{K_{II}}{\sqrt{2\pi r}}\sin\left(\frac{\theta}{2}\right)\left\{2 - 3\sin^2\left(\frac{\theta}{2}\right)\right\}$$

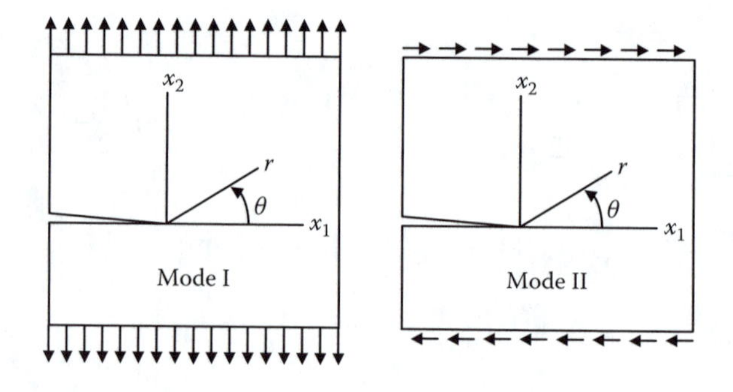


**FIGURE 2.2**
Opening mode (or mode I, left figure) and shearing or sliding mode (or mode II, right figure) loading of a cracked plate.

Symmetric and antisymmetric responses are obtained when the cracked plate is loaded symmetrically and antisymmetrically, respectively, as shown in Figure 2.2. When the load is applied in a direction perpendicular to the crack surface, as shown on the left-hand side of Figure 2.2, then that loading is called opening mode or mode I loading. It is called opening mode because the applied load tries to open the crack. When the load is applied parallel to the crack surface, as shown on the right-hand side of Figure 2.2, then the applied load tries to shear off the crack surface; that loading is called shearing or sliding mode or mode II loading. If the applied load has both opening mode and shearing mode components, then that loading is called mixed mode loading.

### 2.2.4 Strain and Displacement Fields

From the computed stress fields, the strain and displacement fields can be obtained as described in the following sections.

#### 2.2.4.1 *Plane Stress Problems*

For mode I loading from equations (2.37) and (2.31), one obtains

$$E\varepsilon_{rr} = \sigma_{rr} - \nu\sigma_{\theta\theta} = \frac{K_I}{4\sqrt{2\pi r}}\left[\left\{5\cos\left(\frac{\theta}{2}\right) - \cos\left(\frac{3\theta}{2}\right)\right\} - \nu\left\{3\cos\left(\frac{\theta}{2}\right) + \cos\left(\frac{3\theta}{2}\right)\right\}\right]$$

$$\Rightarrow E\frac{\partial u_r}{\partial r} = \frac{K_I}{4\sqrt{2\pi r}}\left[(5-3\nu)\cos\left(\frac{\theta}{2}\right) - (1+\nu)\cos\left(\frac{3\theta}{2}\right)\right]$$

$$\Rightarrow u_r = \frac{2K_I\sqrt{r}}{4E\sqrt{2\pi}}\left[(5-3\nu)\cos\left(\frac{\theta}{2}\right) - (1+\nu)\cos\left(\frac{3\theta}{2}\right)\right]$$

$$= \frac{2(1+\nu)K_I\sqrt{r}}{4E\sqrt{2\pi}}\left[\frac{(5-3\nu)}{(1+\nu)}\cos\left(\frac{\theta}{2}\right) - \cos\left(\frac{3\theta}{2}\right)\right]$$

$$
\begin{aligned}
&= \frac{K_I\sqrt{r}}{4\mu\sqrt{2\pi}}\left[\frac{(5-3\nu)}{(1+\nu)}\cos\left(\frac{\theta}{2}\right)-\cos\left(\frac{3\theta}{2}\right)\right] \\
&= \frac{K_I\sqrt{r}}{4\mu\sqrt{2\pi}}\left[(2\kappa-1)\cos\left(\frac{\theta}{2}\right)-\cos\left(\frac{3\theta}{2}\right)\right]
\end{aligned} \tag{2.39}
$$

where

$$\kappa = \frac{3-\nu}{1+\nu} \text{ for plane stress problems} \tag{2.40}$$

and

$$
\begin{aligned}
E\varepsilon_{\theta\theta} &= E\left(\frac{u_r}{r}+\frac{1}{r}\frac{\partial u_\theta}{\partial\theta}\right) = \sigma_{\theta\theta}-\nu\sigma_{rr} \\
&= \frac{K_I}{4\sqrt{2\pi r}}\left[(3-5\nu)\cos\left(\frac{\theta}{2}\right)+(1+\nu)\cos\left(\frac{3\theta}{2}\right)\right] \\
\Rightarrow \frac{\partial u_\theta}{\partial\theta} &= \frac{K_I\sqrt{r}}{4E\sqrt{2\pi}}\left[(3-5\nu)\cos\left(\frac{\theta}{2}\right)+(1+\nu)\cos\left(\frac{3\theta}{2}\right)\right] \\
&\quad - \frac{K_I\sqrt{r}}{4\mu\sqrt{2\pi}}\left[\left(\frac{5-3\nu}{1+\nu}\right)\cos\left(\frac{\theta}{2}\right)-\cos\left(\frac{3\theta}{2}\right)\right] \\
&= \frac{K_I\sqrt{r}}{8\mu\sqrt{2\pi}}\left[\left(\frac{3-5\nu}{1+\nu}\right)\cos\left(\frac{\theta}{2}\right)+\cos\left(\frac{3\theta}{2}\right)\right] \\
&\quad - \frac{K_I\sqrt{r}}{4\mu\sqrt{2\pi}}\left[\left(\frac{5-3\nu}{1+\nu}\right)\cos\left(\frac{\theta}{2}\right)-\cos\left(\frac{3\theta}{2}\right)\right] \\
&= \frac{K_I\sqrt{r}}{4\mu\sqrt{2\pi}}\left[\left(\frac{3-5\nu-10+6\nu}{2(1+\nu)}\right)\cos\left(\frac{\theta}{2}\right)+\left(\frac{1}{2}+1\right)\cos\left(\frac{3\theta}{2}\right)\right] \\
&= \frac{K_I\sqrt{r}}{4\mu\sqrt{2\pi}}\left[\frac{(-7+\nu)}{2(1+\nu)}\cos\left(\frac{\theta}{2}\right)+\frac{3}{2}\cos\left(\frac{3\theta}{2}\right)\right] \\
\therefore u_\theta &= \frac{K_I\sqrt{r}}{4\mu\sqrt{2\pi}}\left[\frac{2(-7+\nu)}{2(1+\nu)}\sin\left(\frac{\theta}{2}\right)+\frac{2}{3}\times\frac{3}{2}\sin\left(\frac{3\theta}{2}\right)\right] \\
&= \frac{K_I\sqrt{r}}{4\mu\sqrt{2\pi}}\left[-(2\kappa+1)\sin\left(\frac{\theta}{2}\right)+\sin\left(\frac{3\theta}{2}\right)\right]
\end{aligned} \tag{2.41}
$$

In equations (2.39) and (2.41) the definition of $\kappa$ is the same, as defined in equation (2.40).

### 2.2.4.2 Plane Strain Problems

For plane strain problems,

$$\begin{aligned} \varepsilon_{rr} &= \frac{1}{E}(\sigma_{rr} - \nu\sigma_{\theta\theta} - \nu\sigma_{zz}) \\ \varepsilon_{\theta\theta} &= \frac{1}{E}(\sigma_{\theta\theta} - \nu\sigma_{rr} - \nu\sigma_{zz}) \\ \varepsilon_{zz} &= \frac{1}{E}(\sigma_{zz} - \nu\sigma_{\theta\theta} - \nu\sigma_{rr}) = 0 \quad \Rightarrow \sigma_{zz} = \nu(\sigma_{\theta\theta} + \sigma_{rr}) \end{aligned} \tag{2.42}$$

Therefore,

$$E\varepsilon_{rr} = \sigma_{rr} - \nu\sigma_{\theta\theta} - \nu\sigma_{zz} = \sigma_{rr} - \nu\sigma_{\theta\theta} - \nu^2(\sigma_{rr} + \sigma_{\theta\theta}) = (1-\nu^2)\sigma_{rr} - \nu(1+\nu)\sigma_{\theta\theta}$$

$$\begin{aligned} \frac{E}{(1+\nu)}\varepsilon_{rr} &= (1-\nu)\sigma_{rr} - \nu\sigma_{\theta\theta} \\ \frac{E}{(1+\nu)}\varepsilon_{\theta\theta} &= (1-\nu)\sigma_{\theta\theta} - \nu\sigma_{rr} \end{aligned} \tag{2.43}$$

Then, for mode I loading,

$$\begin{aligned} \frac{E}{(1+\nu)}\varepsilon_{rr} = 2\mu\frac{\partial u_r}{\partial r} &= \frac{(1-\nu)K_I}{4\sqrt{2\pi r}}\left(5\cos\frac{\theta}{2} - \cos\frac{3\theta}{2}\right) - \frac{\nu K_I}{4\sqrt{2\pi r}}\left(3\cos\frac{\theta}{2} + \cos\frac{3\theta}{2}\right) \\ &= \frac{K_I}{4\sqrt{2\pi r}}\left((5-5\nu-3\nu)\cos\frac{\theta}{2} - \cos\frac{3\theta}{2}\right) \\ &= \frac{K_I}{4\sqrt{2\pi r}}\left((5-8\nu)\cos\frac{\theta}{2} - \cos\frac{3\theta}{2}\right) \end{aligned}$$

$$\therefore u_r = \frac{K_I\sqrt{r}}{4\mu\sqrt{2\pi}}\left((5-8\nu)\cos\frac{\theta}{2} - \cos\frac{3\theta}{2}\right) = \frac{K_I\sqrt{r}}{4\mu\sqrt{2\pi}}\left((2\kappa-1)\cos\frac{\theta}{2} - \cos\frac{3\theta}{2}\right) \tag{2.44}$$

where, for plane strain problems,

$$\kappa = 3 - 4\nu \tag{2.45}$$

Similarly,

$$\begin{aligned} \frac{E}{(1+\nu)}\varepsilon_{\theta\theta} = 2\mu\left(\frac{u_r}{r} + \frac{1}{r}\frac{\partial u_\theta}{\partial\theta}\right) &= (1-\nu)\sigma_{\theta\theta} - \nu\sigma_{rr} \\ &= (1-\nu)\frac{K_I}{4\sqrt{2\pi r}}\left(3\cos\frac{\theta}{2} + \cos\frac{3\theta}{2}\right) - \frac{\nu K_I}{4\sqrt{2\pi r}}\left(5\cos\frac{\theta}{2} - \cos\frac{3\theta}{2}\right) \end{aligned}$$

$$\Rightarrow \frac{1}{r}\frac{\partial u_\theta}{\partial \theta} = \frac{K_I(1-\nu)}{8\mu\sqrt{2\pi r}}\left(3\cos\frac{\theta}{2} + \cos\frac{3\theta}{2}\right) - \frac{\nu K_I}{8\mu\sqrt{2\pi r}}\left(5\cos\frac{\theta}{2} - \cos\frac{3\theta}{2}\right)$$
$$- \frac{K_I}{4\mu\sqrt{2\pi r}}\left((5-8\nu)\cos\frac{\theta}{2} - \cos\frac{3\theta}{2}\right)$$

$$\Rightarrow \frac{\partial u_\theta}{\partial \theta} = \frac{K_I\sqrt{r}}{8\mu\sqrt{2\pi}}\left((3-3\nu-5\nu-10+16\nu)\cos\frac{\theta}{2} + (1-\nu+\nu+2)\cos\frac{3\theta}{2}\right)$$
$$= \frac{K_I\sqrt{r}}{8\mu\sqrt{2\pi}}\left((-5+8\nu)\cos\frac{\theta}{2} + 3\cos\frac{3\theta}{2}\right) \tag{2.46}$$

$$\therefore u_\theta = \frac{K_I\sqrt{r}}{8\mu\sqrt{2\pi}}\left(2(-5+8\nu)\sin\frac{\theta}{2} + 3\times\frac{2}{3}\sin\frac{3\theta}{2}\right)$$
$$= \frac{K_I\sqrt{r}}{4\mu\sqrt{2\pi}}\left(-(2\kappa-1)\sin\frac{\theta}{2} + \sin\frac{3\theta}{2}\right)$$

Similarly, the displacement fields for mode II loading can be obtained. After obtaining stress and displacement fields in the polar coordinate system, those expressions in the Cartesian coordinate system can be derived by transforming the stress and displacement expressions using the transformation laws.

In summary, near a crack tip where the term corresponding to $n = 1$ dominates, the stress and displacement fields in polar and Cartesian coordinate systems are given by

$$\sigma_{\theta\theta} = \frac{K_I}{\sqrt{2\pi r}}\cos^3\left(\frac{\theta}{2}\right) - \frac{K_{II}}{\sqrt{2\pi r}}3\sin\left(\frac{\theta}{2}\right)\cos^2\left(\frac{\theta}{2}\right)$$
$$= \frac{K_I}{4\sqrt{2\pi r}}\left[3\cos\left(\frac{\theta}{2}\right) + \cos\left(\frac{3\theta}{2}\right)\right] + \frac{K_{II}}{4\sqrt{2\pi r}}\left[-3\sin\left(\frac{\theta}{2}\right) - 3\sin\left(\frac{3\theta}{2}\right)\right]$$

$$\sigma_{rr} = \frac{K_I}{\sqrt{2\pi r}}\cos\left(\frac{\theta}{2}\right)\left[1+\sin^2\left(\frac{\theta}{2}\right)\right] + \frac{K_{II}}{\sqrt{2\pi r}}\sin\left(\frac{\theta}{2}\right)\left[1-3\sin^2\left(\frac{\theta}{2}\right)\right]$$
$$= \frac{K_I}{4\sqrt{2\pi r}}\left[5\cos\left(\frac{\theta}{2}\right) - \cos\left(\frac{3\theta}{2}\right)\right] + \frac{K_{II}}{4\sqrt{2\pi r}}\left[-5\sin\left(\frac{\theta}{2}\right) + 3\sin\left(\frac{3\theta}{2}\right)\right] \tag{2.47a}$$

$$\sigma_{r\theta} = \frac{K_I}{\sqrt{2\pi r}}\sin\left(\frac{\theta}{2}\right)\cos^2\left(\frac{\theta}{2}\right) + \frac{K_{II}}{\sqrt{2\pi r}}\cos\left(\frac{\theta}{2}\right)\left[1-3\sin^2\left(\frac{\theta}{2}\right)\right]$$
$$= \frac{K_I}{4\sqrt{2\pi r}}\left[\sin\left(\frac{\theta}{2}\right) + \sin\left(\frac{3\theta}{2}\right)\right] + \frac{K_{II}}{4\sqrt{2\pi r}}\left[\cos\left(\frac{\theta}{2}\right) + 3\cos\left(\frac{3\theta}{2}\right)\right]$$

$$\sigma_{yy} = \frac{K_I}{\sqrt{2\pi r}}\cos\left(\frac{\theta}{2}\right)\left[1+\sin\left(\frac{\theta}{2}\right)\sin\left(\frac{3\theta}{2}\right)\right]+\frac{K_{II}}{\sqrt{2\pi r}}\frac{1}{2}\sin(\theta)\cos\left(\frac{3\theta}{2}\right)$$

$$\sigma_{xx} = \frac{K_I}{\sqrt{2\pi r}}\cos\left(\frac{\theta}{2}\right)\left[1-\sin\left(\frac{\theta}{2}\right)\sin\left(\frac{3\theta}{2}\right)\right]$$
$$+\frac{K_{II}}{\sqrt{2\pi r}}\sin\left(\frac{\theta}{2}\right)\left[-2-\cos\left(\frac{\theta}{2}\right)\cos\left(\frac{3\theta}{2}\right)\right] \tag{2.47b}$$

$$\sigma_{xy} = \frac{K_I}{\sqrt{2\pi r}}\frac{1}{2}\sin(\theta)\cos\left(\frac{3\theta}{2}\right)+\frac{K_{II}}{\sqrt{2\pi r}}\cos\left(\frac{\theta}{2}\right)\left[1-\sin\left(\frac{\theta}{2}\right)\sin\left(\frac{3\theta}{2}\right)\right]$$

$$u_\theta = \frac{K_I\sqrt{r}}{4\mu\sqrt{2\pi}}\left[-(2\kappa+1)\sin\left(\frac{\theta}{2}\right)+\sin\left(\frac{3\theta}{2}\right)\right]$$
$$+\frac{K_{II}\sqrt{r}}{4\mu\sqrt{2\pi}}\left[-(2\kappa+1)\cos\left(\frac{\theta}{2}\right)+3\cos\left(\frac{3\theta}{2}\right)\right]$$
$$u_r = \frac{K_I\sqrt{r}}{4\mu\sqrt{2\pi}}\left[(2\kappa-1)\cos\left(\frac{\theta}{2}\right)-\cos\left(\frac{3\theta}{2}\right)\right]$$
$$+\frac{K_{II}\sqrt{r}}{4\mu\sqrt{2\pi}}\left[-(2\kappa-1)\sin\left(\frac{\theta}{2}\right)+3\sin\left(\frac{3\theta}{2}\right)\right] \tag{2.48}$$
$$u_y = \frac{K_I\sqrt{r}}{2\mu\sqrt{2\pi}}\left[(\kappa+1)\sin\left(\frac{\theta}{2}\right)-\sin(\theta)\cos\left(\frac{\theta}{2}\right)\right]$$
$$+\frac{K_{II}\sqrt{r}}{2\mu\sqrt{2\pi}}\cos\left(\frac{\theta}{2}\right)\left[(\kappa-1)+2\sin^2\left(\frac{\theta}{2}\right)\right]$$
$$u_x = \frac{K_I\sqrt{r}}{2\mu\sqrt{2\pi}}\left[(\kappa-1)\cos\left(\frac{\theta}{2}\right)+\sin(\theta)\sin\left(\frac{\theta}{2}\right)\right]$$
$$+\frac{K_{II}\sqrt{r}}{2\mu\sqrt{2\pi}}\sin\left(\frac{\theta}{2}\right)\left[(\kappa+1)+2\cos^2\left(\frac{\theta}{2}\right)\right]$$

where $\kappa = \frac{3-\nu}{1+\nu}$ for plane stress problems and $\kappa = 3 - 4\nu$ for plane strain problems.

## 2.3 Stress Intensity Factor and Fracture Toughness

Equation (2.47) implies that for any two-dimensional in-plane (plane stress or plane strain) problems, irrespective of the problem geometry and applied load distribution, the stress field near the crack tip is dominated by the expressions given in equation (2.47) because of the square root singularity in the stress field expressions. Values of the constants $K_I$ and $K_{II}$ depend on the problem geometry and the applied load distribution; however, $r$ and $\theta$ dependence of the functions does not change with the problem geometry or loading. $K_I$ and $K_{II}$ are called the *stress intensity factor* (SIF) for mode I and mode II loadings, respectively. Note that the stress field near the crack tip is linearly dependent on the SIF; in other words, if $K_I$ and $K_{II}$ are multiplied by a factor $n$, then the stress and displacement fields near the crack tip are also multiplied by the same factor $n$, while the stress values at the crack tip remain infinity.

Since the stress values near a crack tip are always very high (and infinity at the tip), the strength-of-material approach of failure prediction that the material fails when the stress exceeds some critical value (ultimate stress or yield stress) cannot be used here. When a cracked plate is subjected to a small load, although the stress field near the crack tip becomes very high, the plate does not fail. However, as the applied load increases to some critical value, the plate fails. In the fracture mechanics approach, instead of comparing the maximum stress value with a critical stress value, the material failure is predicted by comparing the stress intensity factors $K_I$ and $K_{II}$ with some critical value $K_c$. This critical value is called the *critical stress intensity factor* or the *fracture toughness* of the material. It will be shown later that when the applied load is small, $K_I$ and $K_{II}$ values are small. As the applied load increases, $K_I$ and $K_{II}$ values increase proportionately and the structure fails when $K_I$, $K_{II}$ values exceed some critical stress intensity factor or the fracture toughness of the material. Note that $K_c$ is a material property, as the ultimate stress and the yield stress are, while $K_I$ and $K_{II}$ depend on the problem geometry and applied loads, as the stress developed inside a structure is dependent on the problem geometry and applied loads.

## 2.4 Stress and Displacement Fields for Antiplane Problems

When the applied loads act in the $x_3$ direction only but the problem geometry and applied loads are functions of $x_1$ and $x_2$ only (see the Figure of exercise problem 2.2, Figure 2.8), the solution fields are expected to have the following form:

$$
\begin{aligned}
u_1 &= u_2 = 0 \\
u_3 &= u_3(x_1, x_2) = u_3(r, \theta)
\end{aligned}
\tag{2.49}
$$

These problems are called antiplane or out-of-plane problems. Applying Williams' method to antiplane problems, it can be shown (see the solution of exercise problem 2.3) that for such problems the stress and displacement fields are given by

$$\sigma_{\theta 3} = \frac{K_{III}}{\sqrt{2\pi r}}\cos\left(\frac{\theta}{2}\right)$$

$$\sigma_{r3} = \frac{K_{III}}{\sqrt{2\pi r}}\sin\left(\frac{\theta}{2}\right) \tag{2.50}$$

$$u_3 = \frac{K_{III}}{\mu}\sqrt{\frac{2r}{\pi}}\sin\left(\frac{\theta}{2}\right)$$

This mode of loading is called the *tearing mode* or mode III loading. It is known as the tearing mode because this type of loading is applied to tear a paper. $K_{III}$ is the stress intensity factor for mode III loading.

## 2.5 Different Modes of Fracture

From the preceding discussions one can see that three different modes of loading relative to the crack geometry can exist. These are mode I (opening mode), mode II (shearing or sliding mode), and mode III (tearing mode), as shown in Figure 2.3. Note that modes I and II correspond to the in-plane problems and mode III corresponds to the out-of-plane problems. If the crack propagates under any of these three modes, then the failure or the fracture mode is identified with that mode. Under general loading conditions, the applied load may have components of all three modes; this situation is known as mixed mode loading. Failure or crack propagation under mixed mode loading is known as the mixed mode failure or mixed mode fracture.

## 2.6 Direction of Crack Propagation

Erdogan and Sih's (1963) hypothesis of crack propagation is that the crack propagates in the direction perpendicular to the direction of the maximum tensile stress $\sigma_{\theta\theta}$. This hypothesis is identified as the direction of the

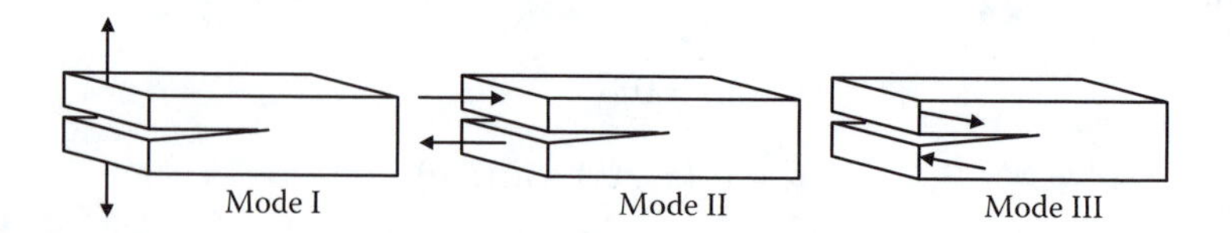


**FIGURE 2.3**
Three modes of loading relative to a crack: mode I (opening mode), mode II (shearing or sliding mode), and mode III (tearing mode).

maximum tensile stress hypothesis. Although other hypotheses, such as that the crack propagates in the maximum strain energy release rate direction, can be adopted to predict the crack propagation direction. For its simplicity, the maximum stress criterion is followed here to predict the crack propagation direction. It can be shown that all these hypotheses give similar results.

From equation (2.47),

$$\sigma_{\theta\theta} = \frac{1}{\sqrt{2\pi r}}\left\{K_I \cos^3\left(\frac{\theta}{2}\right) - 3K_{II}\sin\left(\frac{\theta}{2}\right)\cos^2\left(\frac{\theta}{2}\right)\right\}$$

$$\therefore \frac{\partial\sigma_{\theta\theta}}{\partial\theta} = \frac{1}{\sqrt{2\pi r}}\left\{-\frac{3}{2}K_I\cos^2\left(\frac{\theta}{2}\right)\sin\left(\frac{\theta}{2}\right) - \frac{3}{2}K_{II}\cos^3\left(\frac{\theta}{2}\right)\right. \tag{2.51}$$

$$\left. + 3K_{II}\sin^2\left(\frac{\theta}{2}\right)\cos\left(\frac{\theta}{2}\right)\right\}$$

For maximum value of $\sigma_{\theta\theta}$,

$$\frac{\partial\sigma_{\theta\theta}}{\partial\theta} = \frac{-3}{2\sqrt{2\pi r}}\cos\left(\frac{\theta}{2}\right)\left\{K_I\cos\left(\frac{\theta}{2}\right)\sin\left(\frac{\theta}{2}\right) + K_{II}\cos^2\left(\frac{\theta}{2}\right) - 2K_{II}\sin^2\left(\frac{\theta}{2}\right)\right\} = 0 \tag{2.52}$$

Therefore,

$$\cos\left(\frac{\theta}{2}\right)\left\{K_I\cos\left(\frac{\theta}{2}\right)\sin\left(\frac{\theta}{2}\right) + K_{II}\cos^2\left(\frac{\theta}{2}\right) - 2K_{II}\sin^2\left(\frac{\theta}{2}\right)\right\} = 0$$

$$\Rightarrow \cos\left(\frac{\theta}{2}\right)\left\{K_I\cos\left(\frac{\theta}{2}\right)\sin\left(\frac{\theta}{2}\right) + K_{II}\left[3\cos^2\left(\frac{\theta}{2}\right) - 2\right]\right\} = 0 \tag{2.53}$$

$$\Rightarrow \cos\left(\frac{\theta}{2}\right) = 0 \quad \text{or} \quad K_I\cos\left(\frac{\theta}{2}\right)\sin\left(\frac{\theta}{2}\right) + K_{II}\left[3\cos^2\left(\frac{\theta}{2}\right) - 2\right] = 0$$

Equation (2.53) is satisfied if

$$\cos\left(\frac{\theta}{2}\right) = 0 \Rightarrow \frac{\theta}{2} = \pm\frac{\pi}{2}, \pm\frac{3\pi}{2}, \pm\frac{5\pi}{2}, \ldots \quad \text{or} \quad \theta = \pm\pi, \pm 3\pi, \pm 5\pi, \ldots$$

Note that these values of $\theta$ correspond to the top and bottom crack surfaces where $\sigma_{\theta\theta} = 0$. The crack cannot propagate in that direction since the crack surface already exists there. Therefore, $K_I\cos(\frac{\theta}{2})\sin(\frac{\theta}{2}) + K_{II}[3\cos^2(\frac{\theta}{2}) - 2]$ must be equal to zero. In this expression, after substituting $\cos(\frac{\theta}{2}) = x$ for simplicity, one gets

$$\pm K_I x\sqrt{1-x^2} + K_{II}(3x^2-2) = 0$$
$$\Rightarrow K_{II}(3x^2-2) = \quad K_I x\sqrt{1-x^2} \tag{2.54}$$
$$\Rightarrow \frac{K_{II}}{K_I} = \frac{x\sqrt{1-x^2}}{(3x^2-2)}$$

Note that from equation (2.54) it can be stated that for $\frac{K_{II}}{K_I} = 0$—that is, for mode I loading—$x = \cos(\frac{\theta}{2}) = 0$ or $\pm 1$. Therefore, $\theta = 0, \pm\pi, \pm 2\pi, \pm 3\pi, \ldots$ Since $\theta = \pm\pi, \pm 3\pi, \pm 5\pi, \ldots$ are not of interest for the reasons stated earlier (that the crack cannot propagate in the direction in which crack already exists), we only consider $\theta = 0, \pm 2\pi, \pm 4\pi, \pm 6\pi, \ldots$ Note that all these values correspond to the same plane at $\theta = 0$ located just ahead of the crack tip. Therefore, for mode I loading, the crack propagates along $\theta = 0$ plane.

For $\frac{K_{II}}{K_I} = \infty$ or $\frac{K_I}{K_{II}} = 0$—that is, for mode II loading—equation (2.54) is satisfied when $3x^2 - 2 = 0 \Rightarrow x = \pm\sqrt{\frac{2}{3}} = \cos(\frac{\theta}{2})$. Therefore, $\theta = 2\cos^{-1}(\pm\sqrt{\frac{2}{3}}) = \pm 70.53°, \pm 430.53°$.

One can easily show from equation (2.47) that, for $\theta = +70.53°$, the stress component $\sigma_{\theta\theta}$ is negative or compressive. The crack cannot propagate under the compressive stress. The maximum positive normal stress is obtained for $\theta = -70.53°$. Therefore, the crack propagates along the plane located at $\theta = -70.53°$ under mode II loading.

When both $K_I$ and $K_{II}$ have nonzero positive values, then the crack propagates in the direction between $\theta = 0°$ and $\theta = -70.53°$ because $\sigma_{\theta\theta}$ is positive in this region. Then, in equation (2.53), $\cos\frac{\theta}{2}$ is positive and $\sin\frac{\theta}{2}$ is negative. Therefore, the right-hand side of equation (2.54) should be positive, as shown:

$$\frac{K_{II}}{K_I} = \frac{x\sqrt{1-x^2}}{(3x^2-2)} \tag{2.55}$$

For the mixed mode loading condition $0 \le \frac{K_{II}}{K_I} \le \infty$, the direction of crack propagation or the exact value of $\theta$ can be obtained from equation (2.55) as shown in Table 2.1.

**TABLE 2.1**

Crack Propagation Directions ($\theta$) for Different Values of $K_{II}/K_I$

| $\theta$ | 0° | –10° | –20° | –30° | –40° | –50° | –60° | –70° | –70.4° | –70.53° |
|---|---|---|---|---|---|---|---|---|---|---|
| $K_{II}/K_I$ | 0 | 0.0888 | 0.1880 | 0.3129 | 0.4952 | 0.8252 | 1.7320 | 36.058 | 148.25 | $\infty$ |

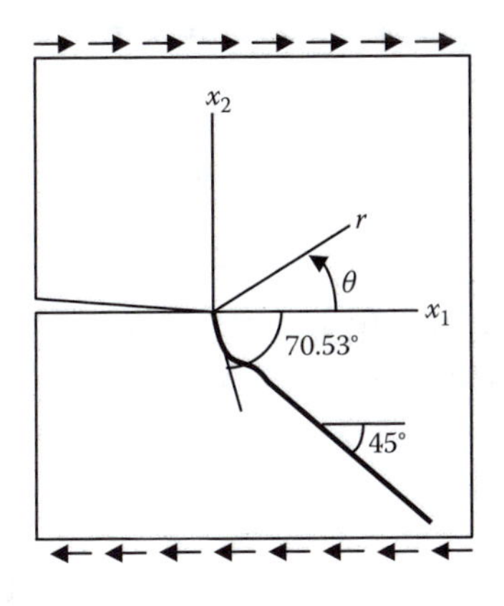

**FIGURE 2.4**
Crack propagation direction (shown by the dark line) under shearing mode (mode II) loading.

For pure shear (mode II) loading, $\frac{K_{II}}{K_I} = \infty$; therefore, the crack starts to propagate at an angle $\theta = -70.53°$ relative to the $x_1$ axis as shown in Figure 2.4. However, as soon as the crack tip propagates a little distance, the applied stress field no longer satisfies the pure mode II condition relative to the crack tip and it becomes a mixed mode loading situation. Under mixed mode loading condition, the crack propagation direction continuously changes and, finally, the crack propagates at an angle 45° relative to the horizontal axis, as shown in Figure 2.4, because this is the direction perpendicular to the maximum tensile stress.

Note that the crack propagation direction has been obtained from the maximum stress hypothesis—that the crack propagates in the direction perpendicular to the maximum $\sigma_{\theta\theta}$ direction. An alternative criterion can be defined based on the potential energy release rate criterion. In this approach the crack front can be extended in different directions by a small amount and the reduction in the total potential energy of the problem geometry can be calculated. The crack should propagate in the direction that gives maximum reduction in the total potential energy. Similar but not necessarily identical results are obtained from the maximum stress and the maximum energy release rate criteria.

## 2.7 Mixed Mode Failure Curve for In-Plane Loading

One can experimentally observe that under mode I loading, the crack starts to propagate along the $\theta = 0$ line when the applied load reaches some critical value that corresponds to some critical stress intensity factor $K_c$.

From equation (2.47) one can compute the maximum $\sigma_{\theta\theta}$ value for crack propagation:

$$\sigma_{\theta\theta}\Big|_{\max} = \frac{K_c}{\sqrt{2\pi r}} \tag{2.56}$$

In the maximum stress hypothesis, it is assumed that under mixed mode loading the crack starts to propagate when $\sigma_{\theta\theta}|_{\max}$ reaches the same critical value of $\frac{K_c}{\sqrt{2\pi r}}$. Therefore, if $\theta = \theta_c$ gives $\sigma_{\theta\theta}|_{\max}$, then the failure criterion can be written as

$$\sigma_{\theta\theta}\Big|_{\max} = \frac{1}{\sqrt{2\pi r}}\left\{K_I \cos^3\left(\frac{\theta_c}{2}\right) - 3K_{II}\sin\left(\frac{\theta_c}{2}\right)\cos^2\left(\frac{\theta_c}{2}\right)\right\} = \frac{K_c}{\sqrt{2\pi r}}$$

$$\therefore K_I \cos^3\left(\frac{\theta_c}{2}\right) - 3K_{II}\sin\left(\frac{\theta_c}{2}\right)\cos^2\left(\frac{\theta_c}{2}\right) = K_c$$

$$\therefore \frac{K_I}{K_c}\left\{\cos^3\left(\frac{\theta_c}{2}\right) - 3\frac{K_{II}}{K_I}\sin\left(\frac{\theta_c}{2}\right)\cos^2\left(\frac{\theta_c}{2}\right)\right\} = 1 \tag{2.57}$$

$$\therefore \frac{K_I}{K_c} = \left\{\cos^3\left(\frac{\theta_c}{2}\right) - 3\frac{K_{II}}{K_I}\sin\left(\frac{\theta_c}{2}\right)\cos^2\left(\frac{\theta_c}{2}\right)\right\}^{-1}$$

To generate the failure curve for mixed mode loading from equation (2.57), the following steps are to be taken:

(1) Select a $\frac{K_{II}}{K_I}$ value.
(2) Obtain $\theta_c$ using the relation (2.55) $\frac{K_{II}}{K_I} = \frac{x\sqrt{1-x^2}}{3x^2-2}$, where $x = \cos\frac{\theta_c}{2}$ and $-\sqrt{1-x^2} = \sin\frac{\theta_c}{2}$.
(3) Obtain $\frac{K_I}{K_c}$ from equation (2.57).
(4) Obtain $\frac{K_{II}}{K_c}$ from steps (1) and (3) since $\frac{K_{II}}{K_c} = \frac{K_{II}}{K_I} \times \frac{K_I}{K_c}$.
(5) Plot $\frac{K_I}{K_c}$ versus $\frac{K_{II}}{K_c}$. The plot will look like the curve shown in Figure 2.5.

The failure criterion developed here and presented in Figure 2.5 is called the maximum stress ($\sigma_{\theta\theta}$) criterion. Similar criterion can be derived from the energy consideration instead. These two criteria are similar, although not identical, as shown in Figure 2.6. Equation of the dashed curve in Figure 2.6 is given by

$$\left(\frac{K_I}{K_c}\right)^2 + 1.56\left(\frac{K_{II}}{K_c}\right)^2 = 1 \tag{2.58}$$

while that for the solid curve is given by equation (2.56).

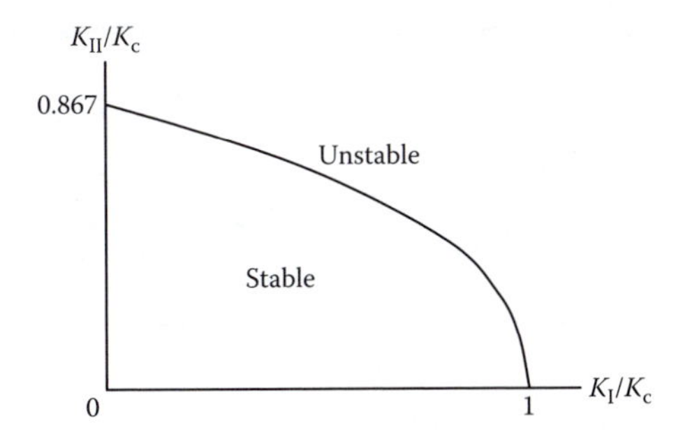


**FIGURE 2.5**
Failure curve for mixed mode loading. $K_c$ is the critical stress intensity factor.

## 2.8 Stress Singularities for Other Wedge Problems

In this chapter we have computed the stress field near a crack tip or a wedge tip in a homogeneous solid. However, if the wedge tip meets one or two interfaces (as shown in Figures 2.7a and 2.7b, respectively), then this problem becomes more complex. The problem of Figure 2.7a has been solved by Suhir (1988) and Hattori, Sakata, and Murakami (1989), while the problem described in Figure 2.7b has been solved by Theocaris (1974) and Carpenter and Byers (1987), among others. Interested readers are referred to these publications.

## 2.9 Concluding Remarks

Fundamentals of linear elastic fracture mechanics are presented in this chapter. Here it is shown that the stress field should have a square root singular behavior near a crack tip. The concept of the stress intensity factor, fracture toughness, failure curve, three modes of fracture, and the direction

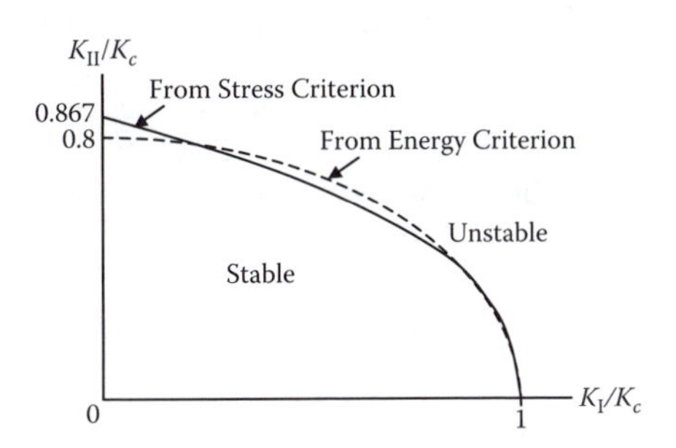


**FIGURE 2.6**
Failure curves for the mixed mode failure from the maximum stress criterion (solid line) and the energy criterion (dashed line). $K_c$ is the critical stress intensity factor.

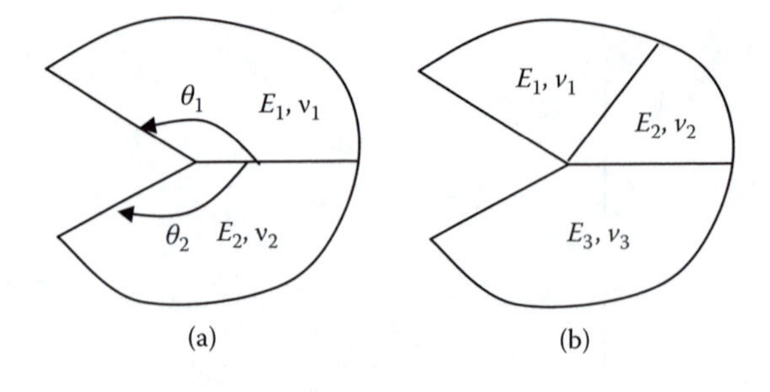


**FIGURE 2.7**
Wedge in inhomogeneous solids. (a) Wedge meeting two different solids; (b) wedge meeting three different solids.

of crack propagation for different fracture modes are introduced here. In a short course on the fundamentals of fracture mechanics, all the concepts discussed here should be covered. Depending on the availability of time and interest of the students, a few more relatively advanced topics from subsequent chapters, like Griffith's energy balance, plasticity correction factor, J-integral, fatigue crack growth, and numerical/analytical fracture mechanics analysis, can be included in a short course. However, in a regular graduate level course on fracture mechanics, all materials discussed in chapters 2–9 should be covered.

## References

Carpenter, W. C. and Byers, C. A path independent integral for computing stress intensities for V-notch cracks in a bi-material. *International Journal of Fracture*, 35, 245, 1987.

Erdogan, F. and Sih, G. C. On the crack extension in plates under plane loading and transverse shear. *ASME Journal of Basic Engineering*, 85D, 519, 1963.

Hattori, T., Sakata, S., and Murakami, G. A stress singularity parameter approach… LSI devices. *ASME Journal of Electronic Packaging*, 3, 243–248, 1989.

Suhir, E. Stress in dual-coated optical fibers. *ASME Journal of Applied Mechanics*, 55, 822, 1988.

Theocaris, P. S. The order of singularity at a multiwedge corner of a composite plate. *International Journal of Engineering Science*, 12, 107, 1974.

Williams, M. L. Stress distribution at the base of a stationary crack. *ASME Journal of Applied Mechanics*, 24, 109–114, 1957.

## Exercise Problems

Problem 2.1: An Airy stress function for an elastic solid without any body force must be biharmonic. Can the following functions be Airy stress functions in absence of body force?

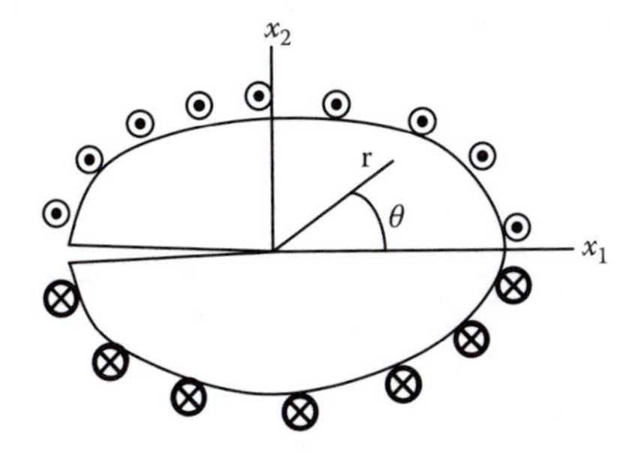


**FIGURE 2.8**
A cracked cylinder subjected to antiplane loading.

(a) $A\ln r$
(b) $Br^2$
(c) $Cr^2\ln r$
(d) $Dr^2\theta$
(e) $E\theta$

If you find some valid Airy stress functions in this list, then state why were these not considered while solving the crack problem using Williams' method? Justify your answer with proper derivation.

Problem 2.2: Let us consider a cracked cylinder that has a shape as shown in Figure 2.8. The cylinder extends to infinity in the $x_3$ direction. External loads are acting only in the $x_3$ direction as shown in the Figure. For this problem geometry $u_1 = u_2 = 0$ and $u_3$ is nonzero. These problems are called antiplane or out-of-plane problems.

(a) Show that the Navier's equation of equilibrium

$$(\lambda + 2\mu)\underline{\nabla}(\underline{\nabla} \cdot \underline{U}) - \mu(\underline{\nabla} \times \underline{\nabla} \times \underline{U}) + \underline{F} = 0$$

(where $\underline{U}$ is the displacement vector and $\underline{F}$ is the body force vector per unit volume) is simplified to the following equation for the antiplane problems in absence of body force:

$$u_{3,11} + u_{3,22} = 0$$

(b) Following Williams's approach, show that the stress and displacement fields near the crack tip ($r \ll 1$) for this problem are

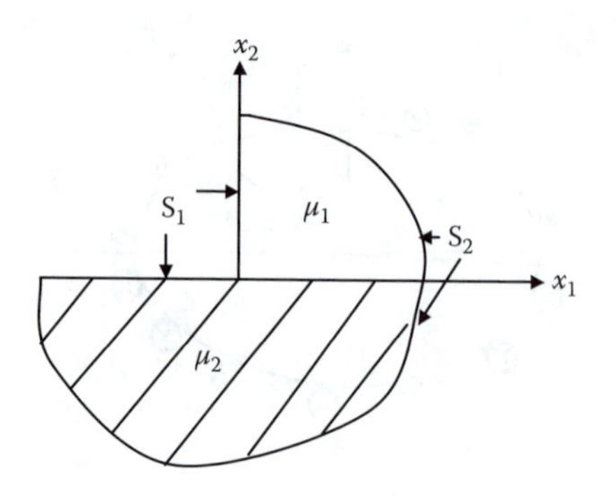


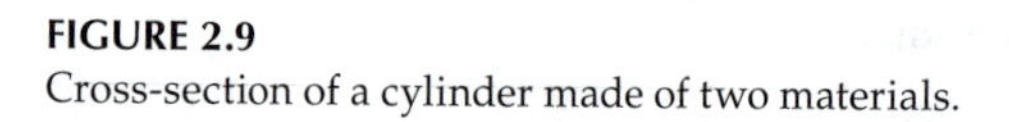

**FIGURE 2.9**
Cross-section of a cylinder made of two materials.

given by the following equations, where $K_{III}$ is a constant:

$$\tau_{\theta 3} = \frac{K_{III}}{\sqrt{2\pi r}} \cos\left(\frac{\theta}{2}\right)$$

$$\tau_{r3} = \frac{K_{III}}{\sqrt{2\pi r}} \sin\left(\frac{\theta}{2}\right)$$

$$u_3 = \frac{K_{III}}{\mu} \sqrt{\frac{2r}{\pi}} \sin\left(\frac{\theta}{2}\right)$$

Problem 2.3: Consider the cross-section of a cylinder as shown in Figure 2.9. It is composed of two materials, 1 and 2. The body is subjected to antiplane stress field on surface $S_2$. The surface $S_1$ is traction free.

(a) Try to determine the character of the stress field in the neighborhood of point O. (Set up the characteristic equation for nontrivial solution but do not solve it.)

(b) Complete all necessary details for the case $\mu_1 = \mu_2$ and obtain the stress field in the neighborhood of point O.

(c) Consider the two limiting cases, (1) $\mu_1 = 0$, $\mu_2$ finite and (2) $\mu_2 = 0$, $\mu_1$ finite, and then obtain the stress fields near the origin for both these cases.

Problem 2.4: Two half spaces are joined together in a circular region of radius $a$ as shown in Figure 2.10. The circular connection between the two half spaces transmits a resultant torque $T_0$. In the circular region at $x_3 = 0$, it is known that the shear stress field

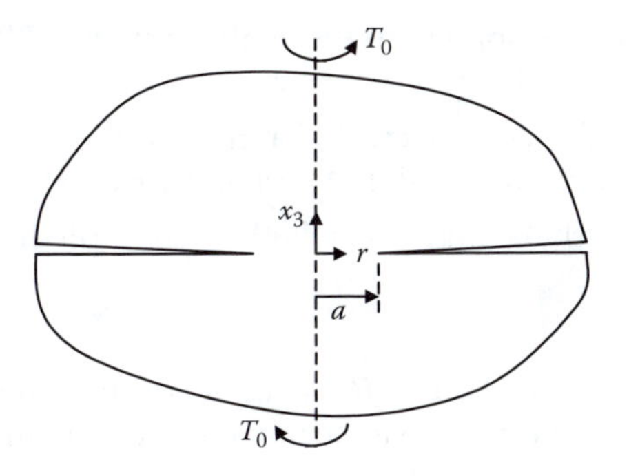


**FIGURE 2.10**
Two half spaces joined in a circular region.

is given by

$$\tau_{\theta 3}(r,\theta,0) = \frac{C_0 r}{\sqrt{a^2 - r^2}} \ldots\ldots. r < a$$

$$\tau_{\theta 3}(r,\theta,0) = 0 \ldots\ldots\ldots\ldots\ldots\ldots r > a$$

(a) Find the crack tip stress intensity factor in terms of torque $T_0$ and the radius of the connected region.

(b) What is the mode of loading (I, II, or III) for this problem?

Problem 2.5: Take a long piece of chalk or any cylindrical brittle material and make three surface cracks with a razor blade as shown in Figure 2.11. All cracks should have same length and depth. Position them as far as possible from one another so that there is no interaction effect.

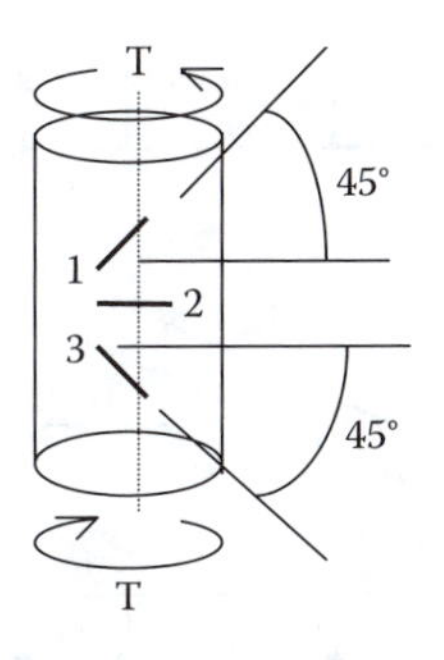


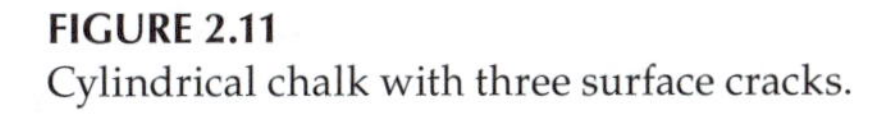

**FIGURE 2.11**
Cylindrical chalk with three surface cracks.

(a) If a torque $T$ is applied to the chalk as shown, identify the stress modes (I, II, or III) for each crack.

(b) Which is the critical crack for this applied loading—in other words, which crack will fail first and why?

(c) Break the chalk and see whether your prediction is right or wrong.

Problem 2.6:

(a) Plot an interaction curve ($K_I/K_c$ along the horizontal axis and $K_{II}/K_c$ along the vertical axis) for mixed mode failure. Take at least six values of $K_{II}/K_I$ to plot this curve.

(b) Let a cracked material ($K_c = 200$ kip.in.$^{-3/2}$) be subjected to different loading conditions that give following combinations of stress intensity factors:

| **Case#** | **1** | **2** | **3** | **4** | **5** | **6** | **7** |
|---|---|---|---|---|---|---|---|
| KI kip.in.$^{-3/2}$ | 180 | 160 | 140 | 120 | 100 | 40 | 0 |
| KII kip.in.$^{-3/2}$ | 0 | 40 | 100 | 120 | 140 | 160 | 180 |

State for each case whether the crack will propagate.

Problem 2.7. Stress intensity factors $K_I$ and $K_{II}$ of a Griffith crack in an infinite plate under opening mode (mode I) and shearing or sliding mode (mode II) are $\sigma_o(\pi a)^{1/2}$ and $\tau_o(\pi a)^{1/2}$ respectively, where $\sigma_o$ and $\tau_o$ are applied normal and shear stress fields, respectively.

(a) Using this information, obtain $K_I$ and $K_{II}$ for the loading shown in Figure 2.12.

(b) What should be the initial and final directions (with respect to the horizontal axis $x$) of crack propagation for this problem geometry if $\theta$ is 30°?

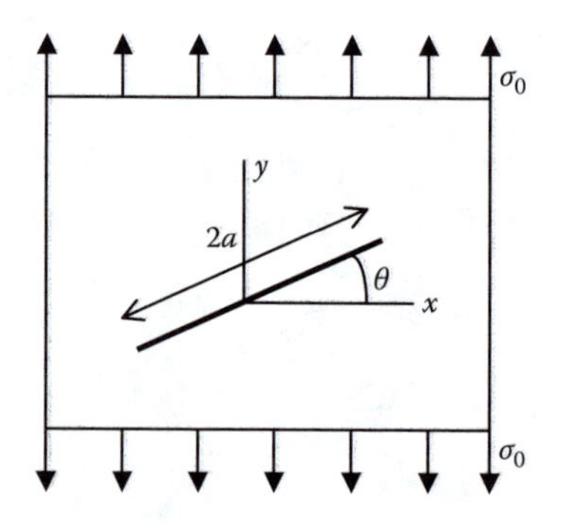


**FIGURE 2.12**
Griffith crack under mixed mode loading.

# 3

# Energy Balance

## 3.1 Introduction

Some fundamental equations of fracture mechanics are derived in this chapter from the principle of conservation of energy. This approach of deriving the fracture mechanics equations from the energy conservation laws was first proposed by Griffith (1921, 1924), who is often considered the father of modern fracture mechanics.

## 3.2 Griffith's Energy Balance

In Figure 3.1 we see a cracked body subjected to some external loading at two different states. In state 0 the body has a crack of surface area $A$. In state 1 the surface area of the crack is $A + \Delta A$. According to Griffith, if the total stored energy in the body in state 0 plus if the work done on the body by the external loads (body force and surface traction) going from state 0 to state 1 is greater than the total energy stored in the body in state 1 then the body moves from state 0 to state 1 because all objects try to achieve the minimum energy state.

Note that the total strain energy in state 0 is given by the following volume integral:

$$U = \int_V U_0 dV \tag{3.1}$$

where $U_0$ is the strain energy density function, and $V$ is the total volume of the body. Work done by the applied loads going from state 0 to state 1 is given by

$$W = \int_V f_i\left(u_i^1 - u_i^0\right)dV + \int_S T_i\left(u_i^1 - u_i^0\right)dS \tag{3.2}$$

where $f_i$ is the body force per unit volume, $T_i$ is the surface traction per unit surface area, and $S$ is the surface area of the body.

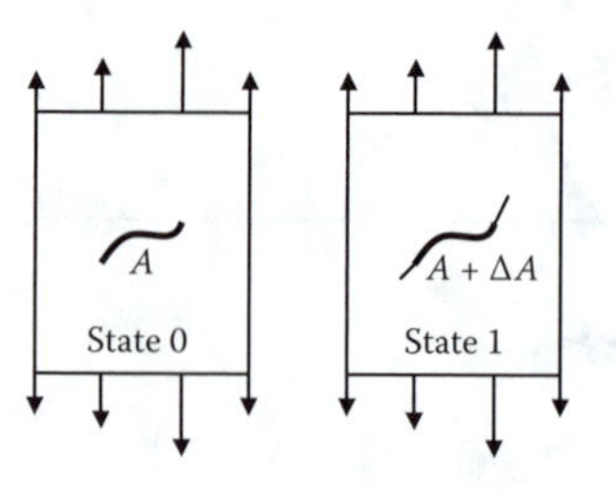


**FIGURE 3.1**
A loaded cracked body with crack surface areas A (state 0) and $A + \Delta A$ (state 1).

Total energy in state 1 can be written as

$$E_1 = \int_V U_1 dV + \int_{\Delta A} \gamma dS + \int_V C(T_1 - T_0)\rho dV + \int_V \frac{1}{2}\rho\left(\frac{du_i^1}{dt}\right)\left(\frac{du_i^1}{dt}\right)dV \tag{3.3}$$

On the right-hand side of equation (3.3) the first term is the strain energy, the second term is the additional surface energy due to the creation of the new crack surface area $\Delta A$, the third term is the increase in the heat energy in the body due to the temperature rise from $T_0$ to $T_1$, and the last term is the kinetic energy in the body. Note that $\gamma$ is the surface energy per unit area of the new crack surface, $C$ is the specific heat, and $\rho$ is the mass density.

If the temperature rise from state 0 to state 1 is not large and the velocity in state 1 is negligible, then one can ignore the last two integrals of equation (3.3) and Griffith's criterion for crack propagation from state 0 to state 1 can be written as

$$\int_V U_0 dV + \int_V f_i\left(u_i^1 - u_i^0\right)dV + \int_S T_i\left(u_i^1 - u_i^0\right)dS \geq \int_V U_1 dV + \int_{\Delta A} \gamma dA \tag{3.4}$$

From equation (3.4),

$$\int_V U_0 dV - \int_V f_i u_i^0 dV - \int_S T_i u_i^0 dS \geq \int_V U_1 dV - \int_V f_i u_i^1 dV - \int_S T_i u_i^1 dS + \int_{\Delta A} \gamma dA \tag{3.5}$$

From the definition of the potential energy, $\Pi$ ($\Pi = U - W$ = strain energy–work done by the applied loads), equation (3.5) can be rewritten as

$$\Pi_0 \geq \Pi_1 + \int_{\Delta A} \gamma dA$$

$$\Rightarrow 0 \geq \Pi_1 - \Pi_0 + \int_{\Delta A} \gamma dA \tag{3.6}$$

$$\Rightarrow \Pi_1 - \Pi_0 + \int_{\Delta A} \gamma dA \leq 0$$

For small $\Delta A$, the preceding equation takes the following form:

$$\begin{aligned}
&\Pi_1 - \Pi_0 + \gamma \Delta A \le 0 \\
&\Rightarrow \frac{\Pi_1 - \Pi_0}{\Delta A} + \gamma \le 0 \\
&\Rightarrow \frac{\Delta \Pi}{\Delta A} + \gamma \le 0 \\
&\Rightarrow \frac{d\Pi}{dA} + \gamma \le 0, \quad as\ \Delta A \to 0
\end{aligned} \tag{3.7}$$

Therefore, according to Griffith, if $\frac{d\Pi}{dA} + \gamma \le 0$, then the crack will propagate and the body will move from state 0 to state 1. However, if $\frac{d\Pi}{dA} + \gamma > 0$, then the body will not achieve state 1 because state 0 is the lower energy state and hence more stable than state 1.

Later, Irwin (1948), Orowan (1955), and others concluded that the failure criterion proposed by Griffith works if the surface energy term $\gamma$ is replaced by a much larger term $G_c$, where $G_c$ is an order of magnitude greater than $\gamma$ because the energy required to form a new crack surface area is much greater than the free surface energy that Griffith suggested. The additional energy is required to take care of other phenomena associated with the crack surface formation such as plastic dissipation, heat generation, etc. With this modification Griffith's criterion for crack propagation takes the following form:

$$\frac{d\Pi}{dA} + G_c \le 0 \tag{3.8}$$

When $\gamma$ is replaced by $G_c$, equation (3.6) takes the following form for small $\Delta A$:

$$\Pi_1 + \Delta A \cdot G_c \le \Pi_0 \tag{3.9}$$

Equation (3.9) is the alternate form of Griffith's criterion for crack propagation.

## 3.3 Energy Criterion of Crack Propagation for Fixed Force and Fixed Grip Conditions

Let us consider a spring-mass system at two states, 0 and 1. The system is given a fixed displacement $\Delta$. The mass contains a crack. The crack length is $a$ in state 0 and $a + \Delta a$ in state 1, as shown in Figure 3.2.

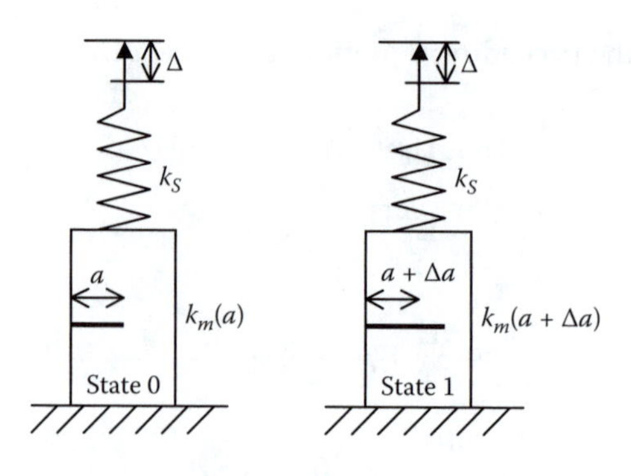


**FIGURE 3.2**
Spring-mass system is given an elongation Δ. Crack length increases from $a$ in state 0 to $a + \Delta a$ in state 1.

Let us denote the displacement of the spring and the mass by $\Delta_S$ and $\Delta_m$, respectively, and let $P$ be the force acting on the system. Since the stiffness of the spring and the mass are $k_s$ and $k_m$, respectively, one can write

$$\begin{aligned} \Delta &= \Delta_S + \Delta_m = \frac{P}{k_s} + \frac{P}{k_m} = P\left(\frac{1}{k_s} + \frac{1}{k_m}\right) \\ \Rightarrow P &= \frac{\Delta}{\left(\frac{1}{k_s} + \frac{1}{k_m}\right)} \end{aligned} \tag{3.10}$$

If the effective stiffness of the spring-mass system is denoted by $k_t$, then

$$P = \Delta k_t \tag{3.11}$$

Comparing equations (3.10) and (3.11),

$$k_t = \frac{1}{\left(\frac{1}{k_s} + \frac{1}{k_m}\right)} \tag{3.12}$$

Strain energies $U_S$ and $U_m$ in the spring and in the mass are given by

$$\begin{aligned} U_S &= \frac{1}{2} P \Delta_S = \frac{P^2}{2k_s} = \frac{\Delta^2}{2k_s\left(\frac{1}{k_s} + \frac{1}{k_m}\right)^2} \\ U_m &= \frac{1}{2} P \Delta_m = \frac{P^2}{2k_m} = \frac{\Delta^2}{2k_m\left(\frac{1}{k_s} + \frac{1}{k_m}\right)^2} \end{aligned} \tag{3.13}$$

Therefore, the total strain energy $U$ in the system

$$U = U_S + U_m = \frac{\Delta^2}{2k_s\left(\frac{1}{k_s}+\frac{1}{k_m}\right)^2} + \frac{\Delta^2}{2k_m\left(\frac{1}{k_s}+\frac{1}{k_m}\right)^2}$$
$$= \frac{\Delta^2}{2}\left(\frac{1}{k_s}+\frac{1}{k_m}\right)\frac{1}{\left(\frac{1}{k_s}+\frac{1}{k_m}\right)^2} = \frac{\Delta^2}{2\left(\frac{1}{k_s}+\frac{1}{k_m}\right)} \tag{3.14}$$

Total strain energies in the spring-mass system in states 0 and 1 are denoted as $U_0$ and $U_1$, respectively, where

$$U_0 = \frac{\Delta^2}{2\left(\frac{1}{k_s}+\frac{1}{k_m(a)}\right)} = \frac{\Delta^2}{2\left(\frac{1}{k_s}+\frac{1}{k_m^0}\right)}$$
$$U_1 = \frac{\Delta^2}{2\left(\frac{1}{k_s}+\frac{1}{k_m(a+\Delta a)}\right)} = \frac{\Delta^2}{2\left(\frac{1}{k_s}+\frac{1}{k_m^1}\right)} \tag{3.15}$$

Equation (3.9) states that the crack propagates when $\Pi_1 + \Delta A.G_c \le \Pi_0$. From Figure 3.2 it is clear that applied force $P$ does not move when the system moves from state 0 to state 1. Therefore, there is no external work done on the system. Hence, $\Pi_1 - \Pi_0 = U_1 - U_0$ and Griffith's criterion of crack propagation becomes

$$U_1 - U_0 + \Delta A \cdot G_c \le 0 \tag{3.16}$$

From equations (3.15) and (3.16),

$$U_1 - U_0 + \Delta A \cdot G_c = \frac{\Delta^2}{2}\left\{\frac{1}{\left(\frac{1}{k_s}+\frac{1}{k_m^1}\right)} - \frac{1}{\left(\frac{1}{k_s}+\frac{1}{k_m^0}\right)}\right\} + G_c t \cdot \Delta a \le 0$$
$$\therefore \frac{\Delta^2}{2t}\frac{d}{da}\left(\frac{1}{\frac{1}{k_s}+\frac{1}{k_m}}\right) + G_c \le 0 \tag{3.17}$$

It should be noted here that in equation (3.17) $t$ is the thickness of the mass. Therefore, the incremental crack length $\Delta a$ and the incremental crack surface area $\Delta A$ are related in the following manner:

$$\Delta A = t \cdot \Delta a \tag{3.18}$$

Let us now specialize equation (3.17) for two special cases: soft spring ($k_s << k_m$) and hard spring ($k_s >> k_m$).

### 3.3.1 Soft Spring Case

For $k_s << k_m$, one can write

$$\frac{1}{\frac{1}{k_s}+\frac{1}{k_m}} = \frac{1}{\frac{1}{k_s}\left(1+\frac{k_s}{k_m}\right)} = k_s\left(1+\frac{k_s}{k_m}\right)^{-1} \approx k_s\left(1-\frac{k_s}{k_m}\right) \tag{3.19}$$

Therefore,

$$\frac{d}{da}\left(\frac{1}{\frac{1}{k_s}+\frac{1}{k_m}}\right) = \frac{d}{da}\left(k_s - \frac{k_s^2}{k_m}\right) = -k_s^2\frac{d}{da}\left(\frac{1}{k_m}\right) \tag{3.20}$$

Combining equations (3.17) and (3.20),

$$\begin{aligned}&\frac{\Delta^2}{2t}\frac{d}{da}\left(\frac{1}{\frac{1}{k_s}+\frac{1}{k_m}}\right)+G_c \le 0\\ &\Rightarrow -\frac{\Delta^2 k_s^2}{2t}\frac{d}{da}\left(\frac{1}{k_m}\right)+G_c \le 0\end{aligned} \tag{3.21}$$

Combining equations (3.10) and (3.19),

$$P = \frac{\Delta}{\left(\frac{1}{k_s}+\frac{1}{k_m}\right)} \approx \Delta\cdot k_s\left(1-\frac{k_s}{k_m}\right) \approx \Delta\cdot k_s \tag{3.22}$$

Both $\Delta$ and $k_s$ remain unchanged when the system moves from state 0 to state 1. Therefore, from equation (3.22) one can conclude that the applied force $P$ remains unchanged. For this reason this case is also known as the *fixed force* case.

Combining equations (3.21) and (3.22),

$$-\frac{\Delta^2 k_s^2}{2t}\frac{d}{da}\left(\frac{1}{k_m}\right)+G_c = -\frac{P^2}{2t}\frac{d}{da}\left(\frac{1}{k_m}\right)+G_c = -\frac{1}{t}\frac{d}{da}\left(\frac{P^2}{2k_m}\right)+G_c \le 0 \tag{3.23}$$

Substituting equations (3.13) and (3.18) into equation (3.23),

$$-\frac{1}{t}\frac{d}{da}\left(\frac{P^2}{2k_m}\right)+G_c = -\frac{d}{dA}(U_m)+G_c = -\frac{dU_m}{dA}+G_c \le 0 \tag{3.24}$$

or

$$\frac{dU_m}{dA} \ge G_c \tag{3.25}$$

### 3.3.2 Hard Spring Case

For $k_s >> k_m$, one can write

$$\frac{1}{\frac{1}{k_s}+\frac{1}{k_m}} = \frac{1}{\frac{1}{k_m}\left(1+\frac{k_m}{k_s}\right)} = k_m\left(1+\frac{k_m}{k_s}\right)^{-1} = k_m\left(1-\frac{k_m}{k_s}+\cdots\right) \approx k_m \tag{3.26}$$

Therefore,

$$\frac{d}{da}\left(\frac{1}{\frac{1}{k_s}+\frac{1}{k_m}}\right) \approx \frac{dk_m}{da} \tag{3.27}$$

Combining equations (3.17) and (3.27),

$$\frac{\Delta^2}{2t}\frac{d}{da}\left(\frac{1}{\frac{1}{k_s}+\frac{1}{k_m}}\right) + G_c \le 0$$
$$\Rightarrow \frac{\Delta^2}{2t}\frac{dk_m}{da} + G_c \le 0 \tag{3.28}$$

Combining equations (3.13) and (3.26),

$$U_m = \frac{\Delta^2}{2k_m\left(\frac{1}{k_s}+\frac{1}{k_m}\right)^2} = \frac{\Delta^2 k_m^2}{2k_m} = \frac{\Delta^2 k_m}{2} \tag{3.29}$$

From equations (3.28) and (3.29),

$$\frac{\Delta^2}{2t}\frac{dk_m}{da} + G_c = \frac{1}{t}\frac{d}{da}\left(\frac{\Delta^2 k_m}{2}\right) + G_c = \frac{dU_m}{dA} + G_c \le 0 \tag{3.30}$$

or

$$-\frac{dU_m}{dA} \ge G_c \tag{3.31}$$

From equations (3.10) and (3.26), one can write

$$P = \frac{\Delta}{\left(\frac{1}{k_s}+\frac{1}{k_m}\right)} \approx \Delta k_m \tag{3.32}$$

but

$$P = \Delta_m k_m \tag{3.33}$$

Therefore, $\Delta_m = \Delta$. Since $\Delta$ is constant, $\Delta_m$ should also be constant. For this reason this case is called the *fixed grip* case. Note that as the spring-mass system moves from state 0 to state 1, although the displacement $\Delta_m$ remains constant, the force $P$ changes because $k_m$ changes. Thus, this is not a fixed force situation.

### 3.3.3 General Case

For the general case when the spring stiffness $k_s$ and the mass stiffness $k_m$ are of the same order, then

$$\frac{d}{da}\left(\frac{1}{\frac{1}{k_s}+\frac{1}{k_m}}\right) = -\frac{1}{\left(\frac{1}{k_s}+\frac{1}{k_m}\right)^2}\left(-\frac{1}{k_m^2}\right)\frac{dk_m}{da} = \frac{1}{k_m^2}\frac{1}{\left(\frac{1}{k_s}+\frac{1}{k_m}\right)^2}\frac{dk_m}{da} \tag{3.34}$$

Therefore, equation (3.17) becomes

$$\frac{\Delta^2}{2t}\frac{d}{da}\left(\frac{1}{\frac{1}{k_s}+\frac{1}{k_m}}\right) + G_c = \frac{\Delta^2}{2t}\frac{1}{k_m^2\left(\frac{1}{k_s}+\frac{1}{k_m}\right)^2}\frac{dk_m}{da} + G_c \leq 0 \tag{3.35}$$

From equations (3.13) and (3.35),

$$\begin{aligned} &\frac{U_m}{tk_m}\frac{dk_m}{da} + G_c \leq 0 \\ &\Rightarrow -\frac{U_m}{tk_m}\frac{dk_m}{da} \geq G_c \end{aligned} \tag{3.36}$$

Note that the stiffness of the mass decreases as the crack length increases. Therefore, $\frac{dk_m}{da} < 0$. For this reason the negative sign appears on the left-hand side of equation (3.36).

## 3.4 Experimental Determination of $G_c$

In 1964 Strawley, Jones, and Gross described an experimental technique to determine the critical strain energy release rate for a given material. They fabricated a number of double cantilever specimens with different crack lengths. A typical double-cantilever specimen is shown in Figure 3.3. When two opposing forces ($P$) are applied at the free end, the crack opens by an amount $2\Delta$. Therefore, the work done on the system that goes into the material as the strain energy is given by

$$U = \frac{1}{2}P\Delta + \frac{1}{2}P\Delta = P\Delta \tag{3.37}$$

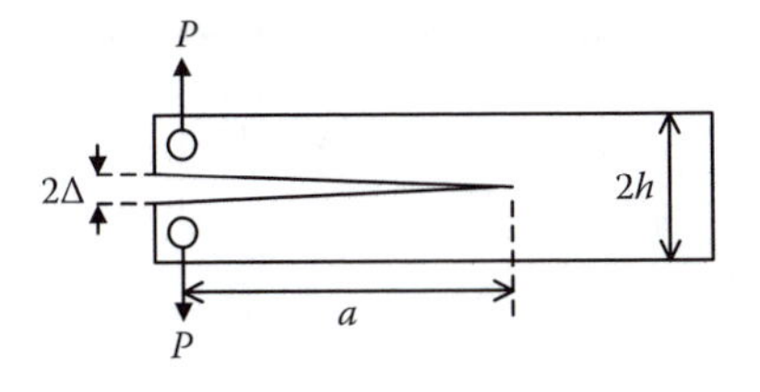


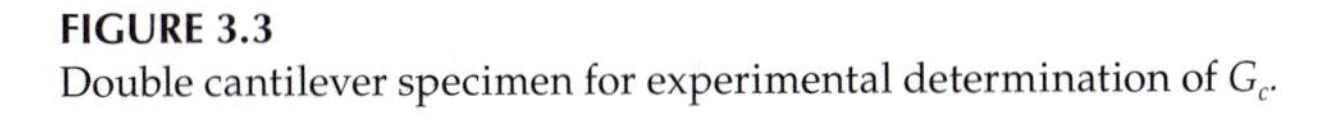
**FIGURE 3.3**
Double cantilever specimen for experimental determination of $G_c$.

In terms of compliance $(C = \frac{\Delta}{P})$, the preceding equation can be written as

$$U = P\Delta = P^2C = \frac{\Delta^2}{C} \tag{3.38}$$

For the fixed force case, since $P$ is constant, one can write from equation (3.38):

$$\frac{dU}{dA} = \frac{P^2}{t}\frac{dC}{da} \tag{3.39}$$

For the fixed grip case, since $\Delta$ is constant, it is possible to write from equation (3.38):

$$\frac{dU}{dA} = -\frac{\Delta^2}{tC^2}\frac{dC}{da} \tag{3.40}$$

Clearly, to evaluate strain energy release rate for given $P$ or $\Delta$ one needs to know the compliance $C$ and its rate of change with the crack length $a$. For this purpose a number of double cantilever specimens, as shown in Figure 3.3, with different crack lengths are fabricated. Deflection ($\Delta$) versus load ($P$) curves are obtained experimentally for different specimens. A typical $\Delta$–$P$ curve is shown in Figure 3.4. Note that in this plot the slope of the linear part is the compliance $C$.

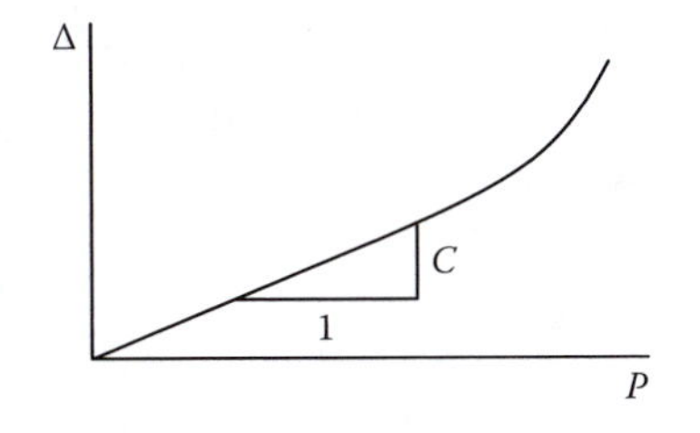

**FIGURE 3.4**
Typical Δ-P curve of double cantilever specimens. Slope of this curve is the compliance.

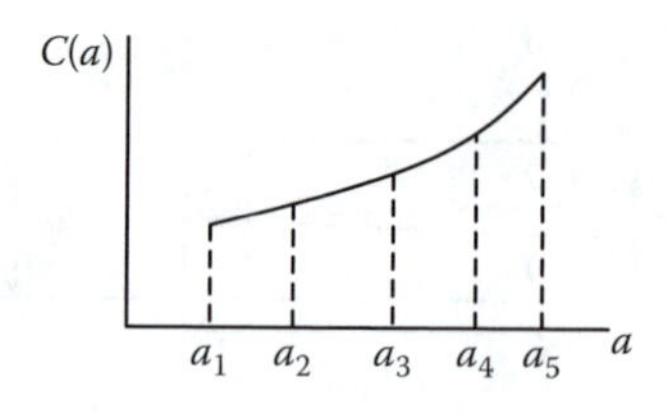


**FIGURE 3.5**
Experimentally obtained crack length versus compliance curve.

After measuring compliance of different specimens, compliance versus the crack length curve is obtained, as shown in Figure 3.5.

### 3.4.1 Fixed Force Experiment

A cracked specimen as shown in Figure 3.3 is loaded by a pair of force $P$. Then the load $P$ is continuously increased until the crack starts to propagate. If the crack propagation starts at load $P = P_c$, then the critical strain energy release rate $G_c$ is obtained from equation (3.39):

$$G_c = \left.\frac{dU}{dA}\right|_{critical} = \frac{P_c^2}{t}\frac{dC}{da} \tag{3.41}$$

$\frac{dC}{da}$ is obtained from Figure 3.5. Note that once the crack starts to propagate at load $P = P_c$, the applied load is not changed during the crack propagation. Therefore, in this case the crack propagation occurs under fixed force condition.

### 3.4.2 Fixed Grip Experiment

If the crack propagation occurs under fixed force condition, as discussed in the previous section, then once the crack starts to propagate, the specimen fails. Alternately, the experiment can be carried out under fixed grip conditions. In this setup, the specimen is subjected to a specified displacement $\Delta$ and then this $\Delta$ is continuously increased until the crack starts to propagate for $\Delta = \Delta_c$. After propagating a little distance, the crack stops when $\Delta$ is kept constant at $\Delta_c$ since the compliance of the specimen increases as the crack length increases. In this case the critical strain energy release rate $G_c$ is obtained from equation (3.40):

$$G_c = -\left.\frac{dU}{dA}\right|_{critical} = \frac{\Delta_c^2}{tC^2}\frac{dC}{da} \tag{3.42}$$

### 3.4.3 Determination of $G_c$ from One Specimen

If the crack length of the double cantilever specimen, shown in Figure 3.3, is much greater than the depth of the beam ($h$), then one can use the formula obtained from the beam theory to compute the crack opening $\Delta$, as shown:

$$\Delta = \frac{Pa^3}{3EI} = \frac{12Pa^3}{3Eth^3} = \frac{4Pa^3}{Eth^3}$$
$$\therefore C = \frac{\Delta}{P} = \frac{4a^3}{Eth^3} \tag{3.43}$$
$$\therefore \frac{dC}{da} = \frac{12a^2}{Eth^3}$$

Therefore, for fixed force experiments combining equations (3.41) and (3.43), one obtains

$$G_c = \frac{P_c^2}{t}\frac{dC}{da} = \frac{12a^2P_c^2}{Et^2h^3} \tag{3.44}$$

and for fixed grip experiments combining equations (3.42) and (3.43), one gets

$$G_c = \frac{\Delta_c^2}{tC^2}\frac{dC}{da} = \frac{\Delta_c^2}{t} \times \frac{E^2t^2h^6}{16a^6} \times \frac{12a^2}{Eth^3} = \frac{3\Delta_c^2Eh^3}{4a^4} \tag{3.45}$$

Equations (3.44) and (3.45) give less than 10% error when $\frac{a}{h} > 3$.

## 3.5 Relation between Strain Energy Release Rate ($G$) and Stress Intensity Factor ($K$)

Two crack propagation criteria have been defined earlier. These two criteria are:

(1) The crack propagates when $\sigma_{\theta\theta}\big|_{\max} \geq \frac{K_c}{\sqrt{2\pi r}}$. For opening mode loading, this criterion implies that when the stress intensity factor $K$ exceeds the critical stress intensity factor $K_c$, the crack propagates.

(2) The crack propagates when the strain energy release rate $(G = \pm \frac{dU}{dA})$ exceeds the critical strain energy release rate $G_c$. Fixed force and fixed grip conditions determine whether the positive or the negative sign of $(\frac{dU}{dA})$ should be taken.

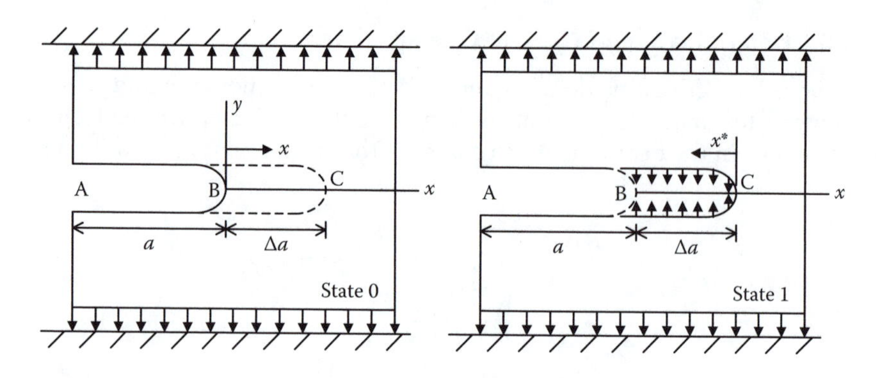


**FIGURE 3.6**
Cracked body under fixed-grip loading. Crack length is $a$ for state 0 and $a + \Delta a$ for state 1.

$G$ and $K$ (or $G_c$ and $K_c$) cannot be completely independent of each other. In this section, the relation between these two parameters is investigated. For this purpose consider a cracked body being stressed and held under fixed grip condition as shown in Figure 3.6.

The original crack length is $a$ in state 0 while it is $a + \Delta a$ in state 1. However, in state 1 a closing force is applied, as shown in the Figure, to close BC part of the crack whose length is $\Delta a$, to make state 1 identical to state 0. To close the crack length BC, one needs to do some work, which increases the strain energy of the body. Therefore, the strain energy in state 0 is greater than the strain energy in state 1 before applying any closing force. Since the problem is linear, the difference in the strain energy between these two states is simply half of the work done by the closing forces going through the closing displacements.

Note that the closing forces in BC in state 1 should be identical to the stress field in state 0 along line BC because, after the crack length BC is closed, states 0 and 1 must be identical. Therefore, the closing force amplitude at position $x(=\Delta a - x^*)$ over an incremental length $dx$ is given by $\frac{K_I}{\sqrt{2\pi x}}tdx$ on the top and bottom surfaces of the crack along line BC. The plate thickness is denoted by $t$. The displacement traveled by this closing force to bring the two surfaces of the crack together along line BC is obtained from the $u_y$ expression given in equation (2.48). Note that for the displacement computation, the crack tip is at point C. Therefore, the radial distance of the point of interest from the crack tip is $x^*$ and the angle measure is $+\pi$ for the top surface and $-\pi$ for the bottom surface. Thus, the displacement of the closing forces on the two faces of the crack is given by

$$u_y = \pm\frac{K_I\sqrt{x^*}}{2\mu\sqrt{2\pi}}(\kappa+1) = \pm\frac{K_I\sqrt{(\Delta a - x)}}{E\sqrt{2\pi}}(1+\nu)(\kappa+1) \tag{3.46}$$

Therefore, the strain energy stored in the body by the closing forces going through the closing displacements is given by

$$\Delta U = 2 \times \frac{1}{2} \int_0^{\Delta a} \frac{K_I}{\sqrt{2\pi x}} \frac{K_I \sqrt{(\Delta a - x)}}{E\sqrt{2\pi}} (1+\nu)(\kappa+1) t dx$$
$$= (1+\nu)(\kappa+1) \frac{tK_I^2}{2\pi E} \int_0^{\Delta a} \frac{\sqrt{(\Delta a - x)}}{\sqrt{x}} dx \tag{3.47}$$

The integral can be solved in closed form by substituting

$$x = \Delta a \sin^2 \theta$$
$$dx = 2\Delta a \sin\theta \cos\theta d\theta \tag{3.48}$$

Then,

$$\int_0^{\Delta a} \frac{\sqrt{(\Delta a - x)}}{\sqrt{x}} dx = \int_0^{\pi/2} \frac{\sqrt{\Delta a} \cos\theta}{\sqrt{\Delta a} \sin\theta} 2\Delta a \sin\theta \cos\theta d\theta = 2\Delta a \int_0^{\pi/2} \cos^2\theta d\theta$$
$$= \Delta a \int_0^{\pi/2} (1+\cos 2\theta) d\theta = \Delta a \left[\theta + \frac{1}{2}\sin 2\theta\right]_0^{\pi/2} = \frac{\pi \Delta a}{2} \tag{3.49}$$

Substituting equation (3.49) into equation (3.47),

$$\Delta U = (1+\nu)(\kappa+1) \frac{tK_I^2}{2\pi E} \int_0^{\Delta a} \frac{\sqrt{(\Delta a - x)}}{\sqrt{x}} dx = (1+\nu)(\kappa+1) \frac{tK_I^2}{2\pi E} \frac{\pi \Delta a}{2}$$
$$= \frac{(1+\nu)(\kappa+1)}{4} \frac{t\Delta a K_I^2}{E} \tag{3.50}$$
$$\therefore \frac{\Delta U}{t\Delta a} = \alpha \frac{K_I^2}{E}$$
$$\Rightarrow \frac{\Delta U}{\Delta A} = \alpha \frac{K_I^2}{E}$$

Note that for the plane stress problems,

$$\alpha = \frac{1}{4}(1+\nu)(\kappa+1) = \frac{1}{4}(1+\nu)\left(\frac{3-\nu}{1+\nu}+1\right) = \frac{1}{4}(1+\nu)\left(\frac{3-\nu+1+\nu}{1+\nu}\right) = 1 \tag{3.51}$$

and for the plane strain problems,

$$\alpha = \frac{1}{4}(1+\nu)(\kappa+1) = \frac{1}{4}(1+\nu)(3-4\nu+1) = \frac{1}{4}(1+\nu)4(1-\nu) = 1-\nu^2 \tag{3.52}$$

In the limiting case, as $\Delta A \to 0$, equation (3.50) becomes

$$G = \frac{dU}{dA} = \alpha \frac{K_I^2}{E} \tag{3.53}$$

Similarly, the critical stress intensity factor and the critical strain energy release rates are related:

$$G_c = \alpha \frac{K_c^2}{E} \tag{3.54}$$

One can also show that for mixed mode loading the strain energy release rate is related to the three stress intensity factors in the following manner (Broek 1997):

$$G = \frac{1-\nu^2}{E}\left( K_I^2 + K_{II}^2 + \frac{K_{III}^2}{1-\nu} \right) \tag{3.55}$$

## 3.6 Determination of Stress Intensity Factor ($K$) for Different Problem Geometries

In this section stress intensity factors (SIFs) of different problem geometries are obtained by applying equation (3.53) to the elasticity solutions. In some occasions when the exact elasticity solutions are not available, engineering judgments are applied to estimate the SIF.

### 3.6.1 Griffith Crack

A crack of finite length (say $2a$) in an infinite plate, as shown in Figure 3.7, is known as the Griffith crack. Note that a Griffith crack does not interact with another crack, inclusion, or the problem boundary since none of them is present in the neighborhood of the crack. When the Griffith crack is subjected to the uniaxial stress field $\sigma_0$ at the far field, then the original problem can be decomposed into two problems, I and II, as shown in Figure 3.7.

Note that problem I does not have any crack. The crack position of the original problem is shown by the dashed line in problem I. Clearly, in

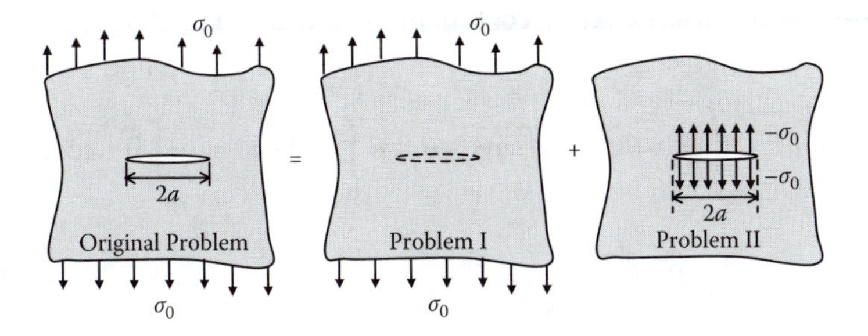


**FIGURE 3.7**
Cracked plate or cross-section of a circular crack in an infinite solid subjected to an axial stress (original problem) can be decomposed into two basic problems, as shown.

problem I along the dashed line the stress field is $\sigma_{22} = \sigma_0$ and $\sigma_{12} = 0$. However, in the original problem, on the crack surface $\sigma_{22} = \sigma_{12} = 0$. To satisfy this boundary condition in problem II, the stress field $\sigma_{22} = -\sigma_0$ is added on the crack surface. When problems I and II are added, both far field conditions and the stress-free boundary conditions on the crack surface are satisfied. Therefore, the superimposed solutions of problems I and II should be the solution of the original problem.

For problem I the stress field at all points is given by $\sigma_{22} = \sigma_0$ and $\sigma_{11} = \sigma_{12} = 0$. For problem II the vertical displacement field at the two crack surfaces can be obtained from the theory of elasticity as shown:

$$u_2 = \pm(\kappa + 1)\frac{\sigma_0}{4\mu}\sqrt{a^2 - x_1^2} \tag{3.56}$$

where the origin of the $x_1$–$x_2$ coordinate system coincides with the center of the crack. Kappa is the same as that defined in equation (2.48).

Strain energy stored in problem II geometry by the applied stress going through the displacement of equation (3.56) is given by

$$U = 2\times 2\int_0^a \frac{1}{2}\sigma_0(\kappa + 1)\frac{\sigma_0}{4\mu}\sqrt{a^2 - x_1^2}\,t\,dx_1 \tag{3.57}$$

In this expression $t$ is the thickness of the plate. The multiplying factor 4 in front of the integral appears because the integral is carried out on only one fourth of the crack surface. The preceding integral is simplified further to obtain

$$U = \int_0^a \frac{1}{2}(\kappa + 1)\frac{\sigma_0^2}{\mu}\sqrt{a^2 - x_1^2}\,t\,dx_1 = \frac{(\kappa + 1)\sigma_0^2 t}{2\mu}\int_0^a \sqrt{a^2 - x_1^2}\,dx_1 \tag{3.58}$$

Substitute $x_1 = a\sin\theta$, $dx_1 = a\cos\theta\, d\theta$ in the integrand to obtain

$$\int_0^a \sqrt{a^2 - x_1^2}\,dx_1 = \int_0^{\pi/2} a\cos\theta\sqrt{a^2 - a^2\sin^2\theta}\,d\theta = a^2\int_0^{\pi/2}\cos^2\theta\, d\theta = \frac{a^2}{2}\int_0^{\pi/2}(1+\cos 2\theta)\,d\theta$$

$$= \frac{a^2}{2}\left[\theta + \frac{\sin 2\theta}{2}\right]_0^{\pi/2} = \frac{\pi a^2}{4} \tag{3.59}$$

Substituting equation (3.59) into equation (3.58),

$$U = \frac{(\kappa+1)\sigma_0^2 t}{2\mu}\int_0^a \sqrt{a^2 - x_1^2}\,dx_1 = \frac{\pi a^2(\kappa+1)\sigma_0^2 t}{8\mu} \tag{3.60}$$

Therefore,

$$\frac{\partial U}{\partial A} = \frac{\partial U}{2t\partial a} = \frac{2\pi a(\kappa+1)\sigma_0^2}{2\times 8\mu} = \frac{(\kappa+1)}{8\mu}\sigma_0^2\pi a \tag{3.61}$$

The crack surface area $A = 2at$; therefore, $\partial A = 2t\partial a$. The total strain energy stored in problem I is independent of the crack length. Therefore, the rate of change of strain energy with the crack length variation in the original problem is the same as that of problem II.

Combining equations (3.53) and (3.61),

$$\begin{aligned} &\alpha\frac{K_I^2}{E} = \frac{\partial U}{\partial A} = \frac{(\kappa+1)}{8\mu}\sigma_0^2\pi a = \frac{(\kappa+1)(1+\nu)}{4E}\sigma_0^2\pi a \\ &\therefore K_I = \sqrt{\frac{(\kappa+1)(1+\nu)}{4\alpha}}\,\sigma_0\sqrt{\pi a} \end{aligned} \tag{3.62}$$

Note that for both thin plate (or plane stress) and thick plate (or plane strain) problems, the argument inside the first square root of equation (3.62) is 1 as shown below.

For plane stress problems,

$$\frac{(\kappa+1)(1+\nu)}{4\alpha} = \frac{1}{4}\left(\frac{3-\nu}{1+\nu}+1\right)(1+\nu) = \frac{1}{4}\left(\frac{3-\nu+1+\nu}{1+\nu}\right)(1+\nu) = 1 \tag{3.63a}$$

and for plane strain,

$$\frac{(\kappa+1)(1+\nu)}{4\alpha} = \frac{(3-4\nu+1)(1+\nu)}{4(1-\nu^2)} = \frac{4(1-\nu)(1+\nu)}{4(1-\nu^2)} = 1 \tag{3.63b}$$

From equations (3.62) and (3.63) one can conclude that for both plane stress and plane strain problems, the opening mode stress intensity factor for the Griffith crack is given by

$$K_I = \sigma_0\sqrt{\pi a} \tag{3.64}$$

Equation (3.56) remains unchanged for the biaxial state of stress. Therefore, for this case also, equation (3.64) is valid.

If the far field is subjected to a shear stress field $\sigma_{12} = \tau_0$ instead of the normal stresses ($\sigma_{22}$ and $\sigma_{11}$), then following similar steps one can show that the sliding mode stress intensity factor is given by

$$K_{II} = \tau_0\sqrt{\pi a} \tag{3.65}$$

### 3.6.2 Circular or Penny-Shaped Crack

Let us now consider a penny-shaped crack of radius $a$ in an infinite solid. Figure 3.7 shows the cross-section of a circular (or penny-shaped) crack in an infinite solid. Following the same arguments presented in section 3.6.1 it can be shown that the rate of change of strain energy with the variation of the crack surface area in the original problem is the same as that for problem II. From the theory of elasticity, the vertical displacement of the circular crack surfaces in problem II can be obtained as

$$u_3 = \beta\frac{\sigma_0}{E}\sqrt{a^2 - r^2} \tag{3.66}$$

In the preceding equation the vertical displacement (normal to the crack surface) is denoted as $u_3$, and the radial distance $r$ is measured from the crack center. The factor $\beta$ is defined as

$$\beta = \frac{4}{\pi}(1 - v^2) \tag{3.67}$$

Therefore, the strain energy stored in problem II by the applied stress field on the crack surface is given by

$$U = 2\times\frac{1}{2}\int_0^a \beta\frac{\sigma_0}{E}\sqrt{a^2 - r^2}\cdot\sigma_0 2\pi r dr = \frac{2\pi\beta\sigma_0^2}{E}\int_0^a \sqrt{a^2 - r^2}\cdot r dr \tag{3.68}$$

Substitute $r = a\sin\theta$, $dr = a\cos\theta\, d\theta$ in the integrand to obtain

$$\begin{aligned}\int_0^a \sqrt{a^2 - r^2}.rdr &= \int_0^{\pi/2} a^2\sin\theta\cos\theta\sqrt{a^2 - a^2\sin^2\theta}d\theta = a^3\int_0^{\pi/2}\cos^2\theta\sin\theta\, d\theta \\ &= \frac{a^3}{3}[-\cos^3\theta]_0^{\pi/2} = \frac{a^3}{3}\end{aligned} \tag{3.69}$$

Substituting equation (3.69) into equation (3.68),

$$U = \frac{2\pi\beta\sigma_0^2 a^3}{3E} \tag{3.70}$$

Note that

$$A = \pi a^2, \partial A = 2\pi a \partial a \tag{3.71}$$

Therefore, from equations (3.53), (3.70), and (3.71),

$$\alpha\frac{K_I^2}{E} = \frac{\partial U}{\partial A} = \frac{1}{2\pi a}\frac{\partial U}{\partial a} = \frac{1}{2\pi a}\frac{2\pi\beta\sigma_0^2 a^2}{E} = \frac{\beta\sigma_0^2 a}{E}$$
$$\therefore K_I = \sqrt{\frac{\beta}{\alpha}}\sigma_0\sqrt{a} \tag{3.72}$$

What should be the right value of $\alpha$ for the penny-shaped crack problem? Note that this problem is neither plane stress nor plane strain, but rather axisymmetric. However, axisymmetric problems are closer to plane strain problems than to plane stress problems because under plane strain conditions all movements in the $x_3$ direction are restricted and in axisymmetric conditions movements in the $\theta$ direction are restricted. Therefore, the value of $\alpha$ should be same as that for the plane strain condition. Thus, we get

$$\frac{\beta}{\alpha} = \frac{4}{\pi}\frac{(1-\nu^2)}{(1-\nu^2)} = \frac{4}{\pi} \tag{3.73}$$

Substituting equation (3.73) into equation (3.72),

$$K_I = \sqrt{\frac{\beta}{\alpha}}\sigma_0\sqrt{a} = \sqrt{\frac{4}{\pi}}\sigma_0\sqrt{a} = \frac{2}{\pi}\sigma_0\sqrt{\pi a} \tag{3.74}$$

### 3.6.3 Semi-infinite Crack in a Strip

Knauss (1966) and Rice (1967) solved the problem of a semi-infinite crack in a strip as shown in Figure 3.8. The strip of width $2h$ and thickness $t$ is stretched in the vertical direction by an amount $2\Delta$. Note that as the crack advances by an amount $\Delta x$, region A on the right side of the crack is reduced and transformed to region B. Note that the strain energy of region B is zero because top and bottom segments of region B simply go through the rigid body translation. However, region A is under tension and therefore should have some strain energy.

If segment A is subjected to uniaxial stress $\sigma_{yy} = \sigma_0 = E(\frac{\Delta}{h})$, then the strain energy density in the segment $U_0 = \frac{1}{2}E(\frac{\Delta}{h})^2$. Note that if segment A is subjected to uniaxial strain $\varepsilon_{yy} = \varepsilon_0 = \frac{\Delta}{h}$, then also the segment should have the

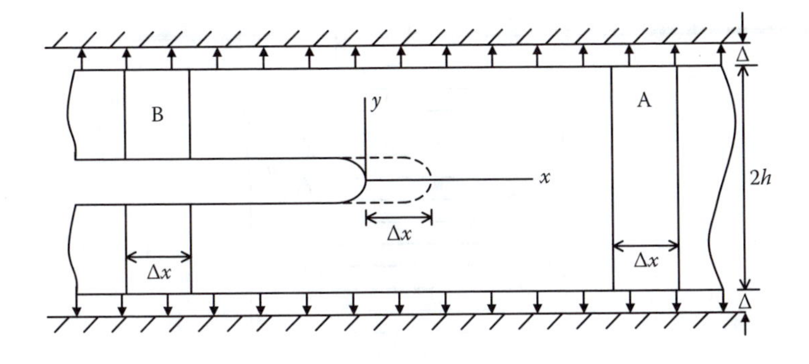


**FIGURE 3.8**
A semi-infinite crack in strip of infinite length.

same strain energy density. For uniaxial stress $\sigma_{xx}$ = 0, but $\varepsilon_{xx} \neq 0$, and for uniaxial strain $\sigma_{xx} \neq 0$, but $\varepsilon_{xx} = 0$, in both cases $\frac{1}{2}\sigma_{xx}\varepsilon_{xx} = 0$.

Strain energy of segment A,

$$\Delta U = \frac{1}{2}E\left(\frac{\Delta}{h}\right)^2 2h\Delta x \cdot t = E\left(\frac{\Delta}{h}\right)^2 h\Delta x \cdot t \tag{3.75}$$

Therefore, the strain energy release rate as the crack propagates is given by

$$\frac{\partial U}{\partial A} = \frac{\Delta U}{t \cdot \Delta x} = \frac{1}{t \cdot \Delta x} E\left(\frac{\Delta}{h}\right)^2 h\Delta x \cdot t = E\left(\frac{\Delta}{h}\right)^2 h$$

$$\therefore \frac{K_I^2}{E} = E\left(\frac{\Delta}{h}\right)^2 h \tag{3.76}$$

$$\therefore K_I = E\left(\frac{\Delta}{h}\right)\sqrt{h} = \sigma_0\sqrt{h}$$

### 3.6.4 Stack of Parallel Cracks in an Infinite Plate

Let us consider a set of parallel cracks of length $2a$ and spacing $2h$ in an infinite sheet as shown in Figure 3.9. If the plate is subjected to uniaxial state of stress $\sigma_{yy} = \sigma_0$, what should be the stress intensity factor? Note that for $h >> a$, the interaction effect between the neighboring cracks can be ignored. In this case the stress intensity factor should be equal to that of the Griffith crack, $K_I = \sigma_0\sqrt{\pi a}$.

On the other hand, for crack length ($a$) much greater than the spacing ($h$), or $\frac{h}{a} << 1$, the cracks may be considered as a stack of semi-infinite cracks, as shown in Figure 3.10, because the crack tips on the right side are not affected by the crack tips on the left side and vice versa.

Note that a strip of width $2h$ (shown in Figure 3.10 by dashed lines) is uniformly stretched to width $2h + 2\Delta$ under the tensile load. It should also be

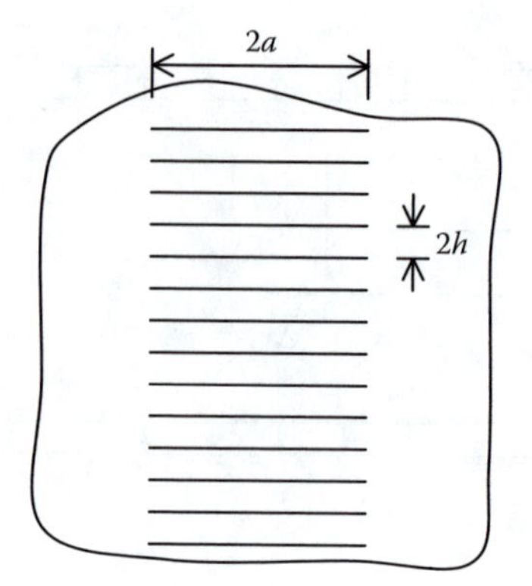


**FIGURE 3.9**
A stack of parallel cracks of length $2a$ and spacing $2h$.

noted here that there is no difference between the geometry of Figure 3.8 and the strip of width $2h$ shown in Figure 3.10 when it is stretched uniformly to the new width $2h + 2\Delta$. The stress intensity factor for this case should be $\sigma_0\sqrt{h}$ as given in equation (3.76).

Therefore, for these two extreme cases, $h >> a$ and $h << a$, the stress intensity factor is given as

$$
\begin{aligned}
K_I &= \sigma_0\sqrt{\pi a} \;\Rightarrow\; \frac{K_I}{\sigma_0\sqrt{\pi a}} = 1 \quad \text{for} \quad \frac{h}{a} >> 1 \\
K_I &= \sigma_0\sqrt{h} \;\Rightarrow\; \frac{K_I}{\sigma_0\sqrt{\pi a}} = \sqrt{\frac{h}{\pi a}} \quad \text{for} \quad \frac{h}{a} << 1
\end{aligned}
\qquad (3.77)
$$

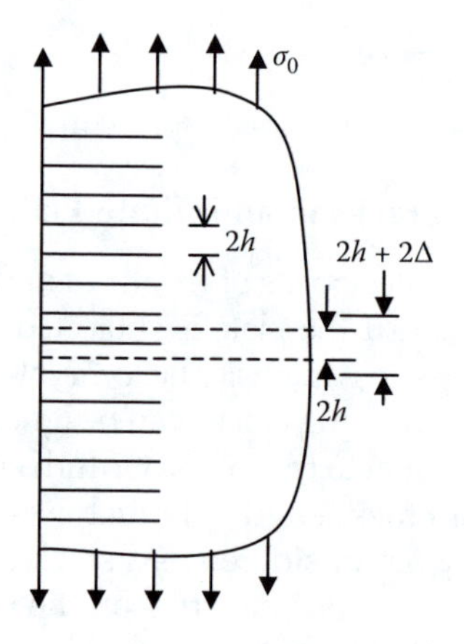


**FIGURE 3.10**
Right half of Figure 3.9. Stack of semi-infinite cracks in a plate subjected to uniaxial tension. Crack length is infinite relative to its spacing.

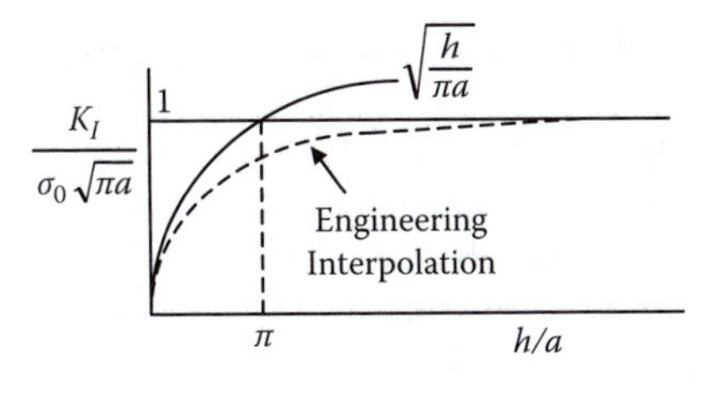


**FIGURE 3.11**
Stress intensity factor variation in a stack of parallel cracks as shown in Figure 3.9 for different $h/a$ ratios.

If we plot the variation of the normalized stress intensity factor (normalized with respect to SIF of the Griffith crack), then we get the plot of Figure 3.11. Note that after knowing the curves for small and large values of $h/a$, it is possible to interpolate the curve for intermediate values of $h/a$, as shown in Figure 3.11 by dashed lines, using our engineering judgment. Stress intensity factor obtained in this manner generally gives less than 5% error.

### 3.6.5 Star-Shaped Cracks

We now consider the geometry of multiple cracks radiating out from a point and thus forming a star-shaped crack system, as shown in Figure 3.12. The cracked plate is subjected to a biaxial state of stress. A total of $n$ cracks are equally spaced. Note that $n$ can be either even or odd. If a brittle plate, such as a glass plate, is struck by an object, then the impact force can form such a star-shaped crack system. Our interest is to obtain the stress intensity factor for this crack system for different values of $n$. Note that if the crack length is $a$, then the diameter of the star formed by $n$ cracks is $2a$.

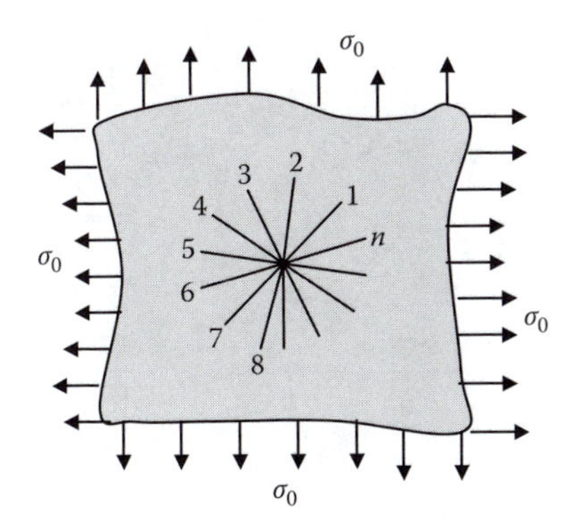


**FIGURE 3.12**
Star-shaped crack system subjected to biaxial state of stress.

For the special case of $n = 2$, a Griffith crack of length $2a$ is formed. For this case we know the stress intensity factor from equation (3.64). For very large value of $n$, the angle $\frac{2\pi}{n}$ between two neighboring cracks is very small. Therefore, in this case it can be approximately assumed that the neighboring cracks are almost parallel to each other. The spacing between two neighboring cracks is $2h = \frac{2\pi a}{n}$. In section 3.6.4, we have seen that the closely packed parallel cracks of spacing $2h$ subjected to an applied stress field of $\sigma_0$ in the direction perpendicular to crack surfaces give a stress intensity factor value of $\sigma_0\sqrt{h}$.

Now the question is for the biaxial state of stress: What should be the applied stress near the crack tip in the direction perpendicular to the crack surfaces? To answer this question, it should be noted here that for a large number of cracks, the star-shaped crack system behaves almost like a circular hole. From the theory of elasticity (see chapter 1, example 1.13) one knows that when a plate containing a circular hole is subjected to a biaxial state of stress, the circumferential stress $\sigma_{\theta\theta} = 2\sigma_0$ at the periphery of the circular hole due to the stress concentration effect, as shown in Figure 3.13. Note that the circumferential stress is perpendicular to the crack surface whose tip touches the periphery of the circle. Therefore, for large $n$ the stress intensity factor of the star-shaped crack system should be equal to $2\sigma_0\sqrt{h} = 2\sigma_0\sqrt{\frac{\pi a}{n}}$. Combining the two special cases: (1) $n = 2$ and (2) $n$ = a large number, the following equation is obtained:

$$
\begin{aligned}
&K_I = \sigma_0\sqrt{\pi a} \;\Rightarrow\; \frac{K_I}{\sigma_0\sqrt{\pi a}} = 1 \quad \text{for} \quad n = 2 \\
&K_I = 2\sigma_0\sqrt{\frac{\pi a}{n}} \;\Rightarrow\; \frac{K_I}{\sigma_0\sqrt{\pi a}} = \frac{2}{\sqrt{n}} \quad \text{for} \quad n = \text{Large Numbers}
\end{aligned}
\tag{3.78}
$$

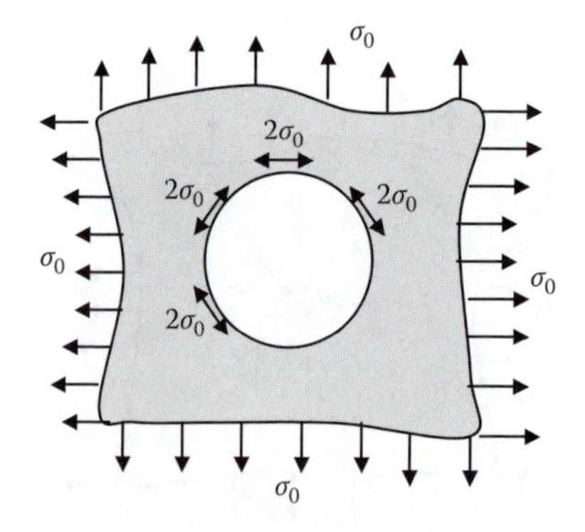


**FIGURE 3.13**
Circumferential stress at the periphery of a circular hole is $2\sigma_0$ when the plate is subjected to a biaxial state of stress.

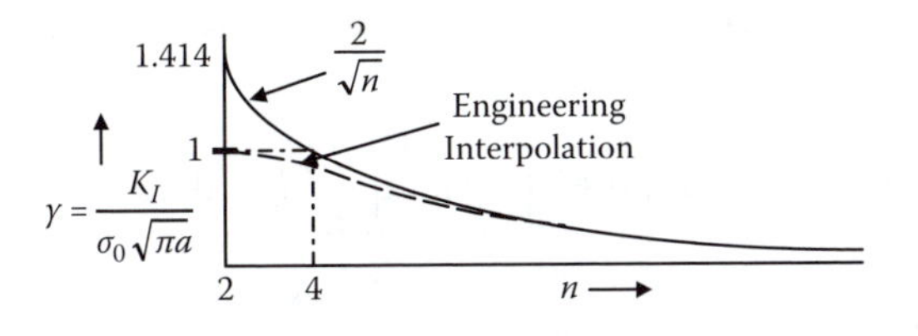


**FIGURE 3.14**
Variation of stress intensity factor for star-shaped crack as the number of cracks $n$ increases from 2 to a large value.

If the normalized stress intensity factor $(\gamma = \frac{K_I}{\sigma_0\sqrt{\pi a}})$ is plotted for the two special cases shown in equation (3.78), we obtain the plot of Figure 3.14.

Similar to Figure 3.11, here also engineering judgment is used to interpolate the curve between $n = 2$ and large values of $n$, and shown as the dashed line in Figure 3.14. Note that for very large $n$, the SIF becomes zero. This is justified because very large $n$ essentially transforms the star-shaped crack into a circular hole. Circular hole causes stress concentration, but the stress field near the periphery of the circular hole is finite. Therefore, SIF must be zero. For a more elaborate study on this problem, the reader is referred to Clark and Irwin (1966).

### 3.6.6 Pressurized Star Cracks

If the star-shaped crack system shown in Figure 3.12 is pressurized instead of having traction-free crack surfaces and the internal pressure inside the crack is not constant, then the problem becomes more complicated. This type of problem appears in mine blast situations. Its solution has been given by Westmann (1964). To solve this problem, all forces acting on a solid segment between two neighboring cracks are first considered. The free body diagram of this segment in Figure 3.15 shows the applied pressures on the two crack surfaces and the internal stress in the solid between two cracks.

Note that for small $\theta$ (i.e., large $n$), the normal stress $\sigma_{\theta\theta}$ can be approximately assumed to be equal to $p(r)$. For computing $\sigma_{rr}$ we apply the force equilibrium condition in the horizontal direction of the solid segment to

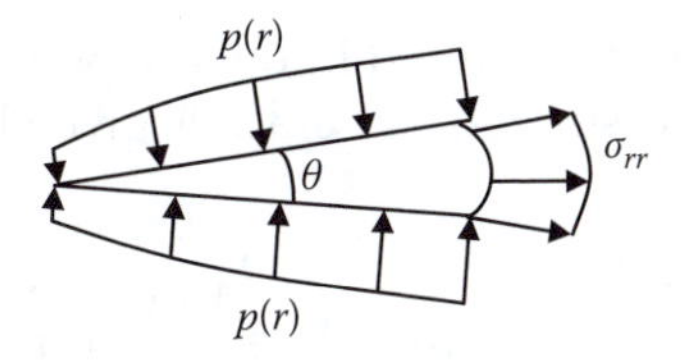

**FIGURE 3.15**
Free body diagram of the solid segment between two neighboring cracks.

obtain

$$\sigma_{rr}\rho\theta + 2\int_0^\rho p(r)\sin\left(\frac{\theta}{2}\right)dr = 0 \tag{3.79}$$

In Figure 3.15, $r$ varies from 0 to $\rho$, and $\rho$ can be any value less than or equal to $a$. Since the pressure field is perpendicular to the crack surface, its horizontal component is obtained by multiplying the pressure $p$ by $\sin(\frac{\theta}{2})$. However, since $\theta$ is small, $\sin(\frac{\theta}{2}) \approx \frac{\theta}{2} = \frac{2\pi}{2n} = \frac{\pi}{n}$. Thus, equation (3.79) becomes

$$\sigma_{rr}\rho\frac{2\pi}{n} + 2\int_0^\rho p(r)\frac{\pi}{n}dr = 0$$

$$\therefore \sigma_{rr}(\rho) = -\frac{2n}{2\pi\rho}\frac{\pi}{n}\int_0^\rho p(r)dr = -\frac{1}{\rho}\int_0^\rho p(r)dr \tag{3.80}$$

$$\Rightarrow \sigma_{rr}(r) = -\frac{1}{r}\int_0^r p(r')dr'$$

Therefore, the radial stress at the periphery of the circle of radius $a$ is $\sigma_{rr}(a) = -\frac{1}{a}\int_0^a p(r')dr'$. Then the stress, strain, and displacement fields outside the star crack region ($r > a$) are very close to a plate with a circular hole of radius $a$ and subjected to a compressive radial stress $\sigma_{rr}(a)$. From the theory of elasticity (see problem II of example 1.13), it can be shown that the radial displacement in a plate with a circular hole subjected to a constant compressive radial stress $\sigma_{rr}(a)$ is given by

$$u_r(r) = -\frac{1+\nu}{E}\sigma_{rr}(a)\frac{a^2}{r}$$
$$\therefore u_r(a) = -\frac{1+\nu}{E}\sigma_{rr}(a)a \tag{3.81}$$

Therefore, the strain energy stored in the region $r \geq a$ in the plate of thickness $t$ containing the pressurized star crack is given by

$$U_0 = -\frac{1}{2}\int_0^{2\pi}\sigma_{rr}(a)u_r(a)ta\,d\theta = \frac{1+\nu}{2E}ta^2\int_0^{2\pi}\sigma_{rr}^2(a)d\theta \tag{3.82}$$

and the strain energy in the region $r < a$, which is the cracked region:

$$U_i = \frac{t}{2E}\int_0^a \left(\sigma_{rr}^2 + \sigma_{\theta\theta}^2 - 2\nu\sigma_{rr}\sigma_{\theta\theta}\right)2\pi r dr = \frac{t\pi}{E}\int_0^a \left(\sigma_{rr}^2 + \sigma_{\theta\theta}^2 - 2\nu\sigma_{rr}\sigma_{\theta\theta}\right) r dr \quad (3.83)$$

For the general case when the pressure $p$ varies as a function of the radial distance $r$, the total strain energy $U = U_0 + U_i$ can be obtained from equations (3.82) and (3.83). Then the stress intensity factor can be derived from equation (3.53).

Let us consider the special case when $p(r) = p_0$, a constant. In this case,

$$\sigma_{rr}(a) = -\frac{1}{a}\int_0^a p(r')dr' = -\frac{1}{a}\int_0^a p_0 dr' = -p_0 \quad (3.84)$$

Then, from equations (3.82),

$$U_0 = \frac{1+\nu}{2E} ta^2 \int_0^{2\pi} p_0^2 d\theta = \frac{1+\nu}{2E} ta^2 p_0^2 2\pi = \pi p_0^2 a^2 t \frac{1+\nu}{E} \quad (3.85)$$

Note that at a radial distance $r$ from the center, $\sigma_{\theta\theta} = -p_0$ (obvious from Figure 3.15), and $\sigma_{rr}(r) = -\frac{1}{r}\int_0^r p(r')dr' = -\frac{1}{r}\int_0^r p_0 dr' = -p_0$; substituting these expressions of $\sigma_{\theta\theta}$ and $\sigma_{rr}$, $U_i$ can be computed as

$$\begin{aligned} U_i &= \frac{t\pi}{E}\int_0^a \left(\sigma_{rr}^2 + \sigma_{\theta\theta}^2 - 2\nu\sigma_{rr}\sigma_{\theta\theta}\right) r dr = \frac{t\pi}{E}\int_0^a (2-2\nu)p_0^2 r dr \\ &= \frac{2(1-\nu)t\pi}{E} p_0^2 \frac{a^2}{2} = \frac{(1-\nu)t\pi p_0^2 a^2}{E} \end{aligned} \quad (3.86)$$

Therefore, the total strain energy in the cracked plate,

$$U = U_0 + U_i = \pi p_0^2 a^2 t \frac{1+\nu}{E} + \pi p_0^2 a^2 t \frac{1-\nu}{E} = \frac{2\pi p_0^2 a^2 t}{E} \quad (3.87)$$

If total crack area is $A$, then

$$A = nat$$

$$\therefore \partial A = nt\partial a$$

Therefore,

$$\frac{\partial U}{\partial A} = \frac{\partial U}{nt\partial a} = \frac{4\pi p_0^2 at}{ntE} = \frac{4\pi p_0^2 a}{nE} = \frac{K_I^2}{E}$$
$$\therefore K_I = \frac{2}{\sqrt{n}} p_0 \sqrt{\pi a} \tag{3.88}$$

### 3.6.7 Longitudinal Cracks in Cylindrical Rods

The next geometry considered is a cylindrical rod with equally spaced longitudinal cracks propagating from the periphery to the central axis of the rod. The cross-section of such a cracked rod is shown in Figure 3.16. One application of this problem can be found in computing the stress intensity factors of cracked fuel element pellets of nuclear reactors.

Consider the case when the spacing between two neighboring crack tips is much smaller than the crack length. In other words, we first investigate the case when $\frac{2\pi}{n} r_1 \ll (R - r_1)$ and the crack is loaded by normal pressure $p(r)$. The free body diagram of a solid segment between two neighboring cracks for $r$ varying between $r_1$ and $R$ is shown in Figure 3.17.

From the force equilibrium in the radial direction for small angle $\theta$, one can write

$$\sigma_{rr} \frac{2\pi r}{n} = 2\int_r^R p(r') \frac{\pi}{n} dr'$$
$$\therefore \sigma_{rr} = \frac{1}{r} \int_r^R p(r') dr' \tag{3.89}$$

Therefore, the tensile stress applied along the periphery ($r = r_1$) of the central region ($r < r_1$) is $\sigma_{rr} = \frac{1}{r_1}\int_r^R p(r')dr'$. From the theory of elasticity we know that

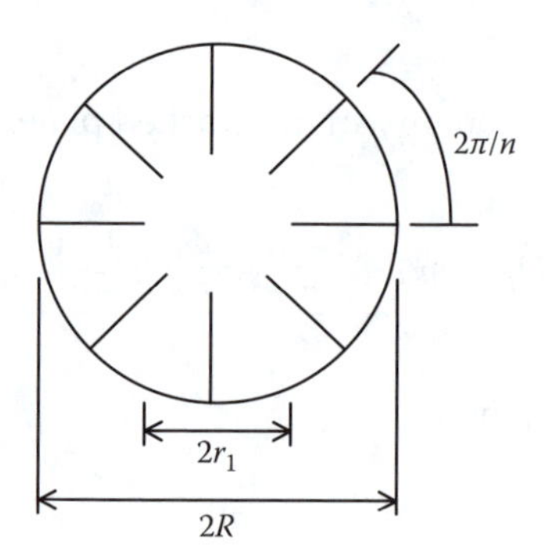


**FIGURE 3.16**
Cross-section of a cracked cylinder.

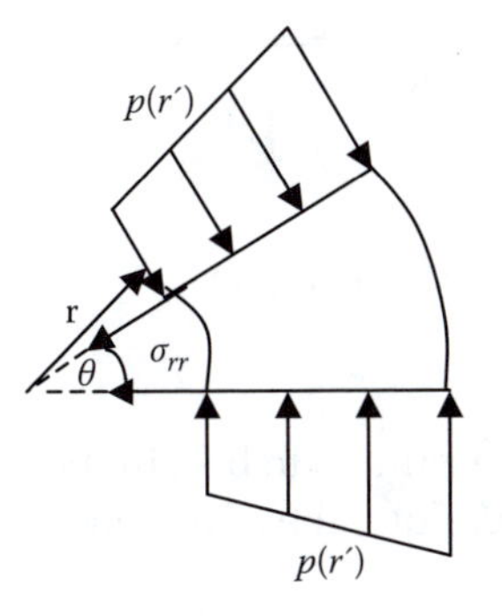


**FIGURE 3.17**
Free body diagram of a solid segment between two neighboring cracks of a cylinder shown in Figure 3.16.

when a circular disk is subjected to a uniform radial stress $\sigma_{rr}$ on its periphery, the state of stress inside the disk is given by

$$\sigma_{\theta\theta} = \sigma_{rr} = \frac{1}{r_1}\int_{r_1}^{R} p(r')\,dr' \tag{3.90}$$

Therefore, the strain energy stored in the inner core ($r < r_1$) is

$$U_i = \frac{1}{2}\int_V (\sigma_{rr}\varepsilon_{rr} + \sigma_{\theta\theta}\varepsilon_{\theta\theta})dV = \frac{1}{2E}\int_V (\sigma_{rr}^2 + \sigma_{\theta\theta}^2 - 2\nu\sigma_{rr}\sigma_{\theta\theta})dV \tag{3.91}$$

Let us consider the special case of uniform pressure $p(r) = p_0$. In this case,

$$\sigma_{rr}(r) = \frac{1}{r}\int_r^R p(r')dr' = \frac{1}{r}\int_r^R p_0 dr = \frac{p_0}{r}(R - r) = p_0\left(\frac{R}{r} - 1\right) \tag{3.92}$$

Therefore, the stress field in the inner core region ($r < r_1$) is given by $\sigma_{\theta\theta} = \sigma_{rr} = p_0(\frac{R}{r_1} - 1)$. Strain energy stored in the inner core,

$$U_i = \frac{1}{2E}\int_0^{r_1} (\sigma_{rr}^2 + \sigma_{\theta\theta}^2 - 2\nu\sigma_{rr}\sigma_{\theta\theta})2\pi rt dr = \frac{\pi t}{E}\int_0^{r_1} (\sigma_{rr}^2 + \sigma_{\theta\theta}^2 - 2\nu\sigma_{rr}\sigma_{\theta\theta})r dr$$

$$= \frac{2(1-\nu)\pi t}{E}\int_0^{r_1} \sigma_{rr}^2 r dr$$

$$
\begin{aligned}
&= \frac{2(1-\nu)\pi t p_0^2}{E}\left(\frac{R}{r_1}-1\right)^2 \int_0^{r_1} r dr = \frac{2(1-\nu)\pi t p_0^2}{E}\left(\frac{R}{r_1}-1\right)^2 \frac{r_1^2}{2} \\
&= \frac{(1-\nu)\pi t p_0^2}{E}\left(R^2 - 2Rr_1 + r_1^2\right)
\end{aligned} \tag{3.93}
$$

The length of the rod is assumed to be $t$. In the outer region ($R < r < r_1$), $\sigma_{\theta\theta} = -p_0$ and $\sigma_{rr} = p_0(\frac{R}{r} - 1)$; therefore, the strain energy stored in the outer region is given by

$$
\begin{aligned}
U_o &= \frac{1}{2E}\int_{r_1}^{R} (\sigma_{rr}^2 + \sigma_{\theta\theta}^2 - 2\nu\sigma_{rr}\sigma_{\theta\theta}) 2\pi r t dr \\
&= \frac{\pi t}{E}\int_{r_1}^{R} p_0^2 \left\{\left(\frac{R}{r}-1\right)^2 + 1 + 2\nu\left(\frac{R}{r}-1\right)\right\} r dr \\
&= \frac{\pi t p_0^2}{E}\int_{r_1}^{R} \left\{\left(\frac{R}{r}\right)^2 - 2\frac{R}{r} + 1 + 1 + 2\nu\left(\frac{R}{r}\right) - 2\nu\right\} r dr \\
&= \frac{\pi t p_0^2}{E}\int_{r_1}^{R} \left\{\left(\frac{R}{r}\right)^2 - 2(1-\nu)\frac{R}{r} + 2(1-\nu)\right\} r dr \\
&= \frac{\pi t p_0^2}{E}\int_{r_1}^{R} \left\{\frac{R^2}{r} - 2(1-\nu)R + 2(1-\nu)r\right\} dr \\
&= \frac{\pi t p_0^2}{E}\left\{R^2 \ln r - 2(1-\nu)Rr + 2(1-\nu)\frac{r^2}{2}\right\}_{r_1}^{R} \\
&= \frac{\pi t p_0^2}{E}\left\{R^2(\ln R - \ln r_1) - 2(1-\nu)R(R-r_1) + (1-\nu)\left(R^2 - r_1^2\right)\right\}
\end{aligned} \tag{3.94}
$$

Therefore, the total strain energy stored in the cylinder is given by

$$
\begin{aligned}
U = U_o + U_i &= \frac{\pi t p_0^2}{E}\left\{R^2(\ln R - \ln r_1) - 2(1-\nu)R(R-r_1) + (1-\nu)\left(R^2 - r_1^2\right)\right\} \\
&\quad + \frac{(1-\nu)\pi t p_0^2}{E}\left(R^2 - 2Rr_1 + r_1^2\right)
\end{aligned} \tag{3.95}
$$

Total crack area $A = nt(R - r_1)$; $\therefore\ \partial A = -nt\partial r_1$, since $R$ does not change but $r_1$ changes when crack lengths change. Therefore,

$$\frac{\partial U}{\partial A} = \frac{\partial U}{\partial r_1}\frac{\partial r_1}{\partial A} = \frac{\pi t p_0^2}{E}\left[\left\{R^2\left(-\frac{1}{r_1}\right) + 2(1-\nu)R - (1-\nu)2r_1\right\} + (1-\nu)(-2R + 2r_1)\right]\left(-\frac{1}{nt}\right)$$

$$= \frac{\pi p_0^2}{nE}\left[\left\{\frac{R^2}{r_1} - 2(1-\nu)(R-r_1)\right\} + 2(1-\nu)(R-r_1)\right] = \frac{\pi p_0^2}{nE}\frac{R^2}{r_1} = \alpha\frac{K_I^2}{E}$$

$$\therefore K_I = p_0 R\sqrt{\frac{\pi}{nr_1}} = p_0\sqrt{\pi R}\sqrt{\frac{R}{nr_1}} \tag{3.96}$$

In the preceding equation, $\alpha = 1$ is taken because plane stress condition is assumed; in other words, two ends of the cylinder are free to expand or shrink.

When $r_1$ is close to $R$ or the crack tips are close to the cylinder surface, the stress intensity factor is given by

$$K_I = 1.12 p_0\sqrt{\pi a} = 1.12 p_0\sqrt{\pi(R - r_1)} \tag{3.97}$$

The factor 1.12 appears because of the presence of the free surface near the crack tip. This point will be discussed in detail later. Equation (3.97) can be written in the following form:

$$K_I = 1.12 p_0\sqrt{\pi(R-r_1)} = 1.12 p_0\sqrt{\pi R}\sqrt{\left(1 - \frac{r_1}{R}\right)}$$

$$\therefore \frac{K_I}{p_0\sqrt{\pi R}} = 1.12\sqrt{\left(1 - \frac{r_1}{R}\right)} \tag{3.98}$$

Stress intensity factors given in equations (3.96) and (3.98) are plotted in Figure 3.18. In this plot the horizontal axis varies from 0 to 1 as the crack length varies from 0 to $R$. Note that the stress intensity factor increases with increasing crack length and decreases as the number of cracks increases. Engineering interpolation functions (curved dashed lines) are used to interpolate the curves between small crack lengths and large crack lengths.

## 3.7 Concluding Remarks

The relation between the stress intensity factor and the strain energy release rate is derived in this chapter from the energy balance principle proposed by Griffith (1921, 1924). Using this relation, the stress intensity factors for

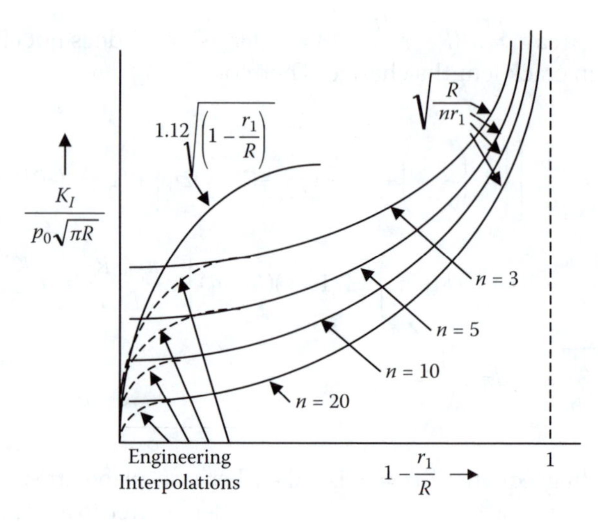


**FIGURE 3.18**
Variations of SIF in a cracked cylinder with variations of crack lengths and number of cracks. See Figure 3.16 for the problem geometry.

various problem geometries are obtained and presented. Stress intensity factors for some more practical problems are given in chapter 7.

## References

Broek, D. *Elementary engineering fracture mechanics,* 4th rev. ed. Dordrecht, the Netherlands: Kluwer Academic Publishers, 1997.

Clark, A. B. J. and Irwin, G. R. Crack propagation behaviors. *Experimental Mechanics,* 6(6), 321–330, 1966.

Griffith, A. A. The phenomena of rupture and flow in solids. *Philosophical Transactions of the Royal Society London,* A221, 163–198, 1921.

——. The theory of rupture. *Proceedings of the First International Congress of Applied Mechanics,* Delft, Biezeno and Burgers eds., Waltman, 55–63, 1924.

Irwin, G. R. Fracture dynamics. In *Fracturing of metals,* ASM, 29th National Metal Congress and Exposition, 147–166, 1948.

Knauss, W. G. Stresses in an infinite strip containing a semi-infinite crack. *ASME Journal of Applied Mechanics,* 53, 356–362, 1966.

Orowan, E. Energy criteria of fracture. *Welding Journal,* 34, 157s–160s, 1955.

Rice, J. R. Discussion on paper by Knauss. *ASME Journal of Applied Mechanics,* 34, 248–250, 1967.

Strawley, J. E., Jones, M. H., and Gross, B. Experimental determination of the dependence of crack extension force on crack length for a single-edge-notch tension specimen. NASA report #TND-2396, 1964.

Westmann, R. A. Pressurized star crack. *Journal of Mathematics and Physics,* 43, 191–198, 1964.

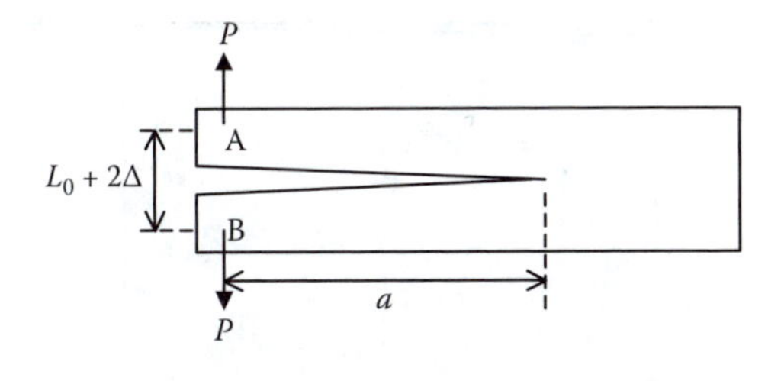


FIGURE 3.19

## Exercise Problems

Problem 3.1: The distance between points A and B is increased from $L_0$ to $L_0 + 2\Delta$ when opposing loads $P$ are applied (see Figure 3.19). Load-deflection ($P$-$\Delta$) curve for the specimen for crack lengths $a$ and $a + \delta a$ ($\delta a << a$) are shown in Figure 3.20 by straight lines OB and OC, respectively. Express in terms of different areas of Figure 3.20 the following parameters:

(a) The strain energy stored in the material with crack length $a$, load $P$, and deflection $\Delta$.

(b) If the crack length is increased from $a$ to $a + \delta a$, keeping $P$ fixed, which area does represent the change in strain energy in the material? Does it give an increase or decrease in the strain energy?

(c) If the crack length is increased from $a$ to $a + \delta a$, keeping $\Delta$ fixed, which area does represent the change in strain energy in the material? Does it give an increase or decrease in the strain energy?

(d) What is the change in the potential energy for cases (b) and (c)? Does the potential energy increase or decrease in cases (b) and (c)?

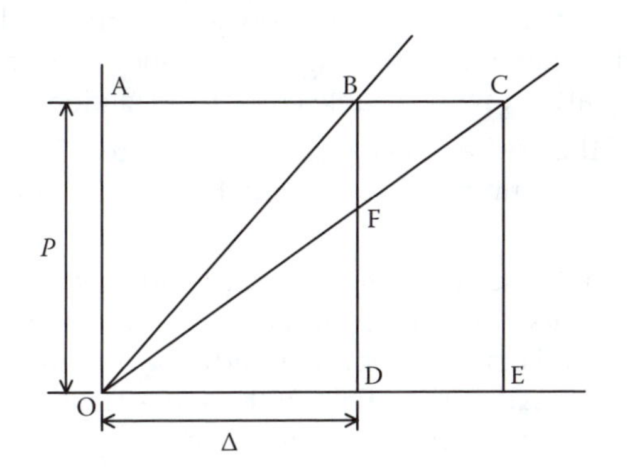


FIGURE 3.20

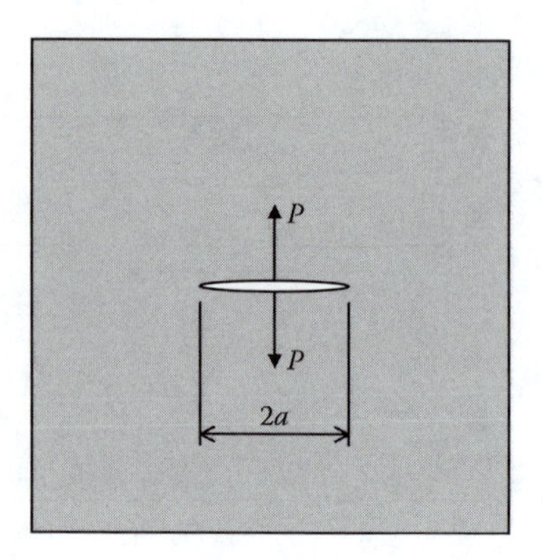


**FIGURE 3.21**

Problem 3.2: A Griffith crack in a large thin plate is loaded in an opening mode by two concentrated loads $P$ as shown in Figure 3.21. It produces a maximum crack opening at the center, given by the expression

$$\frac{4P}{t\pi E}\ln(a)+C$$

where $t$ is the thickness of the plate, $E$ is the Young's modulus, $a$ is the half crack length, and C is a constant.

(a) Calculate the stress intensity factor for this problem geometry.

(b) If the material has critical stress intensity factor $K_c$ = 200 kip. in.$^{-3/2}$, find the load $P$ for which a 2-in. long crack will start to propagate in a 1-in. thick plate.

(c) After the crack starts to propagate, if the load is not reduced, should the crack continue to propagate until the plate fails completely (unstable crack propagation) or should the crack propagation stop after a while (stable crack propagation)?

Problem 3.3: If the critical crack length is equal to 1.5 cm for a Griffith crack in a plate subjected to a biaxial state of stress $\sigma_0$ at the far field,

(a) What should be the critical crack length $2a_c$ when 18 equally spaced cracks of length $2a_c$ intersect at the center points to form a star-shaped crack system with radius $a_c$ and this system is subjected to the same stress field in the same material? Note that the angle between two neighboring cracks is equal to 10°.

(b) What should be the critical crack length $2a$, when only two cracks of length $2a_c$ intersect each other at midpoint at an angle 90° and subjected to the same stress field in the same material?

(c) If a large number of parallel cracks of length of $2a$ and distance $a$ between two neighboring cracks are present in the same material and are subjected to uniaxial tension $\sigma_0$ in the direction perpendicular to the crack axis, then what is the critical crack length $2a_c$?

# 4

# *Effect of Plasticity*

## 4.1 Introduction

In chapter 2 it was shown that the elastic solution gives infinite stress value at the crack tip. It implies that the material very close to the crack tip cannot remain elastic when the cracked body is loaded. This chapter discusses how the plastic zone size in front of the crack tip can be estimated, and what effect, if any, this plastic zone has on the stress computation and failure prediction of a cracked solid.

## 4.2 First Approximation on the Plastic Zone Size Estimation

A cracked plate subjected to tensile stresses is shown in Figure 4.1a. The free body diagram showing all forces on the top half of this linear elastic plate can be seen in Figure 4.1b. Upward and downward forces acting on the free body diagram of Figure 4.1b keep the top half of the plate in equilibrium. However, to keep the plate in equilibrium, the stress very close to the crack tip obtained from the elastic analysis exceeds the yield stress ($\sigma_{YS}$) of the material.

If the solid material is assumed to be an elastic–perfectly plastic material with yield stress $\sigma_{YS}$, then the maximum internal stress at the cut cannot exceed $\sigma_{YS}$. Then the stress field variation along the cut surface should be as shown in Figure 4.2. However, simply chopping off the stresses that are greater than $\sigma_{YS}$, as shown in Figure 4.2a, violates the equilibrium condition because it reduces the total downward force in the free body diagram of Figure 4.2a. The unbalanced force is denoted by the shaded area $A$ in Figure 4.2a. This unbalanced force can be taken care of by simply extending the extent of the plastic zone from $r_p$ to $\alpha r_p$ as shown in Figure 4.2b. Note that the shaded area $A$ must be equal to the additional area obtained from the extension of the plastic zone size, as shown in Figure 4.2b.

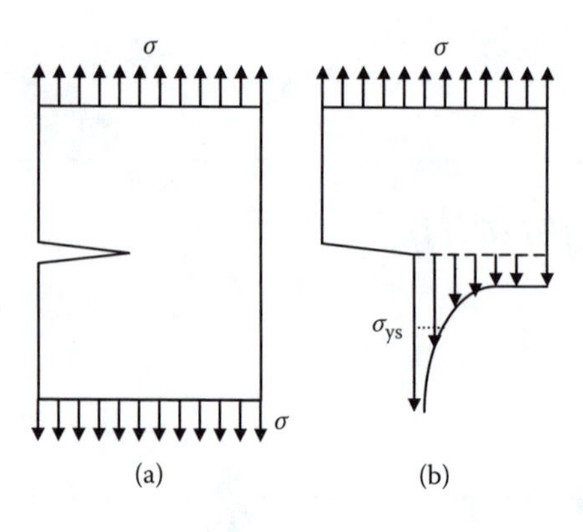


**FIGURE 4.1**
(a) Cracked plate under tensile load; (b) free body diagram of the top half of the plate obtained from elastic analysis.

### 4.2.1 Evaluation of $r_p$

For the opening mode loading, the stress field ahead of the crack tip is given by (see equation 2.47) $\sigma = \frac{K_I}{\sqrt{2\pi r}}$. From this equation and Figure 4.2a one can write

$$\sigma_{YS} = \frac{K_I}{\sqrt{2\pi r_p}}$$
$$\therefore r_p = \frac{1}{2\pi}\left(\frac{K_I}{\sigma_{YS}}\right)^2 \tag{4.1}$$

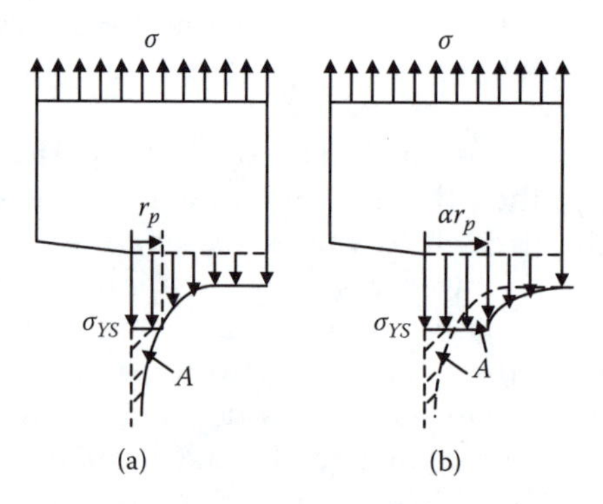


**FIGURE 4.2**
Internal stress field ahead of the crack tip for an elastic–perfectly plastic material. (a) Free body diagram is not in equilibrium—unbalanced force A is shown; (b) free body diagram is in equilibrium.

### 4.2.2 Evaluation of $\alpha r_p$

The unbalanced force area $A$ of Figure 4.2a can be obtained from the following equation:

$$A = \int_0^{r_p} \left( \frac{K_I}{\sqrt{2\pi r}} - \sigma_{YS} \right) dr = \frac{K_I}{\sqrt{2\pi}} 2\sqrt{r_p} - \sigma_{YS} r_p \tag{4.2}$$

Substituting equation (4.1) into equation (4.2),

$$\begin{aligned} A &= \frac{K_I}{\sqrt{2\pi}} 2\sqrt{r_p} - \sigma_{YS} r_p = \frac{K_I}{\sqrt{2\pi}} \frac{2}{\sqrt{2\pi}} \left( \frac{K_I}{\sigma_{YS}} \right) - \sigma_{YS} \frac{1}{2\pi} \left( \frac{K_I}{\sigma_{YS}} \right)^2 \\ &= \frac{2}{2\pi} \frac{K_I^2}{\sigma_{YS}} - \frac{1}{2\pi} \frac{K_I^2}{\sigma_{YS}} = \frac{1}{2\pi} \frac{K_I^2}{\sigma_{YS}} = \sigma_{YS} r_p \end{aligned} \tag{4.3}$$

Note that the chopped off shaded area and the rectangular area shown in Figure 4.3a both have the same area, $A = \sigma_{YS} r_p$. After chopping off the shaded region, the downward force in the free body diagram of Figure 4.3a is reduced to $A + B$, while the total downward force needed for equilibrium is $2A + B$. Therefore, to satisfy the equilibrium condition, the plastic zone must be extended to $\alpha r_p$ as shown in Figure 4.3b. If $\alpha r_p$ is much smaller than the plate width on which the downward force acts, then the total downward force in the elastic region may be approximately assumed to be $B$ in both Figures 4.3a and 4.3b. Adding the downward force ($2A$) of the plastic region, one gets the total downward force = $2A + B$, which is sufficient to satisfy the equilibrium condition. From Figure 4.3b, it is clear that $\alpha r_p = 2r_p$ or $\alpha = 2$.

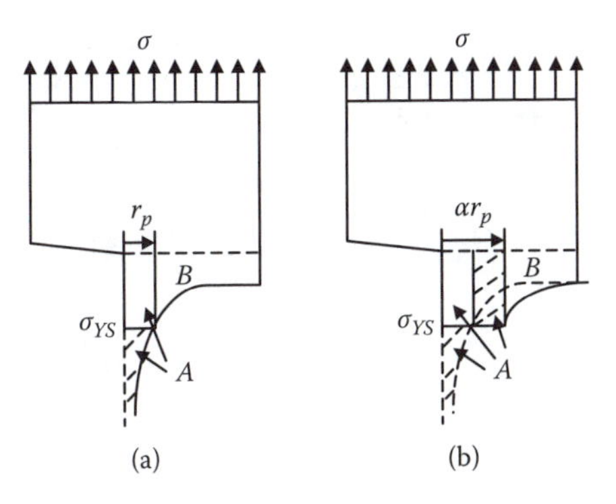


**FIGURE 4.3**
(a) Free body diagram is not in equilibrium—downward force is $A + B$; (b) free body diagram is in equilibrium—downward force is $2A + B$. The elastic region should be much greater than the plastic region.

Therefore, the plastic zone size ahead of the crack tip is given by

$$R = \alpha r_p \approx 2r_p = \frac{1}{\pi}\left(\frac{K_I}{\sigma_{YS}}\right)^2 \tag{4.4}$$

Note that the elastic stress field in Figure 4.3b is obtained by simply moving the elastic stress field of Figure 4.3a toward the right by an amount $r_p$. This field can be obtained by simply moving the crack tip toward the right by an amount $r_p$ or increasing the crack length from $a$ to $a + r_p$. It is a common practice for fracture mechanics analysis to change the crack length from $a$ to $a + r_p$ to take into account the effect of plastic deformation in front of the crack tip.

## 4.3 Determination of the Plastic Zone Shape in Front of the Crack Tip

In the previous section the plastic zone length has been calculated ahead of the crack tip for $\theta = 0$ only. In this section we calculate the extent of the plastic zone for all values of $\theta$. Note that for opening mode loading, the stress field in front of the crack tip is obtained from equation (2.47):

$$\begin{aligned}
\sigma_{rr} &= \frac{K_I}{\sqrt{2\pi r}}\cos\left(\frac{\theta}{2}\right)\left[1+\sin^2\left(\frac{\theta}{2}\right)\right] \\
\sigma_{\theta\theta} &= \frac{K_I}{\sqrt{2\pi r}}\cos^3\left(\frac{\theta}{2}\right) \\
\sigma_{r\theta} &= \frac{K_I}{\sqrt{2\pi r}}\sin\left(\frac{\theta}{2}\right)\cos^2\left(\frac{\theta}{2}\right)
\end{aligned} \tag{4.5}$$

Then the principal stresses can be obtained from the two-dimensional stress transformation law or from Mohr's circle analysis:

$$\sigma_{1,2} = \frac{\sigma_{rr}+\sigma_{\theta\theta}}{2} \pm \sqrt{\left(\frac{\sigma_{rr}-\sigma_{\theta\theta}}{2}\right)^2 + (\sigma_{r\theta})^2} \tag{4.6}$$

Substituting relations (4.5) into equation (4.6) gives

$$\begin{aligned}
\sigma_1 &= \frac{K_I}{\sqrt{2\pi r}}\cos\left(\frac{\theta}{2}\right)\left[1+\sin\left(\frac{\theta}{2}\right)\right] \\
\sigma_2 &= \frac{K_I}{\sqrt{2\pi r}}\cos\left(\frac{\theta}{2}\right)\left[1-\sin\left(\frac{\theta}{2}\right)\right]
\end{aligned} \tag{4.7}$$

$$\sigma_3 = 0 \quad \text{for plane stress condition}$$

$$= \nu(\sigma_1 + \sigma_2) = 2\nu \frac{K_I}{\sqrt{2\pi r}} \cos\left(\frac{\theta}{2}\right) \quad \text{for plane strain condition} \tag{4.8}$$

Von Mises and Tresca's yield criteria, given in equations (4.9) and (4.10), respectively, can be used to calculate the plastic zone size:

$$(\sigma_1 - \sigma_2)^2 + (\sigma_2 - \sigma_3)^2 + (\sigma_3 - \sigma_1)^2 = 2\sigma_{YS}^2 \tag{4.9}$$

$$\text{Max}(|\sigma_1 - \sigma_2|, |\sigma_2 - \sigma_3|, |\sigma_3 - \sigma_1|) = \sigma_{YS} \tag{4.10}$$

Substituting three principal stresses in Von Mises' yield criterion (equation 4.9), one gets for plane stress problems:

$$\begin{aligned}
&\left(\frac{K_I}{\sqrt{2\pi r_p}}\right)^2 \left[4\cos^2\left(\frac{\theta}{2}\right)\sin^2\left(\frac{\theta}{2}\right) + \cos^2\left(\frac{\theta}{2}\right)\left\{1 + \sin\left(\frac{\theta}{2}\right)\right\}^2\right. \\
&\quad \left. + \cos^2\left(\frac{\theta}{2}\right)\left\{1 - \sin\left(\frac{\theta}{2}\right)\right\}^2\right] = 2\sigma_{YS}^2 \\
&\therefore \frac{K_I^2}{2\pi r_p}\cos^2\left(\frac{\theta}{2}\right)\left[4\sin^2\left(\frac{\theta}{2}\right) + 1 + \sin^2\left(\frac{\theta}{2}\right) + 2\sin\left(\frac{\theta}{2}\right)\right. \\
&\quad \left. + 1 + \sin^2\left(\frac{\theta}{2}\right) - 2\sin\left(\frac{\theta}{2}\right)\right] = 2\sigma_{YS}^2 \\
&\therefore \frac{K_I^2}{2\pi r_p}\cos^2\left(\frac{\theta}{2}\right)\left[6\sin^2\left(\frac{\theta}{2}\right) + 2\right] = 2\sigma_{YS}^2 \\
&\therefore r_p = \frac{K_I^2}{4\pi\sigma_{YS}^2}\cos^2\left(\frac{\theta}{2}\right)\left[6\sin^2\left(\frac{\theta}{2}\right) + 2\right] \\
&\quad = \frac{K_I^2}{4\pi\sigma_{YS}^2}\left[2\cos^2\left(\frac{\theta}{2}\right) + 6\sin^2\left(\frac{\theta}{2}\right)\cos^2\left(\frac{\theta}{2}\right)\right] \\
&\quad = \frac{K_I^2}{4\pi\sigma_{YS}^2}\left[2\cos^2\left(\frac{\theta}{2}\right) + \frac{3}{2}\sin^2\theta\right]
\end{aligned} \tag{4.11}$$

For plane strain problems, one gets

$$\begin{aligned}
&\left(\frac{K_I}{\sqrt{2\pi r_p}}\right)^2 \left[4\cos^2\left(\frac{\theta}{2}\right)\sin^2\left(\frac{\theta}{2}\right) + \cos^2\left(\frac{\theta}{2}\right)\left\{1 + \sin\left(\frac{\theta}{2}\right) - 2\nu\right\}^2\right. \\
&\quad \left. + \cos^2\left(\frac{\theta}{2}\right)\left\{1 - \sin\left(\frac{\theta}{2}\right) - 2\nu\right\}^2\right] = 2\sigma_{YS}^2
\end{aligned}$$

$$\therefore \frac{K_I^2}{2\pi r_p}\cos^2\left(\frac{\theta}{2}\right)\left[4\sin^2\left(\frac{\theta}{2}\right)+\left\{(1-2\nu)+\sin\left(\frac{\theta}{2}\right)\right\}^2+\left\{(1-2\nu)-\sin\left(\frac{\theta}{2}\right)\right\}^2\right]=2\sigma_{YS}^2$$

$$\therefore \frac{K_I^2}{2\pi r_p}\cos^2\left(\frac{\theta}{2}\right)\left[4\sin^2\left(\frac{\theta}{2}\right)+2(1-2\nu)^2+2\sin^2\left(\frac{\theta}{2}\right)\right]=2\sigma_{YS}^2$$

$$\therefore r_p=\frac{K_I^2}{4\pi\sigma_{YS}^2}\cos^2\left(\frac{\theta}{2}\right)\left[6\sin^2\left(\frac{\theta}{2}\right)+2(1-2\nu)^2\right] \tag{4.12}$$

$$=\frac{K_I^2}{4\pi\sigma_{YS}^2}\left[2(1-2\nu)^2\cos^2\left(\frac{\theta}{2}\right)+6\sin^2\left(\frac{\theta}{2}\right)\cos^2\left(\frac{\theta}{2}\right)\right]$$

$$=\frac{K_I^2}{4\pi\sigma_{YS}^2}\left[2(1-2\nu)^2\cos^2\left(\frac{\theta}{2}\right)+\frac{3}{2}\sin^2\theta\right]$$

Since the actual plastic zone size $R$ is approximately two times $r_p$ (see equation 4.4), one can write

$$R(\theta)=2r_p(\theta)=\frac{K_I^2}{2\pi\sigma_{YS}^2}\left[2\cos^2\left(\frac{\theta}{2}\right)+\frac{3}{2}\sin^2\theta\right]$$

for plane stress problems

$$R(\theta)=2r_p(\theta)=\frac{K_I^2}{2\pi\sigma_{YS}^2}\left[2(1-2\nu)^2\cos^2\left(\frac{\theta}{2}\right)+\frac{3}{2}\sin^2\theta\right] \tag{4.13}$$

for plane strain problems

The plastic zone shapes given in equation (4.13) are shown in Figure 4.4.

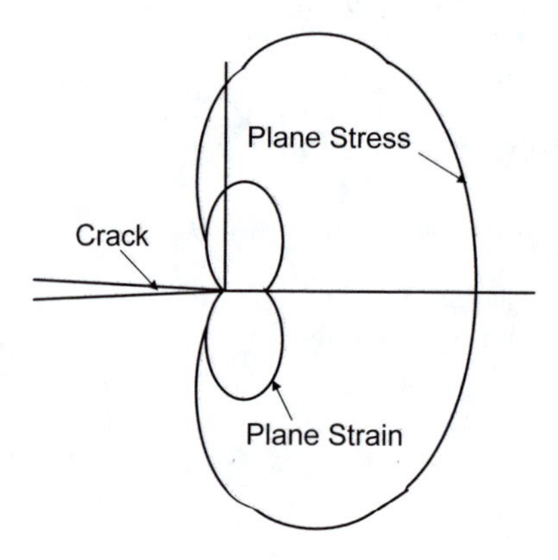


**FIGURE 4.4**
Shape of the plastically deformed regions in front of the crack tip for mode I loading obtained from Von Mises' yield criterion for plane stress and plane strain conditions.

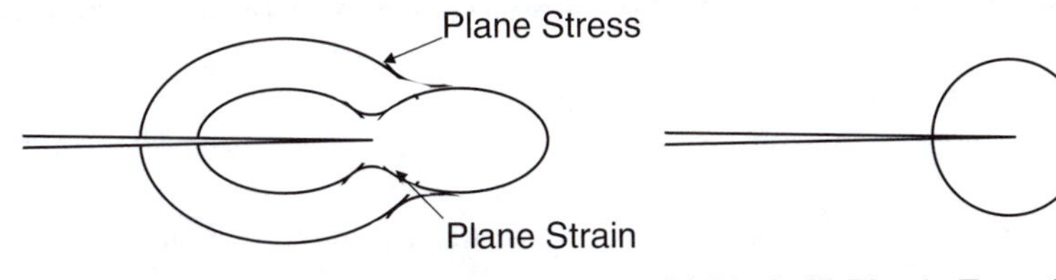


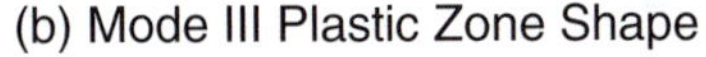


**FIGURE 4.5**
Plastic zone shapes for (a) mode II and (b) mode III loading.

Note that the plane strain plastic zone size is significantly smaller than that for the plane stress condition. Along the $\theta = 0$ line (in front of the crack), one can obtain from equation (4.13):

$$\left[\frac{R\big|_{\text{plane stress}}}{R\big|_{\text{plane strain}}}\right]_{\theta=0} = \left[\frac{r_p\big|_{\text{plane stress}}}{r_p\big|_{\text{plane strain}}}\right]_{\theta=0} = \left[\frac{2\cos^2\left(\frac{\theta}{2}\right) + \frac{3}{2}\sin^2\theta}{2(1-2\nu)^2\cos^2\left(\frac{\theta}{2}\right) + \frac{3}{2}\sin^2\theta}\right]_{\theta=0} = \frac{1}{(1-2\nu)^2} \tag{4.14}$$

Clearly, this ratio is 6.25 for $\nu = 0.3$ and it is 9 for $\nu = 0.33$. Thus, it is very sensitive to the Poisson's ratio of the material.

For mode II and mode III loadings, the plastic zone shapes can be obtained in the same manner considering appropriate stress field expressions. These shapes are shown in Figure 4.5. For more detailed discussion on the plastic zone shapes, readers are referred to McClintock and Irwin (1965). In Figure 4.5 one can see that for mode II loading, plane stress and plane strain plastic zone sizes are identical at $\theta = 0$. This is because, under mode II loading at $\theta = 0$, the stress field from equation (2.47) is obtained as

$$\begin{aligned} \sigma_{xx} &= \sigma_{yy} = 0 \\ \sigma_{xy} &= \frac{K_{II}}{\sqrt{2\pi r}} \end{aligned} \tag{4.15}$$

For the stress field given in equation (4.15), one can easily show from Mohr's circle that two principal stresses in the $xy$ plane are

$$\sigma_1 = \frac{K_{II}}{\sqrt{2\pi r}}, \quad \sigma_2 = -\frac{K_{II}}{\sqrt{2\pi r}} \tag{4.16}$$

The third principal stress in the $z$-direction is 0 for plane stress and $\sigma_3 = \nu(\sigma_1 + \sigma_2) = 0$ for plane strain conditions. Since all three principal stress components match for plane stress and plane strain problems along the $\theta = 0$ line, the plastic zone size on this line is identical for plane stress and plane strain problems.

It is well known that when the plate thickness is much smaller than the characteristic dimensions of the plate, the problem is a plane stress problem; when the plate thickness is much greater than the characteristic dimensions, the problem is a plane strain problem. Length and width of the plate and various dimensions of defects in the plate (such as the radius of a circular hole in the plate, if such a hole exists) are considered as the characteristic dimensions. A crack has two dimensions: its length and width. However, since the crack width is infinitesimally small, the plate thickness is always much larger than the crack width. For deciding whether the plane stress or the plane strain condition dominates in a cracked plate, the plate thickness is compared with the plastic zone size $R$ in front of the crack tip instead of the crack width. If the plate thickness is much smaller than $R$, then it is a plane stress problem; if it is much greater than $R$, then the plane strain condition dominates.

One major difference between a crack-free plate problem and a cracked plate problem should be mentioned here. When a crack-free plate is subjected to in-plane stresses only and is free to expand or contract in the out-of-plane direction, the plate is subjected to a pure plane-stress condition since no out-of-plane stress is developed in the plate. This is because no restrictions are imposed on its movement in the out-of-plane direction. However, in a loaded cracked plate, the plastic zone is formed in front of the crack tip and the plastically deformed zone has different Poisson's ratio (close to 0.5) than

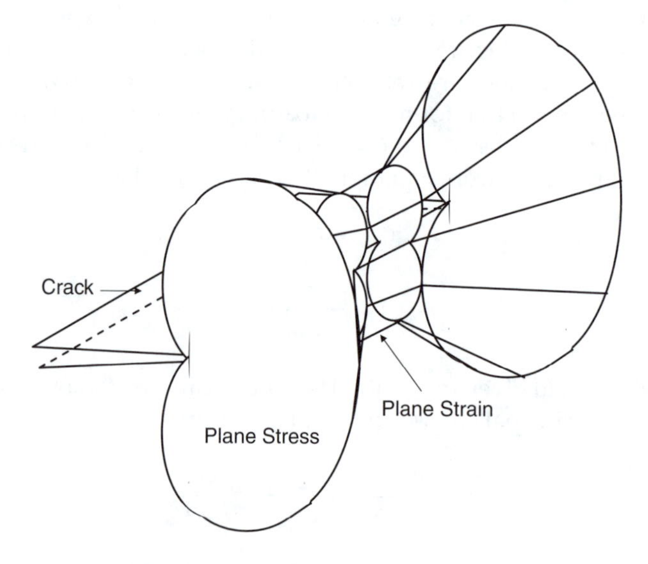


**FIGURE 4.6**
Plastic zone shape in front of the crack tip in a thick plate.

the material in the elastic region. Therefore, the plastically deformed region is not free to move in the out-of-plane direction. For this reason, when a cracked thick plate is loaded, the central part of the plate is subjected to the plane strain condition since the out-of-plane normal stress is developed in this region; the sections close to the two free surfaces of the plate exhibit the plane stress condition because, on the plate surfaces, out-of-plane normal stress and shear stresses are zero. Therefore, under mode I loading, the plastic zone shape in a thick plate in front of the crack tip should be as shown in Figure 4.6.

## 4.4 Plasticity Correction Factor

For a cracked plate subjected to an applied stress $\sigma$ in the far field (far away from the crack), where $\frac{\sigma}{\sigma_{YS}} < 0.5$, the plastic zone size is very small. The small plastic zone has negligible effect on the computed stress field. For $0.5 < \frac{\sigma}{\sigma_{YS}} < 0.7$, the plastic zone affects the computed stress field. Its effect can be taken into account by incorporating the plasticity correction factor. When the applied stress field is in this region, the elastic analysis after plasticity correction can correctly model the problem. However, when the applied stress is too large, $\frac{\sigma}{\sigma_{YS}} > 0.7$, the elastic analysis with plasticity correction is not adequate to correctly model the problem. In this situation complete elastoplastic analysis must be carried out to solve the problem.

Plasticity correction is taken into account following Irwin's suggestion, simply by increasing the crack length from $a$ to $a + r_p$ or from $2a$ to $2(a + r_p)$. Justification for this modification is given in section 4.2. Let us now investigate how it affects the stress intensity factor of a Griffith crack.

Since the modified crack length after plasticity correction is $2a' = 2(a + r_p)$, the modified stress intensity factor is

$$K = \sigma\sqrt{\pi a'} = \sigma\sqrt{\pi(a + r_p)} \tag{4.17}$$

Substituting equation (4.1) into equation (4.17),

$$K^2 = \sigma^2\pi(a + r_p) = \sigma^2\pi\left(a + \frac{1}{2\pi}\left[\frac{K}{\sigma_{YS}}\right]^2\right) = \sigma^2\left(\pi a + \frac{K^2}{2\sigma_{YS}^2}\right)$$

$$\therefore K^2\left\{1 - \frac{1}{2}\left(\frac{\sigma}{\sigma_{YS}}\right)^2\right\} = \sigma^2\pi a \tag{4.18}$$

$$\therefore K = \frac{\sigma\sqrt{\pi a}}{\sqrt{1 - \frac{1}{2}\left(\frac{\sigma}{\sigma_{YS}}\right)^2}}$$

**TABLE 4.1**

Critical Crack Lengths and Plastic Zone Sizes in Different Materials for Applied Stress = 50% of Yield Stress

| Material | $K_{Ic}$ (ksi $\sqrt{\text{in}}$ ) | $\sigma_{YS}$ (ksi) | $\sigma_{ult}$ (ksi) | $a_c$ (in.) | $r_p$ (in.) |
|---|---|---|---|---|---|
| 4340 Steel | 42 | 214 | 264 | 0.0429 | 0.0061 |
| 7075-T6 Al | 30 | 73 | 81 | 0.1882 | 0.0269 |
| Maraging steel | 82 | 250 | 268 | 0.1199 | 0.0171 |

Equation (4.18) can be used to calculate the critical crack length $a_c$ in different materials whose critical stress intensity factor $K_c$ is known.

$$K_c = \frac{\sigma\sqrt{\pi a_c}}{\sqrt{1-\frac{1}{2}\left(\frac{\sigma}{\sigma_{YS}}\right)^2}}$$

$$\therefore a_c = \frac{K_c^2}{\pi\sigma^2}\left\{1-\frac{1}{2}\left(\frac{\sigma}{\sigma_{YS}}\right)^2\right\} \tag{4.19}$$

If applied stress $\sigma = \frac{\sigma_{YS}}{2}$, then equation (4.19) gives

$$a_c = \frac{K_c^2}{\pi\sigma^2}\left\{1-\frac{1}{2}\left(\frac{1}{2}\right)^2\right\} = \frac{7K_c^2}{8\pi\sigma^2} \tag{4.20}$$

From equation (4.20), critical crack lengths for $\sigma = \frac{\sigma_{YS}}{2}$ in different materials can be computed as shown in Table 4.1.

For more complex expressions of stress intensity factors (SIFs), the modified SIF may not be obtained in closed form such as the one given in equation (4.18). In such situations modified SIF can be obtained through iterative calculations, as shown in equation (4.21). The iterative steps shown in equation (4.21) generally converge within a few iterations:

$$K^{(0)} = f(a^{(0)})$$

$$a^{(1)} = a^{(0)} + r_p^{(0)} = a^{(0)} + \frac{1}{2\pi}\left(\frac{K^{(0)}}{\sigma_{YS}}\right)^2$$

$$K^{(1)} = f(a^{(1)})$$

$$a^{(2)} = a^{(1)} + r_p^{(1)} = a^{(1)} + \frac{1}{2\pi}\left(\frac{K^{(1)}}{\sigma_{YS}}\right)^2$$

$$K^{(2)} = f(a^{(2)})$$

$$\dots$$

$$\dots$$

$$\dots \tag{4.21}$$

$$a^{(n)} = a^{(n-1)} + r_p^{(n-1)} = a^{(n-1)} + \frac{1}{2\pi}\left(\frac{K^{(n-1)}}{\sigma_{YS}}\right)^2$$

$$K^{(n)} = f\left(a^{(n)}\right)$$

## 4.5 Failure Modes under Plane Stress and Plane Strain Conditions

### 4.5.1 Plane Stress Case

Equations (4.7) and (4.8) give principal stress expressions near a crack tip for both plane stress and plane strain conditions. Note that under the plane stress condition $\sigma_1 > \sigma_2 > \sigma_3$. Therefore, from Mohr's circle (Figure 4.7a), it is clear that the maximum shear stress occurs at a plane that bisects $\sigma_1$ and $\sigma_3$ directions, as shown in Figure 4.7b. In Figure 4.7b, $\sigma_1$ and $\sigma_2$ are shown in the $xy$ plane in two mutually perpendicular directions. However, it should be noted here that in the $xz$ plane just ahead of the crack tip ($\theta = 0°$), $\sigma_1$ and $\sigma_2$ coincide with $y$ and $x$ directions, respectively, for the opening mode loading.

We have seen earlier that if it is assumed that the crack propagates in the direction perpendicular to the maximum normal stress, then the crack should propagate along $\theta = 0°$ plane. This is true when the material behaves like a brittle material. This theory is known as the brittle fracture theory.

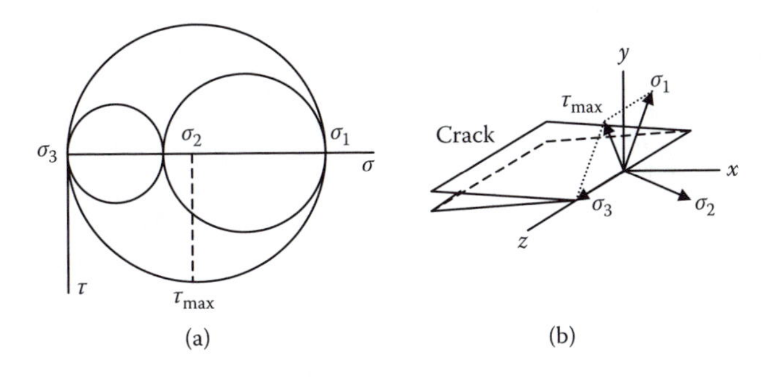


**FIGURE 4.7**
For plane stress condition, (a) Mohr's circle and (b) principal stress directions and maximum shear directions near a crack tip.

However, when a large plastic zone is formed in front of the crack tip, as in the case of the plane stress loading condition, then the material in front of the crack tip is ductile, not brittle. This is because plastically deformed materials show ductile behavior and are weak in shear. For this reason, under the plane stress loading condition the material in front of the crack fails along the maximum shear plane. Figure 4.7b clearly shows that the maximum shear plane forms an angle with the crack surface (*xz* plane); therefore, the failure plane should make an angle relative to the crack surface if the ductile failure occurs.

### 4.5.2 Plane Strain Case

Poisson's ratio is close to 0.5 in the plastically deformed region. Therefore, for plane strain condition $\sigma_3 = 2\nu \frac{K_I}{\sqrt{2\pi r}} \cos(\frac{\theta}{2}) \approx \frac{K_I}{\sqrt{2\pi r}} \cos(\frac{\theta}{2})$. Thus, in this case, from equations (4.7) and (4.8) one concludes that $\sigma_1 > \sigma_3 > \sigma_2$. Mohr's circle and the maximum shear direction for the plane strain case are shown in Figure 4.8. Therefore, if the plastically deformed zone size is large, resulting in ductile failure, then the failure plane should propagate along the maximum shear direction as shown in Figure 4.8b. However, under plane strain conditions, since the plastic zone size is small, the material exhibits mostly brittle failure—that is, the failure occurs not in the maximum shear stress direction but in the direction perpendicular to the maximum normal stress, as discussed in chapter 2.

As discussed before, a cracked plate shows ductile failure under the plane stress condition and brittle failure under the plane strain condition. Therefore, the failure surface and the crack propagation direction show significant difference between plane stress and plane strain situations, as illustrated in Figure 4.9. The failure surfaces in Figure 4.9 are presented for the opening mode loading for three different plate thicknesses: thin, medium, and thick. Note that the stress intensity factor at failure is significantly higher under the plane stress condition. To be on the safe side, the critical stress intensity factor of a material is defined as the stress intensity factor at failure under

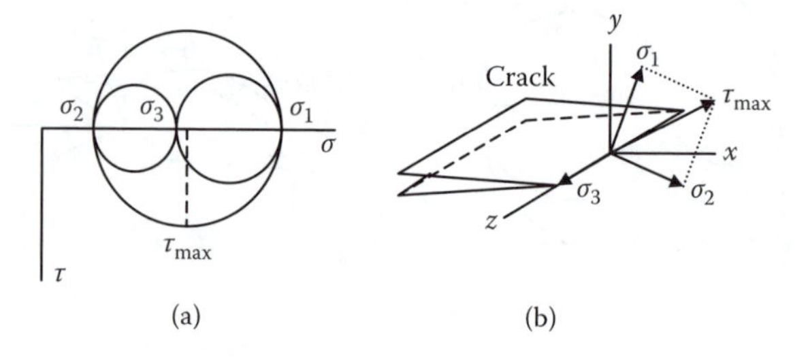


**FIGURE 4.8**
For plane strain condition, (a) Mohr's circle and (b) principal stress directions and maximum shear directions near a crack tip.

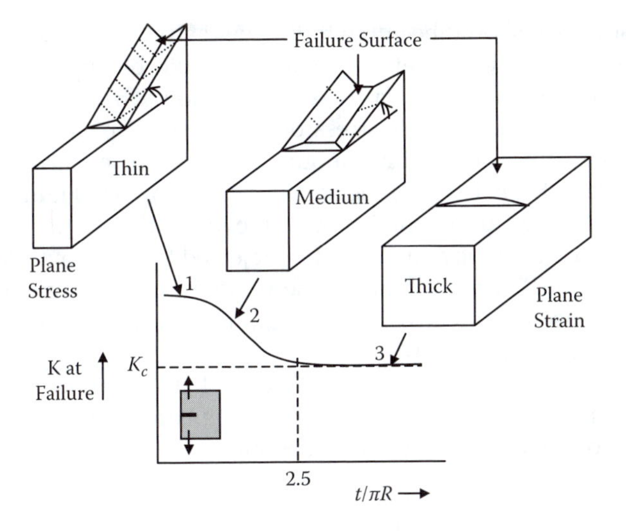


**FIGURE 4.9**
Critical stress intensity factor variation with plate thickness ($t$). Failure surfaces for different plate thicknesses are also shown.

plane strain condition, which is achieved when the plate thickness exceeds $2.5\pi R$, where $R$ is the plastic zone size ahead of the crack tip as defined in equations (4.4) and (4.13).

## 4.6 Dugdale Model

Dugdale (1960) proposed a simple elastoplastic analysis to compute the extent of plastic zone in front of a crack tip. He assumed the plastic zone size to be a thin strip BC in front of the crack AB as shown in Figure 4.10a. The material in region BC inside the thin strip whose boundary is marked by the dashed line in Figure 4.10a is plastically deformed. For the analysis presented next the material is assumed to be elastic–perfectly plastic. Then the stress level

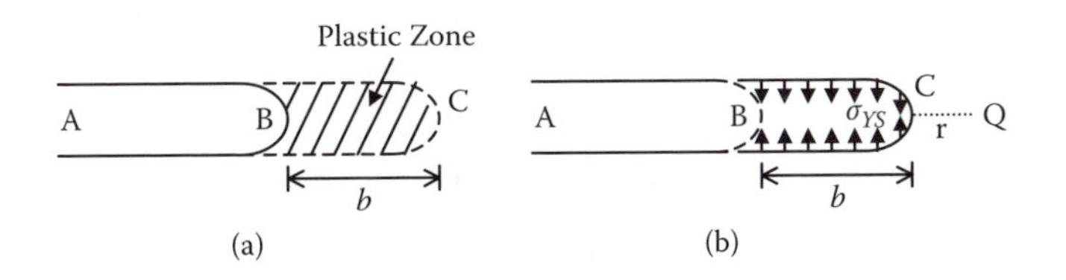


**FIGURE 4.10**
(a) Dugdale model of plastic zone in front of the crack tip; (b) effect of the plastic zone on the elastic material, $\sigma_{YS}$ is the yield stress.

in the plastic zone should be equal to the yield stress $\sigma_{YS}$. The force applied by the plastic region on the elastic material should be opposite to what the plastic region experiences from the elastic region. Therefore, if the plastic region is subjected to a tensile stress of $\sigma_{YS}$, then the plastic region should apply a closing stress of the same amount on the elastic material ahead of the crack tip B along the elastic–plastic boundary, as shown in Figure 4.10b. If the plastically deformed thin zone is now removed and only the elastic part is analyzed, then the elastic region will be subjected to additional closing stress $\sigma_{YS}$, as shown in Figure 4.10b along the removed plastic zone boundary. It will also increase the effective crack length by an amount $b$ since the crack tip will advance from point B to C.

One can compute the stress field near the crack tip C at point Q at a distance $r$ from the tip (see Figure 4.10b) by adding the contributions of all applied loads and those of the closing forces. Therefore,

$$\sigma = \frac{K_I}{\sqrt{2\pi r}} + \frac{K_I^*}{\sqrt{2\pi r}} \tag{4.22}$$

In equation (4.22) $K_I$ is the stress intensity factor for the crack ABC in absence of the closing forces and $K_I^*$ is the stress intensity factor for the crack ABC when it is subjected to only the closing stress $\sigma_{YS}$.

The SIF of a semi-infinite crack subjected to two concentrated forces as shown in Figure 4.11 is given by

$$K = P\sqrt{\frac{2}{\pi\alpha}} \tag{4.23}$$

From equation (4.23), the stress intensity factor $K_I^*$ of equation (4.22) can be obtained by simple superposition:

$$K_I^* = -\sigma_{YS}\sqrt{\frac{2}{\pi}}\int_0^b \frac{d\alpha}{\sqrt{\alpha}} = -\sigma_{YS}\sqrt{\frac{2}{\pi}}\left[2\sqrt{\alpha}\right]_0^b = -\sigma_{YS}\sqrt{\frac{8b}{\pi}} \tag{4.24}$$

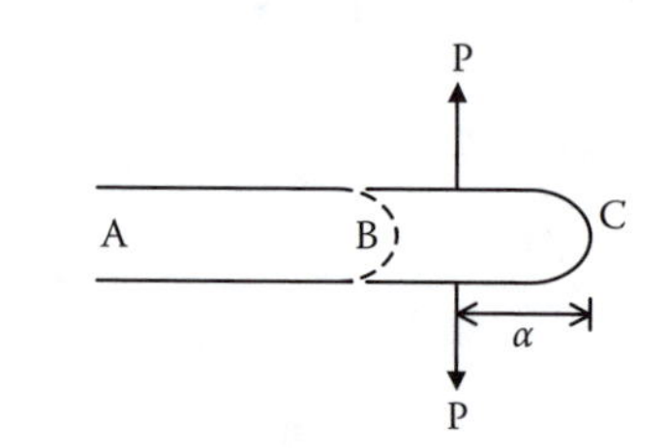


**FIGURE 4.11**
A semi-infinite crack in a linear elastic material, subjected to two opening forces $P$ at a distance $\alpha$ from the crack tip.

From equations (4.22) and (4.24), the stress at a point Q in front of the crack tip C can be obtained:

$$\sigma = \frac{K_I}{\sqrt{2\pi r}} + \frac{K_I^*}{\sqrt{2\pi r}} = \frac{K_I}{\sqrt{2\pi r}} - \sigma_{YS}\sqrt{\frac{8b}{\pi}}\frac{1}{\sqrt{2\pi r}} = \frac{1}{\sqrt{2\pi r}}\left(K_I - \sigma_{YS}\sqrt{\frac{8b}{\pi}}\right) \quad (4.25)$$

However, point Q is in the elastic region. Therefore, the stress at point Q must be finite. Therefore,

$$K_I - \sigma_{YS}\sqrt{\frac{8b}{\pi}} = 0$$
$$\therefore b = \frac{\pi}{8}\left(\frac{K_I}{\sigma_{YS}}\right)^2 \quad (4.26)$$

Equation (4.26) gives the length of the plastic zone obtained from the Dugdale model. Note that the plastic zone size $R$ (equation 4.4) obtained from the $r_p$ model, described in section 4.2, and the plastic zone size $b$ (equation 4.26) obtained from the Dugdale model are different but are of the same order. It should be noted here that Barenblatt (1962) also solved this problem in a slightly different manner. For this reason sometimes this model is called the Dugdale–Barenblatt model.

## 4.7 Crack Tip Opening Displacement

When a plastic zone is formed in front of a crack tip, as shown in Figure 4.10a, the original crack tip at point B opens up as the cracked structure is loaded. This opening displacement of the crack tip is called the crack tip opening displacement (CTOD). Failure of the structure can be predicted from the CTOD value; when the CTOD reaches a critical value $CTOD_c$, the crack starts to propagate.

In Figure 4.10b the opening displacement at point B can be estimated in the following manner. Considering the crack tip at point C, the displacement at point B due to external loads can be computed from equation (2.48) after substituting $r = b$ and $\theta = \pm\pi$:

$$CTOD^{(1)} = 2\frac{K_I\sqrt{b}}{2\mu\sqrt{2\pi}}(\kappa+1) = \frac{2K_I(1+\nu)\sqrt{b}}{E\sqrt{2\pi}}(\kappa+1) \quad (4.27)$$

Substituting $K_I$ from equation (4.26) in the preceding equation,

$$CTOD^{(1)} = \frac{2K_I(1+\nu)\sqrt{b}}{E\sqrt{2\pi}}(\kappa+1) = \frac{2\sigma_{YS}(1+\nu)\sqrt{b}}{E\sqrt{2\pi}}(\kappa+1)\sqrt{\frac{8b}{\pi}} = \frac{4\sigma_{YS}b}{\pi E}(1+\nu)(1+\kappa) \quad (4.28)$$

However, true CTOD at point B is smaller than CTOD[(1)] because the closing stresses shown in Figure 4.10b try to reduce the opening displacement at point B. A pure elastic analysis in absence of the plastically deformed materials in region BC sandwiched in between the closing stresses predicts a closing displacement CTOD[(2)] (due to closing stresses only) at point B equal to negative of CTOD[(1)], as shown:

$$CTOD^{(2)} = -2\frac{\sqrt{b}}{2\mu\sqrt{2\pi}}(\kappa+1)\int dK_I^* = -\frac{2(1+\nu)\sqrt{b}}{E\sqrt{2\pi}}(1+\kappa)\int_0^b \sigma_{YS}\sqrt{\frac{2}{\pi\alpha}}d\alpha$$

$$= -\frac{2\sigma_{YS}(1+\nu)\sqrt{b}}{\pi E}(1+\kappa)\int_0^b \frac{d\alpha}{\sqrt{\alpha}} = -\frac{2\sigma_{YS}\sqrt{b}}{\pi E}(1+\nu)(1+\kappa)[2\sqrt{\alpha}]_0^b \tag{4.29}$$

$$= -\frac{4\sigma_{YS}b}{\pi E}(1+\nu)(1+\kappa)$$

In equation (4.29) the negative sign implies the closing force. Note that the magnitude of CTOD[(2)] is reduced when the plastically deformed material is introduced in region BC, as shown in Figure 4.10a. In presence of the plastically deformed material, the closing stresses cannot move the crack surfaces inward freely. After incorporating all these factors, one can show that the true CTOD (at point B of Figure 4.10a) predicted by Dugdale model is

$$CTOD = \frac{2\sigma_{YS}b}{\pi E}(1+\nu)(1+\kappa) = \frac{2\sigma_{YS}}{\pi E}\frac{\pi}{8}\left(\frac{K_I}{\sigma_{YS}}\right)^2(1+\nu)(1+\kappa)$$

$$= \frac{(1+\nu)(1+\kappa)}{4}\frac{K_I^2}{E\sigma_{YS}} = \frac{\alpha K_I^2}{E\sigma_{YS}} \tag{4.30}$$

where $\alpha = 1$ for plane stress problems and $\alpha = 1 - \nu^2$ for plane strain problems.

From equation (4.30) it is easy to see that CTOD is related to the stress intensity factor as well as the strain energy release rate in the following manner:

$$CTOD = \frac{\alpha K_I^2}{E\sigma_{YS}} = \frac{G}{\sigma_{YS}} \tag{4.31}$$

For more elaborate discussion on CTOD, readers are referred to Burdekin and Stone (1966).

The crack tip opening displacement from the $r_p$ model of plasticity can be computed in the same manner, substituting $r = r_p$ and $\theta = \pm\pi$ in equation (2.48):

$$CTOD = 2\frac{K_I\sqrt{r_p}}{2\mu\sqrt{2\pi}}(\kappa+1) = \frac{2K_I\sqrt{r_p}}{E\sqrt{2\pi}}(1+\nu)(1+\kappa)$$

$$= \frac{2K_I}{E\sqrt{2\pi}}(1+\nu)(1+\kappa)\frac{1}{\sqrt{2\pi}}\left(\frac{K_I}{\sigma_{YS}}\right) \tag{4.32}$$

$$= \frac{(1+\nu)(1+\kappa)}{\pi E}\left(\frac{K_I^2}{\sigma_{YS}}\right) = \frac{4}{\pi}\frac{(1+\nu)(1+\kappa)}{4}\frac{K_I^2}{E\sigma_{YS}} = \frac{4}{\pi}\frac{\alpha K_I^2}{E\sigma_{YS}} = \frac{4}{\pi}\frac{G}{\sigma_{YS}}$$

CTOD and the plastic zone size for in-plane problems (from both the $r_p$ model and the Dugdale model) and out-of-plane problems are given in Table 4.2.

So far we have seen that the crack propagation can be predicted by comparing the stress intensity factor ($K$), the strain energy release rate ($G$), and the crack tip opening displacement (CTOD) with their critical values $K_c$, $G_c$, and $CTOD_c$, respectively. Relations between these three parameters are given in Table 4.2. Other parameters as listed below are also used for predicting the crack propagation:

- critical stress intensity factor, $K_c$
- critical strain energy release rate, $G_c$
- critical crack tip opening displacement, $CTOD_c$
- critical plastic zone size, $R_c$
- critical strain intensity factor, $K_c^{strain}$
- critical J-integral value, $J_c$

Most of these parameters work very well in brittle fracture theory, when the plastic zone size is small (for $\frac{\sigma}{\sigma_{YS}} < 0.5$) and work reasonably well when it is not too large (for $0.7 > \frac{\sigma}{\sigma_{YS}} > 0.5$). However, for ductile fracture condition (for $\frac{\sigma}{\sigma_{YS}} > 0.7$), many of the preceding parameters often do not work

**TABLE 4.2**

Plastic Zone Size and the Crack Tip Opening Displacement for In-Plane and Antiplane Problems

| | $r_p$ Model | Dugdale Model | Antiplane Problems |
|---|---|---|---|
| Plastic zone size | $\frac{1}{\pi}\left(\frac{k_I}{\sigma_{YS}}\right)^2$ | $\frac{\pi}{8}\left(\frac{k_I}{\sigma_{YS}}\right)^2$ | $\frac{1}{\pi}\left(\frac{K_{III}}{\tau_{YS}}\right)^2$ |
| CTOD | $\frac{4}{\pi}\frac{\alpha K_I^2}{E\sigma_{YS}} = \frac{4}{\pi}\frac{G}{\sigma_{YS}}$ | $\frac{\alpha K_I^2}{E\sigma_{YS}} = \frac{G}{\sigma_{YS}}$ | $\frac{2}{\pi}\frac{K_{III}^2}{G\tau_{YS}}$ |

well to predict the failure. Note that under brittle fracture and ductile fracture conditions, the failure modes that give rise to different failure surfaces and crack propagation directions under these two conditions are different, as illustrated in Figure 4.9. Therefore, it is not uncommon to consider different parameters as the governing or critical parameters for predicting the crack propagation under brittle fracture and ductile fracture conditions.

It should be also noted here that when the material in front of the crack tip is plastically deformed, the strain at the crack tip can be unbounded for finite stress value if the material shows elastic–perfectly plastic behavior. From this singular strain field, a strain intensity factor can be defined and its critical value can be used as a parameter for predicting the crack propagation. The J-integral will be discussed in the next chapter.

## 4.8 Experimental Determination of $K_c$

To obtain $K_c$ for brittle fracture theory one needs to make sure that the failure occurs under a plane strain condition that produces a relatively small plastic zone and a smaller value of $K_c$ compared to the $K_c$ value under the plane stress condition (see Figure 4.9). If the specimen thickness is not large enough to produce the plane strain condition, then the specimen fails under the plane stress condition and the $K_c$ value is overpredicted for the brittle fracture theory. The experiment should be conducted on a crack with a sharp flat front and the crack should propagate at a stress level that is not too close to the yield stress. All these constraints should be satisfied if the experiment is conducted following the ASTM guidelines as described next.

### 4.8.1 Compact Tension Specimen

The diagram of the compact tension specimen with all its dimensions is shown in Figure 4.12. To ensure the plane strain failure and correct $K_c$ measurement, the following constraint conditions must be satisfied according to ASTM:

$$
\begin{gathered}
\frac{W}{4} \leq B \leq \frac{W}{2} \\
0.45W \leq a \leq 0.55W \\
a,\ B \geq 2.5\pi R = 2.5\left(\frac{K_c}{\sigma_{YS}}\right)^2
\end{gathered}
\tag{4.33}
$$

A step-by-step testing procedure is given next.

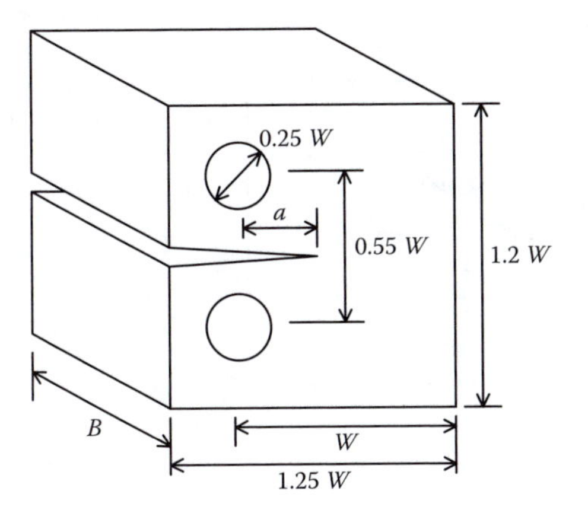


**FIGURE 4.12**
Compact tension specimen.

#### 4.8.1.1 *Step 1: Crack Formation*

(1) Cut or machine a notch (or blunt crack of finite thickness) in the pattern shown in Figure 4.13a.

(2) Apply fatigue loading to grow a sharp fatigue crack from the machined notch as shown in Figure 4.13b. During this crack growth process, the crack front should be almost straight and

$$K_{max} \leq 0.60K_c \tag{4.34}$$

Of course there is no guarantee that the preceding two constraint conditions (crack front being straight and satisfaction of equation 4.34) are satisfied during the fatigue crack formation process. We will check later if these conditions are satisfied.

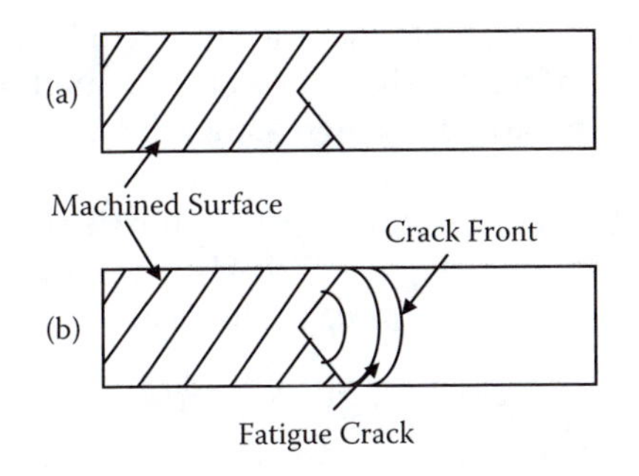


**FIGURE 4.13**
(a) Shape of the machined notch and (b) the fatigue crack developed from the machined notch.

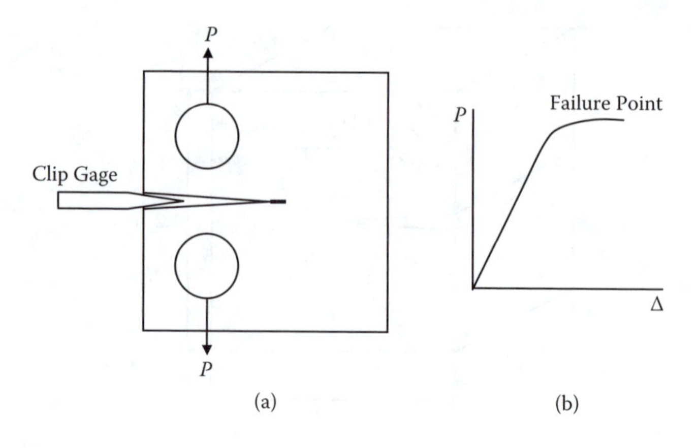


**FIGURE 4.14**
(a) Loading of compact tension specimen; (b) $P$–Δ curve obtained experimentally by loading the specimen to failure.

#### 4.8.1.2 Step 2: Loading the Specimen

Load the specimen as shown in Figure 4.14a. Measure the applied load ($P$) and the crack opening displacement (Δ). The crack opening displacement can be measured by a clip gage as shown in the Figure. The $P$–Δ curve shows a linear behavior in the beginning and then it becomes nonlinear before failure. At this step the constraint condition that must be satisfied is that the nonlinear region in the $P$–Δ curve should be relatively small. If it shows a large nonlinear behavior before failure, then that is an indication of a large, plastically deformed zone developed during the plane stress failure. In that case specimen thickness must be increased and steps 1 and 2 will have to be carried out again.

#### 4.8.1.3 Step 3: Checking Crack Geometry in the Failed Specimen

Observing the surface finish of the failed specimen, identify the crack front of the fatigue crack that was present before the monotonically increasing load $P$ was applied in step 2. Note that the surface finish of the fatigue crack is different from the surface finish of the crack formed during the unstable crack growth under monotonically increasing load $P$:

(1) Measure crack lengths $a_1$, $a_2$, and $a_3$ along the thickness of the plate at $B/4$ intervals as shown in Figure 4.15. Also measure the crack lengths $a_{S1}$ and $a_{S2}$ along the two surfaces of the plate.

(2) Compute the average crack length $a_{av} = \frac{1}{3}(a_1 + a_2 + a_3)$.

(3) Make sure that the following constraint conditions are satisfied:

(a) $|a_i - a_{av}| \le 0.05a_{av}$.

(b) $|a_{Si} - a_{av}| \le 0.1a_{av}$, $i = 1, 2$.

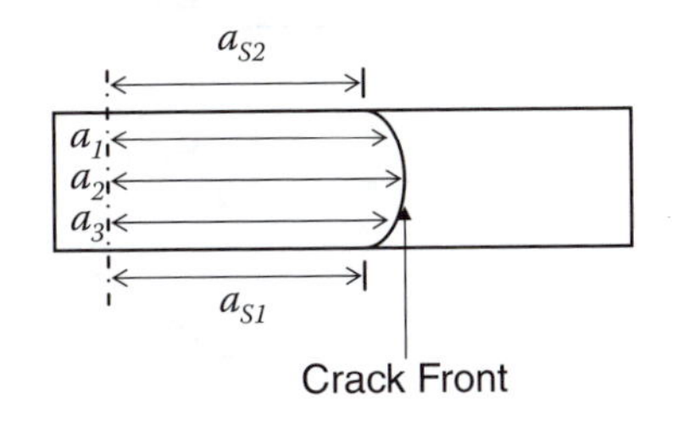


**FIGURE 4.15**
(a) Failure surface of the compact tension specimen.

(c) All parts of the crack front must be at a minimum distance of $0.05a_{av}$ or 1.3 mm, whichever is smaller, from the machined notch.

If any of the preceding three constraint conditions are violated, then steps 1, 2, and 3 will have to be carried out with a new specimen.

#### *4.8.1.4 Step 4: Computation of Stress Intensity Factor at Failure*

The SIF at failure ($K_F$) can be computed from the failure load $P_F$ using the following formula:

$$K_F = \frac{P_F}{B\sqrt{W}} f\left(\frac{a}{W}\right) \tag{4.35}$$

where

$$f\left(\frac{a}{W}\right) = 29.6\left(\frac{a}{W}\right)^{\frac{1}{2}} - 185.5\left(\frac{a}{W}\right)^{\frac{3}{2}} + 655.7\left(\frac{a}{W}\right)^{\frac{5}{2}} - 1017.0\left(\frac{a}{W}\right)^{\frac{7}{2}} + 638.9\left(\frac{a}{W}\right)^{\frac{9}{2}} \tag{4.36}$$

Equation (4.36) is valid only in the region $0.45 \leq \frac{a}{W} \leq 0.55$. For this reason the second constraint condition of equation (4.33) is necessary. Variation of the function $f(\frac{a}{W})$ in this range is shown in Figure 4.16. Clearly, for $0.45 \leq \frac{a}{W} \leq 0.55$, the function value variation is given by $8.34 \leq f(\frac{a}{W}) \leq 11.26$.

Srawley (1976) proposed an alternate expression for the function $f(\frac{a}{W})$ of equation (4.35). Srawley's expression is given in equation (4.37). It covers a wider range: $0.2 \leq \frac{a}{W} \leq 1$. In this range the maximum error of the stress intensity factor obtained from equation (4.37) is less than 0.5%:

$$f\left(\frac{a}{W}\right) = \frac{\left(2+\frac{a}{W}\right)\left[0.886 + 4.64\left(\frac{a}{W}\right) - 13.32\left(\frac{a}{W}\right)^2 + 14.72\left(\frac{a}{W}\right)^3 - 5.6\left(\frac{a}{W}\right)^4\right]}{\left(1-\frac{a}{W}\right)^{\frac{3}{2}}} \tag{4.37}$$

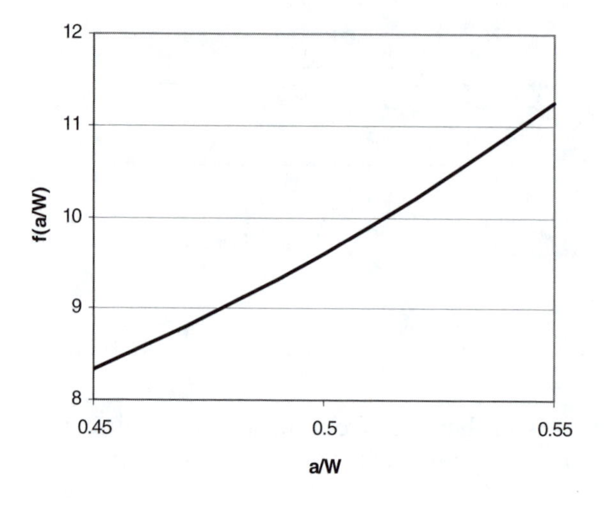


**FIGURE 4.16**
Variation of the function given in equation (4.36).

### 4.8.1.5 *Step 5: Final Check*

$K_F$ computed in step 4 is the critical stress intensity factor $K_c$ of the specimen material if the following two constraint conditions are satisfied:

(1) $a_{av}, B \geq 2.5(\frac{K_F}{\sigma_{YS}})^2$

(2) During the fatigue crack growth described in step 1, the maximum stress intensity factor did not exceed 60% of the stress intensity factor at failure, or $K_{max}|_{fatigue} \leq 0.60K_F$.

If the preceding two conditions are satisfied, then $K_F = K_c$. If these conditions or the conditions stated in step 3 are not satisfied, then consider the $K_F$ obtained in step 4 as the first estimate of $K_c$ and design a new specimen satisfying the constraint conditions stated in equation (4.33). Then repeat steps 1–5 to obtain $K_c$.

## 4.8.2 Three-Point Bend Specimen

The diagram of the three-point bend specimen with all its dimensions is shown in Figure 4.17. To ensure plane strain failure and correct $K_c$ measurement, the following constraint conditions must be satisfied according to ASTM:

$$\frac{W}{4} \leq B \leq \frac{W}{2}$$

$$0.45W \leq a \leq 0.55W \tag{4.38}$$

$$a,\ B \geq 2.5\pi R = 2.5\left(\frac{K_c}{\sigma_{YS}}\right)^2$$

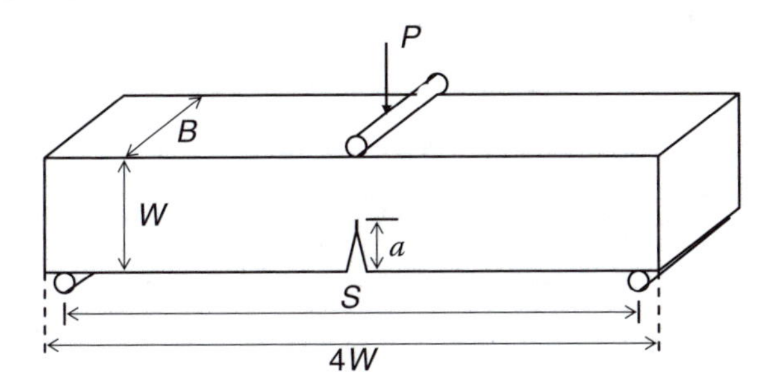


**FIGURE 4.17**
Three-point bend specimen.

Steps 1–3 for the three-point bend specimen are identical to the corresponding steps for the compact tension specimen described in sections 4.8.1.1–4.8.1.3 and are not repeated here. Step 4, which involves the SIF calculation at the failure load, differs from section 4.8.1.4 because the $K_F$ formula for the bend specimen is different from equation (4.35):

$$K_F = \frac{P_F S}{BW^{3/2}} f\left(\frac{a}{W}\right) \tag{4.39}$$

where

$$f\left(\frac{a}{W}\right) = 2.9\left(\frac{a}{W}\right)^{\frac{1}{2}} - 4.6\left(\frac{a}{W}\right)^{\frac{3}{2}} + 21.8\left(\frac{a}{W}\right)^{\frac{5}{2}} - 37.6\left(\frac{a}{W}\right)^{\frac{7}{2}} + 38.7\left(\frac{a}{W}\right)^{\frac{9}{2}} \tag{4.40}$$

Equation (4.40) is valid only in the region $0.45 \le \frac{a}{W} \le 0.55$. For this reason the second constraint condition of equation (4.38) is necessary. Variation of the function $f(\frac{a}{W})$ in this range is shown in Figure 4.18. Clearly, for $0.45 \le \frac{a}{W} \le 0.55$, the function value variation is given by $2.28 \le f(\frac{a}{W}) \le 3.15$.

Srawley (1976) proposed an alternate expression for the function $f(\frac{a}{W})$ of equation (4.39). Srawley's expression is given in equation (4.41). It covers the entire range of $\frac{a}{W}$. In this range the maximum error of the stress intensity factor obtained from equations (4.39) and (4.41) is less than 0.5%:

$$f\left(\frac{a}{W}\right) = \frac{3\sqrt{\frac{a}{W}}\left[1.99 - \left(\frac{a}{W}\right)\left(1 - \frac{a}{W}\right)\left(2.15 - 3.93\frac{a}{W} + 2.7\frac{a^2}{W^2}\right)\right]}{2\left(1 + 2\frac{a}{W}\right)\left(1 - \frac{a}{W}\right)^{\frac{3}{2}}} \tag{4.41}$$

Step 5, described in section 4.8.1.5, is then repeated to obtain the crucial stress intensity factor $K_c$ for the specimen material.

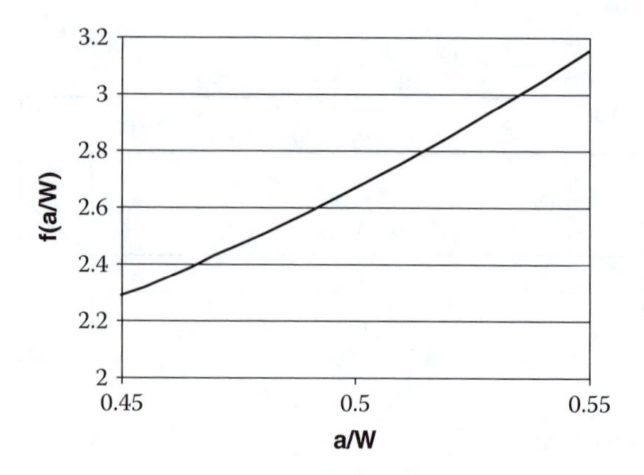


**FIGURE 4.18**
Variation of the function given in equation (4.40).

### 4.8.3 Practical Examples

Let us investigate what size of specimen is needed to test 7075 aluminum and reactor steel A533B.

#### 4.8.3.1 *7075 Aluminum*

For 7075 aluminum, $K_c = 27$ kip-in.$^{-3/2}$, $\sigma_{YS} = 79$ ksi; $\therefore 2.5\pi R = 2.5(\frac{27}{79})^2 = 0.29''$. Therefore, different dimensions ($a$, $B$, etc.) of the specimen should be greater than 0.29 in. This condition is easy to satisfy.

#### 4.8.3.2 *A533B Reactor Steel*

For reactor steel $K_c = 180$ kip-in.$^{-3/2}$, $\sigma_{YS} = 50$ ksi; $\therefore 2.5\pi R = 2.5(\frac{180}{50})^2 = 32.4''$. Therefore, different dimensions ($a$, $B$, etc.) of the specimen should be greater than 32.4 in. This condition is not easy to satisfy for either compact tension specimen or three-point bend specimen, as illustrated next.

##### 4.8.3.2.1 *Compact Tension Specimen*

Let us take $B = 33$ in. (Note that $B$ must be greater than 32.4 in.) Then, $W = 2B = 66$ in.:

$$\therefore \frac{a}{W} = 0.5;\ f\left(\frac{a}{W}\right) \approx 10$$

From equation (4.35),

$$K_F = \frac{P_F}{B\sqrt{W}} f\left(\frac{1}{W}\right) = \frac{10P_F}{33\sqrt{66}}$$

$$\therefore P_F = \frac{180 \times 33\sqrt{66}}{10} = 4.8 \times 10^6 \text{ pounds}$$

This failure load is too high.

*4.8.3.2.2 Three-Point Bend Specimen*

Let us take $B = 33$ in. (Note that $B$ must be greater than 32.4 in.) Then $W = 2B = 66$ in., $S \approx 4W = 132''$.

$$\therefore \frac{a}{W} = 0.5; \; g\left(\frac{a}{W}\right) \approx 2.75$$

From equation (4.39),

$$K_F = \frac{P_F S}{BW^{3/2}} g\left(\frac{a}{W}\right) = \frac{P_F \times 264}{33 \times 66\sqrt{66}} \times 2.75$$

$$\therefore P_F = \frac{180 \times 33\sqrt{66}}{4 \times 2.75} = 4.4 \times 10^6 \text{ pounds}$$

This failure load is too high.

## 4.9 Concluding Remarks

A complete analysis of fracture mechanics problems after ignoring the theory of plasticity is not possible because some plastic deformation always occurs near the crack tip. Whether the effect of this plastic deformation is negligible or not is discussed in this chapter. When it is negligible, the straightforward application of the linear elastic fracture mechanics (LEFM) analysis is permitted. When the plastic zone size is small but not necessarily negligible, LEFM analysis with some corrections for plastic deformation gives good results also. However, when the plastically deformed region is large, LEFM no longer works and complete elastoplastic analysis must be carried out. This chapter discussed what parameters and criteria decide whether the plastic zone size is (1) negligibly small so that straightforward LEFM is applicable; (2) not negligible but still small, which requires LEFM analysis with plasticity correction factor; or (3) large, for which complete elastoplastic analysis is required.

In chapter 1 the background knowledge on the theory of elasticity that is necessary to understand the linear elastic fracture mechanics was presented. However, no fundamental knowledge on the theory of plasticity, such as the derivation of Von Mises and Tresca's yield criteria, used here was given in chapter 1. Readers who lack the fundamental knowledge on plasticity are referred to basic text books on the theory of plasticity (Hill 1950; Mendelson 1968).

## References

Barenblatt, G. I. The mathematical theory of equilibrium of cracks in brittle fracture. *Advances in Applied Mechanics*, 7, 55–129, 1962.

Burdekin F. M. and Stone, D. E. W. The crack opening displacement approach to fracture mechanics in yielding materials. *Journal of Strain Analysis*, 1, 145–153, 1966.

Dugdale, D. S. Yielding of steel sheets containing slots. *Journal of the Mechanics and Physics of Solids*, 8, 100–104, 1960.

Hill, R. *The mathematical theory of plasticity*. New York: Oxford University Press, 1950.

McClintock, F. and Irwin, G. R. Plasticity aspects of fracture mechanics. *ASTM STP*, 381, 84–113, 1965.

Mendelson, A. *Plasticity: Theory and applications*. Huntington, NY: Robert E. Krieger Publishing Company, 1968.

Srawley, J. E. Wide-range stress intensity factor expressions for ASTM E-399 standard fracture toughness specimens. *International Journal of Fracture*, 12, 475–476, 1976.

## Exercise Problems

Problem 4.1:

(a) Using Tresca yield criterion, determine $r_p(\theta)$ (first estimate of the plastic zone in front of the crack tip for mode I loading) for a state of plane stress and plane strain (Poisson's ratio, $\upsilon = 0.3$).

(b) Plot $r_p(\theta)$ for the case of plane stress in front of the crack tip showing the plastic zone shape.

(c) Determine the ratio of $r_p(0)$ for plane stress to that for plane strain.

Problem 4.2: Assume a center-cracked tensile specimen as shown in Figure 4.19 and let the crack tip stress intensity factor be given by

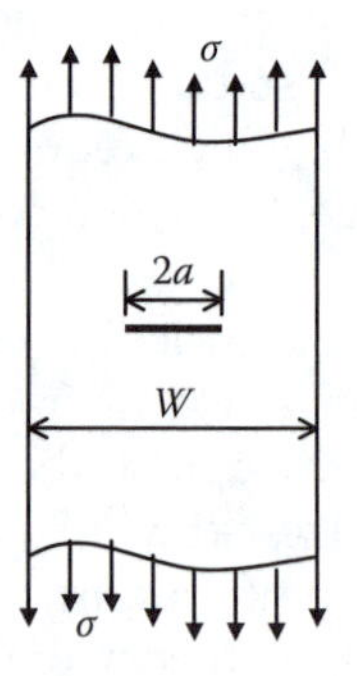


FIGURE 4.19

$$K^2 = \sigma^2 \pi a \sec\left(\frac{\pi a}{W}\right)$$

Assume $a = W/4$. If $r_p = R/2 = a/10$, then find the ratio of the average net section stress $\sigma_N$ [$\sigma_N = \sigma W/(W-2a)$] to the yield stress $\sigma_{YS}$ (a) ignoring plasticity correction and (b) considering plasticity correction.

Problem 4.3: A specimen is being tested to determine $K_c$. Unfortunately the recorder broke during the test and nobody noticed, so the value of the load was not recorded. However, just prior to failure it was estimated (from the dimple location) that the plastic zone size ahead of the crack tip was about $R = 0.2$ in. If $\sigma_{YS}$ is 50 ksi, give an estimate of $K_c$ based on (a) $r_p$ model and (b) Dugdale model.

# 5

# J-Integral

## 5.1 Introduction

An integral expression proposed by James R. Rice (1968) can compute the strain energy release rate for a cracked elastic solid in a different and simpler manner. This integral, named after its inventor, is known as the J-integral. The J-integral value also helps one to predict when a crack should propagate, as discussed in this chapter.

## 5.2 Derivation of J-Integral

Consider a cracked plate of thickness $t$ containing a crack. An area $A$ with boundary $S$ contains the crack tip as shown in Figure 5.1. If the control volume of Figure 5.1 experiences a surface traction **T** along the boundary S and no body force, then the potential energy in the control volume is given by

$$\Pi = \left\{ \int_A U dA - \int_S \mathbf{T} \cdot \mathbf{u} \, dS \right\} t = \left\{ \int_A U dA - \int_S T_i u_i dS \right\} t \tag{5.1}$$

In equation (5.1) $U$ is the strain energy density and **u** is the displacement vector. If the crack is extended by an amount $\Delta a$, then the potential energy in the control volume should change because both integrands of equation (5.1) would change. If the potential energy in the control volume is denoted as $\Pi_1$ before the crack extension and as $\Pi_2$ after the crack extension, then the potential energy release rate can be written as

$$-\frac{\partial \Pi}{\partial A} = -\lim_{\Delta a \to 0} \left\{ \frac{\Pi_2 - \Pi_1}{t \Delta a} \right\} \tag{5.2}$$

Note that for a crack in an infinite plate the potential energies $\Pi_1$ and $\Pi_2$ can be computed by extending the crack toward the right, as shown in Figure 5.2a, or keeping the crack length fixed but moving the control volume

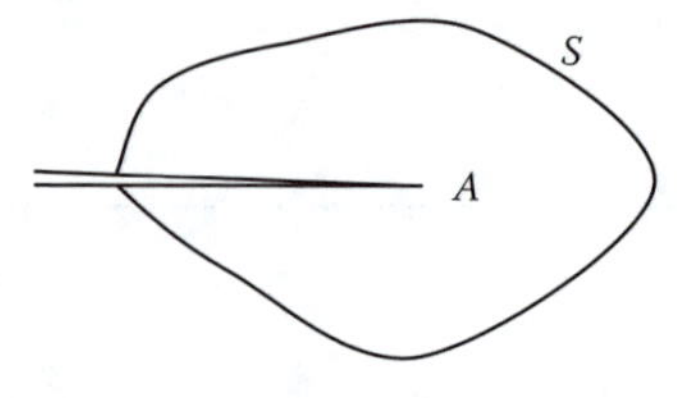


**FIGURE 5.1**

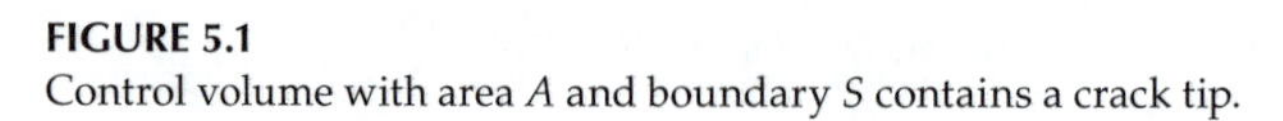
Control volume with area $A$ and boundary $S$ contains a crack tip.

toward the left by the same amount, as shown in Figure 5.2b. In Figure 5.2b, boundary $S_1$ (marked by the solid line) and $S_2$ (marked by the dashed line) are the same boundary after it is shifted toward the left by amount $\Delta a$. $A_1$ and $A_2$ are areas enclosed by boundaries $S_1$ and $S_2$, respectively.

From equation (5.2) and Figure 5.2b it is possible to write

$$
\begin{aligned}
-\frac{\partial \Pi}{\partial A} &= -\lim_{\Delta a \to 0} \left\{ \frac{\Pi_2 - \Pi_1}{t\Delta a} \right\} \\
&= -\lim_{\Delta a \to 0} \frac{1}{t\Delta a} \left\{ \left( \int_{A_2} U dA - \int_{S_2} T_i u_i dS \right) - \left( \int_{A_1} U dA - \int_{S_1} T_i u_i dS \right) \right\} t \\
&= -\lim_{\Delta a \to 0} \frac{1}{\Delta a} \left\{ \int_{A_2 - A_1} U dA - \int_{S_2} T_i u_i dS + \int_{S_1} T_i u_i dS \right\} \\
&= \lim_{\Delta a \to 0} \frac{1}{\Delta a} \left\{ \int_{A_1 - A_2} U dA + \int_{S_2} T_i u_i dS - \int_{S_1} T_i u_i dS \right\}
\end{aligned}
\tag{5.3}
$$

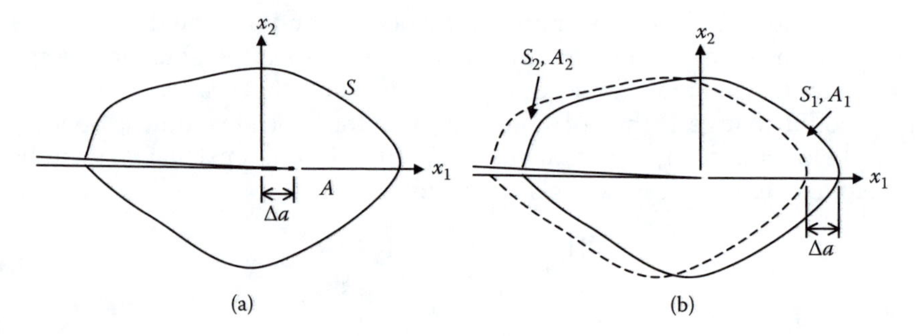


**FIGURE 5.2**
Crack extending toward the right (a), and the control volume moving toward the left (b), should have the same effect in computing the expression of equation (5.2).

In the limiting case $\Delta a$ approaching zero, one can assume that the displacement field $u_i$ is varying linearly between boundaries $S_1$ and $S_2$. Note that over small regions the displacement field can always be assumed to be varying linearly, even for nonlinear displacement fields, as long as there is no jump in the displacement field between $S_1$ and $S_2$. For linear displacement fields, the strain, stress, and traction fields should be constant. Therefore, equation (5.3) can be written as

$$
\begin{aligned}
-\frac{\partial \Pi}{\partial A} &= \lim_{\Delta a \to 0} \frac{1}{\Delta a}\left\{ \int_{A_1 - A_2} U dA + \int_{S_2} T_i u_i dS - \int_{S_1} T_i u_i dS \right\} \\
&= \lim_{\Delta a \to 0} \frac{1}{\Delta a}\left\{ \int_{A_1 - A_2} U dA - \int_{S} T_i \left( u_i^{(1)} - u_i^{(2)} \right) dS \right\}
\end{aligned}
\tag{5.4}
$$

In Figure 5.3 one can clearly see that the two boundaries $S_1$ and $S_2$ are separated by a horizontal distance $\Delta a$; therefore, the displacement fields on these two boundaries are related in the following manner:

$$
u_i^{(1)} = u_i^{(2)} + \frac{\partial u_i}{\partial x_1} \Delta a \tag{5.5}
$$

From equations (5.4) and (5.5),

$$
\begin{aligned}
-\frac{\partial \Pi}{\partial A} &= \lim_{\Delta a \to 0} \frac{1}{\Delta a}\left\{ \int_{A_1 - A_2} U dA - \int_{S} T_i \left( u_i^{(1)} - u_i^{(2)} \right) dS \right\} \\
&= \lim_{\Delta a \to 0} \frac{1}{\Delta a}\left\{ \int_{A_1 - A_2} U dA - \int_{S} T_i \frac{\partial u_i}{\partial x_1} \Delta a dS \right\}
\end{aligned}
\tag{5.6}
$$

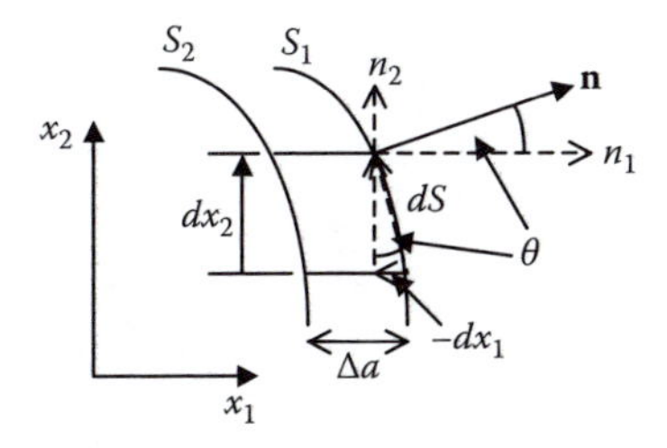


**FIGURE 5.3**
Two boundaries $S_1$ and $S_2$ separated by a horizontal distance $\Delta a$.

In Figure (5.3) one can clearly see that the elemental area $dA = \Delta a \cdot dx_2$. Substituting it in equation (5.6),

$$-\frac{\partial \Pi}{\partial A} = \lim_{\Delta a \to 0} \frac{1}{\Delta a}\left\{ \int_{A_1-A_2} U dA - \int_S T_i \frac{\partial u_i}{\partial x_1} \Delta a dS \right\}$$

$$= \lim_{\Delta a \to 0} \frac{1}{\Delta a}\left\{ \int_S U \Delta a dx_2 - \int_S T_i \frac{\partial u_i}{\partial x_1} \Delta a dS \right\} \tag{5.7}$$

$$= \int_S U dx_2 - \int_S T_i \frac{\partial u_i}{\partial x_1} dS = \int_S \left( U dx_2 - T_i \frac{\partial u_i}{\partial x_1} dS \right)$$

Integral expression of equation (5.7) is known as the J-integral over boundary $S$.

$$J = \int_S \left( U dx_2 - T_i \frac{\partial u_i}{\partial x_1} dS \right) \tag{5.8}$$

From the preceding derivation it is clear that, for an elastic solid, the J-integral is another way of expressing the potential energy release rate.

## 5.3 J-Integral over a Closed Loop

Note that the line $S$ of Figure 5.1 is not a closed loop because of the presence of the crack. Two ends of line $S$ meet the top and bottom surfaces of the crack. In absence of a crack when two ends of line meet, as shown in Figure 5.4, a closed loop is formed. We are interested in computing the J-integral over such a closed loop.

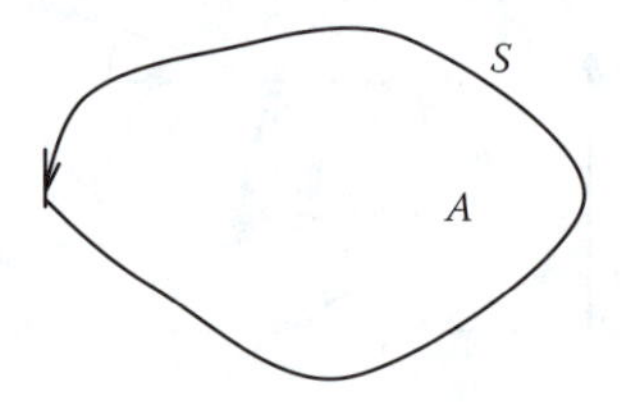


**FIGURE 5.4**
Closed loop $S$ on which J-integral is to be computed.

In Figure 5.3 it is easy to see that

$$n_1 = \cos\theta = \frac{dx_2}{dS}$$
$$n_2 = \sin\theta = -\frac{dx_1}{dS} \tag{5.9}$$

Therefore,

$$T_1 = \sigma_{11}n_1 + \sigma_{12}n_2 = \sigma_{11}\frac{dx_2}{dS} - \sigma_{12}\frac{dx_1}{dS}$$
$$T_2 = \sigma_{21}n_1 + \sigma_{22}n_2 = \sigma_{21}\frac{dx_2}{dS} - \sigma_{22}\frac{dx_1}{dS} \tag{5.10}$$

Substituting equation (5.10) into equation (5.8),

$$\begin{aligned} J &= \int_S \left( U dx_2 - T_i \frac{\partial u_i}{\partial x_1} dS \right) = \int_S \left( U dx_2 - T_1 \frac{\partial u_1}{\partial x_1} dS - T_2 \frac{\partial u_2}{\partial x_1} dS \right) \\ &= \int_S \left\{ U dx_2 - \left( \sigma_{11}\frac{dx_2}{dS} - \sigma_{12}\frac{dx_1}{dS} \right)\frac{\partial u_1}{\partial x_1} dS - \left( \sigma_{21}\frac{dx_2}{dS} - \sigma_{22}\frac{dx_1}{dS} \right)\frac{\partial u_2}{\partial x_1} dS \right\} \\ &= \int_S \left\{ U dx_2 - \left( \sigma_{11}\frac{\partial u_1}{\partial x_1} dx_2 - \sigma_{12}\frac{\partial u_1}{\partial x_1} dx_1 \right) - \left( \sigma_{21}\frac{\partial u_2}{\partial x_1} dx_2 - \sigma_{22}\frac{\partial u_2}{\partial x_1} dx_1 \right) \right\} \\ &= \int_S \left\{ \left( U - \sigma_{11}\frac{\partial u_1}{\partial x_1} - \sigma_{21}\frac{\partial u_2}{\partial x_1} \right) dx_2 + \left( \sigma_{12}\frac{\partial u_1}{\partial x_1} + \sigma_{22}\frac{\partial u_2}{\partial x_1} \right) dx_1 \right\} \end{aligned} \tag{5.11}$$

Applying Green's theorem,

$$\int_S (M dx_1 + N dx_2) = \int_A \left( \frac{\partial N}{\partial x_1} - \frac{\partial M}{\partial x_2} \right) \tag{5.12}$$

into equation (5.11), the line integral of equation (5.11) can be converted to the area integral as shown:

$$\begin{aligned} J &= \int_S \left\{ \left( U - \sigma_{11}\frac{\partial u_1}{\partial x_1} - \sigma_{21}\frac{\partial u_2}{\partial x_1} \right) dx_2 + \left( \sigma_{12}\frac{\partial u_1}{\partial x_1} + \sigma_{22}\frac{\partial u_2}{\partial x_1} \right) dx_1 \right\} \\ &= \int_A \left\{ \frac{\partial}{\partial x_1}\left( U - \sigma_{11}\frac{\partial u_1}{\partial x_1} - \sigma_{21}\frac{\partial u_2}{\partial x_1} \right) - \frac{\partial}{\partial x_2}\left( \sigma_{12}\frac{\partial u_1}{\partial x_1} + \sigma_{22}\frac{\partial u_2}{\partial x_1} \right) \right\} \end{aligned} \tag{5.13}$$

Note that the strain energy density $U$ can be expressed in terms of strain components. For two-dimensional problems,

$$U = U(\varepsilon_{11}, \varepsilon_{22}, \gamma_{12}) \tag{5.14}$$

Applying the chain rule,

$$\frac{\partial U}{\partial x_1} = \frac{\partial U}{\partial \varepsilon_{11}}\frac{\partial \varepsilon_{11}}{\partial x_1} + \frac{\partial U}{\partial \varepsilon_{22}}\frac{\partial \varepsilon_{22}}{\partial x_1} + \frac{\partial U}{\partial \gamma_{12}}\frac{\partial \gamma_{12}}{\partial x_1} = \sigma_{11}u_{1,11} + \sigma_{22}u_{2,21} + \sigma_{12}(u_{1,21} + u_{2,11}) \tag{5.15}$$

In equation (5.15) we have used the relation $\sigma_{ij} = \frac{\partial U}{\partial \varepsilon_{ij}}$. This relation is true for elastic material only. Therefore, the region $A$ bounded by $S$ must be elastic for the preceding equation to be applicable.

$$\begin{aligned}
J &= \int_A \left\{ \frac{\partial}{\partial x_1}\left( U - \sigma_{11}\frac{\partial u_1}{\partial x_1} - \sigma_{21}\frac{\partial u_2}{\partial x_1} \right) - \frac{\partial}{\partial x_2}\left( \sigma_{12}\frac{\partial u_1}{\partial x_1} + \sigma_{22}\frac{\partial u_2}{\partial x_1} \right) \right\} \\
&= \int_A \{ \sigma_{11}u_{1,11} + \sigma_{22}u_{2,21} + \sigma_{12}u_{1,21} + \sigma_{12}u_{2,11} - \sigma_{11}u_{1,11} - \sigma_{11,1}u_{1,1} - \sigma_{21}u_{2,11} \\
&\quad - \sigma_{21,1}u_{2,1} - \sigma_{12}u_{1,12} - \sigma_{12,2}u_{1,1} - \sigma_{22}u_{2,12} - \sigma_{22,2}u_{2,1} \} \\
&= \int_A \{ -\sigma_{11,1}u_{1,1} - \sigma_{21,1}u_{2,1} - \sigma_{12,2}u_{1,1} - \sigma_{22,2}u_{2,1} \} \\
&= \int_A \{ -u_{1,1}(\sigma_{11,1} + \sigma_{12,2}) - u_{2,1}(\sigma_{21,1} + \sigma_{22,2}) \} = 0
\end{aligned} \tag{5.16}$$

In equation (5.16) the equilibrium equation in two dimensions in absence of any body force has been used. From the derivation presented previously, one can conclude that the J-integral value over a closed loop is zero if the region inside the loop is elastic and has zero body force.

## 5.4 Path Independence of J-Integral

In the previous section it has been shown that the J-integral value over any closed loop is zero as long as the region inside the closed loop is elastic and does not have any body force. This property of J-integral can be used to prove that its value should be the same on different paths $S_1$, $S_2$, and $S_3$ shown in Figure 5.5.

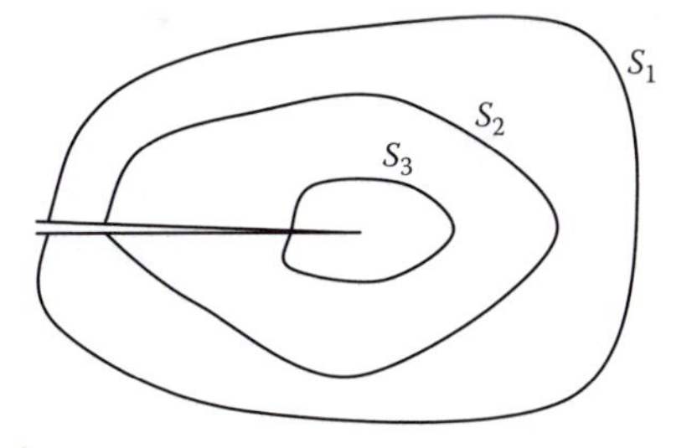


**FIGURE 5.5**
Three different contours for J-integral should give the same J-integral value.

To prove the path independence of J-integral, let us consider the J-integral contour as shown in Figure 5.6. Let the total path S be the union of four paths $S_1$, $S_2$, $S_3$, and $S_4$ ($S = S_1 \cup S_2 \cup S_3 \cup S_4$). Directions of integration in the four segments are shown by the arrows in the Figure. Note that $S$ is a closed contour and the material inside the contour is elastic, although a plastically deformed region may exist in front of the crack tip, as shown in Figure 5.6. Therefore,

$$J = \int_S = \int_{S_1} + \int_{S_2} + \int_{S_3} + \int_{S_4} = J_1 + J_2 + J_3 + J_4 = 0 \tag{5.17}$$

Note that $J_3$ and $J_4$ must be zero because on paths $S_3$ and $S_4$ the traction $T_i = 0$ and $dx_2 = 0$. Therefore, both terms of J-integral given in equation 5.8 are zero on these two paths. Substituting these zero values in equation (5.17):

$$J_1 + J_2 = 0 \tag{5.18}$$

Note that the integration direction is counterclockwise on path $S_1$ and clockwise on path $S_2$. If both integrals are carried out in the same direction, then $J_1 = J_2$.

The main advantage of the path independence property of the J-integral is that it can be computed by choosing a path of our liking. For example, instead

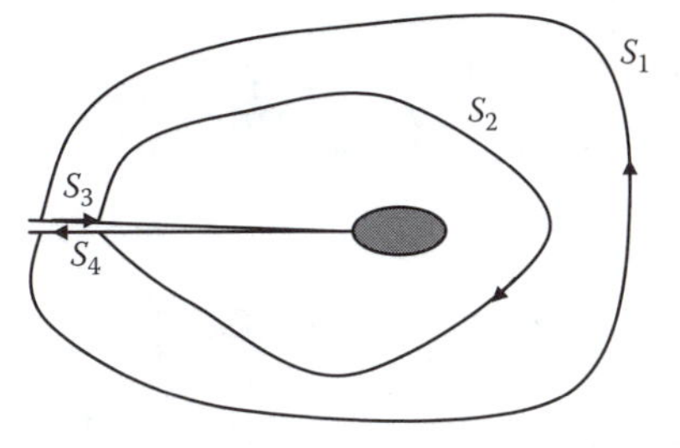


**FIGURE 5.6**
J-integral contour $S = S_1 \cup S_3 \cup S_2 \cup S_4$. Plastic zone shape ahead of the crack tip is shown by the gray region.

of taking the path very close to the crack tip, where it is difficult to compute the stress and displacement fields accurately because of the stress singularities or the presence of the plastic zone near the crack tip, it is now possible to take the path along the boundary and/or lines of symmetry on which displacement, traction, and strain energy can be computed relatively easily.

## 5.5 J-Integral for Dugdale Model

Let us now compute the J-integral on a path surrounding the plastic zone predicted by the Dugdale plasticity model. The J-integral is carried out on path $S$ (=$S^- \cup S^+$) as shown in Figure 5.7. Note that the path independence property of the J-integral is valid as long as the material between two J-integral paths is elastic.

Since the material enclosed by the J-integral path shown in Figure 5.7 is not elastic, the concept of the elastic strain energy release rate ($G$) does not exist here. Thus, in this case we cannot say that $J = G$. However, it is still possible to compute the J-integral value for this problem geometry and equate it to another important parameter, the crack tip opening displacement (CTOD), as shown:

$$J = \int_S \left( U dx_2 - T_i \frac{\partial u_i}{\partial x_1} dS \right) = \int_{S^-} \left( U dx_2 - T_i \frac{\partial u_i}{\partial x_1} dS \right) + \int_{S^+} \left( U dx_2 - T_i \frac{\partial u_i}{\partial x_1} dS \right) \quad (5.19)$$

From Figure 5.8 it is clear that, on the lower path ($S^-$), $T_2 = -\sigma_{YS}$, $dS = dx_1$ and, on the upper path ($S^+$), $T_2 = \sigma_{YS}$, $dS = dx_1$; on both paths, $dx_2 = 0$. Therefore, equation (5.19) can be written as

$$J = \int_0^b \left( U \times 0 - T_i^- \frac{\partial u_i^-}{\partial x_1} dx_1 \right) + \int_b^0 \left( U \times 0 - T_i^+ \frac{\partial u_i^+}{\partial x_1} (-dx_1) \right)$$

$$= \int_0^b \left( -T_2^- \frac{\partial u_2^-}{\partial x_1} dx_1 \right) + \int_0^b \left( -T_2^+ \frac{\partial u_2^+}{\partial x_1} dx_1 \right)$$

$$= \int_0^b \left( \sigma_{YS} \frac{\partial u_2^-}{\partial x_1} dx_1 \right) + \int_0^b \left( -\sigma_{YS} \frac{\partial u_2^+}{\partial x_1} dx_1 \right)$$

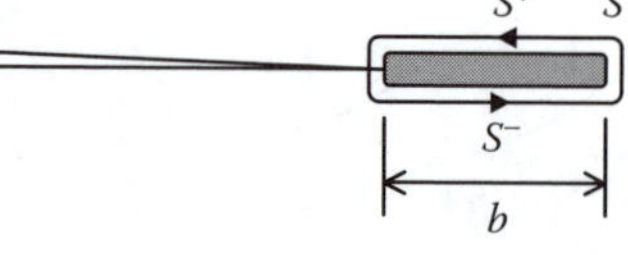


**FIGURE 5.7**
J-integral contour around the plastic zone predicted by the Dugdale model.

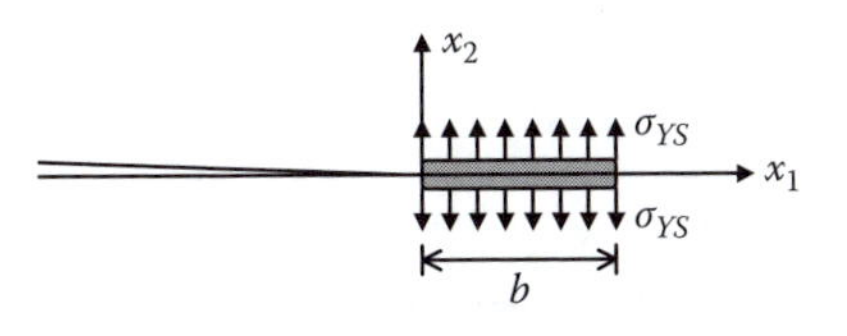


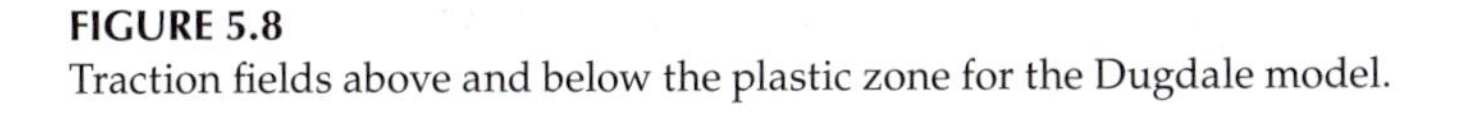

**FIGURE 5.8**
Traction fields above and below the plastic zone for the Dugdale model.

$$= \sigma_{YS}\left[\int_0^b \left(\frac{\partial u_2^-}{\partial x_1} dx_1\right) - \int_0^b \left(\frac{\partial u_2^+}{\partial x_1} dx_1\right)\right] = \sigma_{YS}\left[\int_0^b \frac{\partial (u_2^- - u_2^+)}{\partial x_1} dx_1\right] \tag{5.20}$$

$$= \sigma_{YS}\left[u_2^- - u_2^+\right]_0^b = -\sigma_{YS}\left[u_2^+ - u_2^-\right]_0^b = -\sigma_{YS}(0 - CTOD) = \sigma_{YS} \times CTOD$$

In equation (5.20), $CTOD = [u_2^+ - u_2^-]_{x_1=0}$ = crack tip opening displacement. At the end of the plastic zone, at $x_1 = b$, $[u_2^+ - u_2^-]_{x_1=b} = 0$.

Therefore, we see that for the Dugdale model one can calculate the crack tip opening displacement simply by dividing the J-integral value by the yield stress ($\sigma_{YS}$) of the material. Therefore, the J-integral value is the strain energy release rate as well as the crack tip opening displacement multiplied by the yield stress of the material. Clearly, for a material, if there is a critical value of strain energy release rate ($G_c$) that governs the crack propagation, then there must be a critical value for the J-integral ($J_c$), as well as a critical value of the crack tip opening displacement ($CTOD_c$) governing the crack propagation phenomenon.

## 5.6 Experimental Evaluation of Critical J-Integral Value, $J_c$

Since the J-integral is related to the strain-energy release rate and crack tip opening displacement—the two parameters that can be experimentally evaluated—it is possible to evaluate the J-integral value at failure or the critical J-integral value ($J_c$) experimentally. Begely and Landis (1972) described the experimental evaluation technique for critical J-integral. The test specimen needed for this experiment is shown in Figure 5.9.

If the force $P$ versus the displacement $\Delta$ is plotted, then for small values of $P$, the curve should show a linear variation and for larger values of $P$ it should show a nonlinear variation, as shown in Figure 5.10. If the crack length $a$ of Figure 5.9 is then changed to $a + \Delta a$, then the $P$–$\Delta$ curve of Figure 5.10 should also change, showing a greater displacement $\Delta$ for the same value of $P$, as shown in Figure 5.11.

For the fixed force or fixed grip loading, the strain energy stored in the specimen can be obtained from Figure 5.12, drawn in the linear region. Note

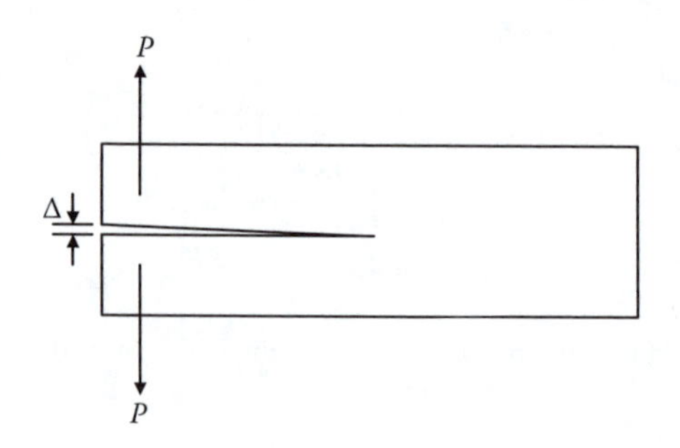


**FIGURE 5.9**
A double cantilever specimen subjected to two opening forces $P$ producing a crack opening Δ.

that the strain energy stored in the specimen with crack length $a$ for applied load $P_0$ is the triangular area OAC. For another specimen with crack length $a + \Delta a$, for the same applied load $P_0$ the strain energy stored in the specimen is the triangular area OBD. Therefore, for the fixed force ($P_0$), loading the increase in the strain energy as the crack length increases from $a$ to $a + \Delta a$, is equal to the triangular area OAB. Note that both the height (AC) and the base length (AB) of triangle OAB are proportional to displacement Δ. Therefore, the area of triangle OAB is proportional to $\Delta^2$:

$$J = G = \frac{dU}{dA} = \frac{1}{t\Delta a} \times \frac{AB \times AC}{2} = k\Delta^2 \tag{5.21}$$

where $k$ of equation (5.21) is the proportionality constant.

Instead of a fixed force condition, if a fixed grip condition is maintained and the displacement (Δ) is kept constant at OC, then the load will decrease from AC (or $P_0$) to LC, resulting in a decrease in the strain energy in the material from area OAC to area OLC. Therefore, the strain energy released in the process is equal to the area AOL. Note that in this case also the base AL and height OC of the triangle AOL are proportional to displacement Δ.

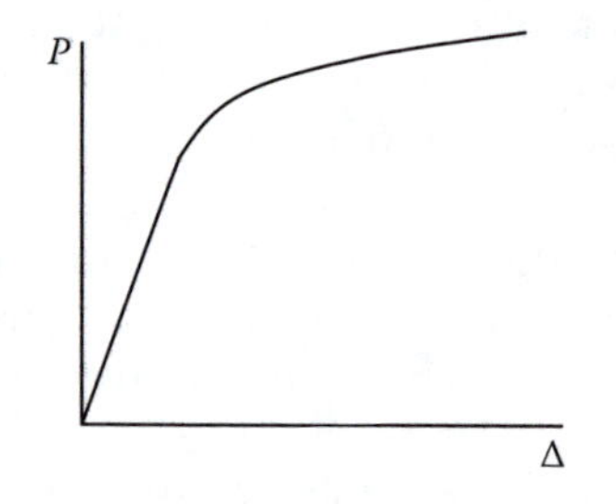


**FIGURE 5.10**
Force ($P$) versus displacement (Δ) curve for the double cantilever specimen shown in Figure 5.9.

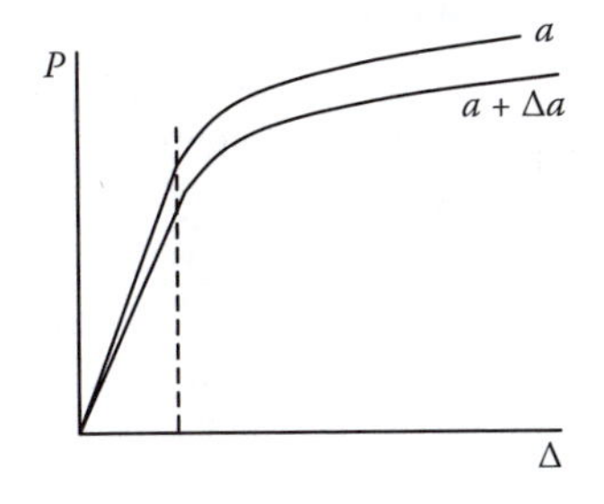


**FIGURE 5.11**
Force ($P$) versus displacement ($\Delta$) curve for the double cantilever specimen shown in Figure 5.9 for two different crack lengths. Dashed vertical line separates the linear and nonlinear regions.

Therefore, the area of triangle AOL is proportional to $\Delta^2$:

$$J = G = -\frac{dU}{dA} = \frac{1}{t\Delta a} \times \frac{AL \times OC}{2} = k\Delta^2 \tag{5.22}$$

where $k$ of equation (5.22) is the proportionality constant.

If the strain energy release rate is now computed in the nonlinear range of the $P$–$\Delta$ curve (Figure 5.11), then one needs to refer to Figure 5.13, instead of Figure 5.12, to obtain the strain energy release rate as the crack length is increased from $a$ to $a + \Delta a$ under fixed grip conditions. The shaded area of Figure 5.13 represents the strain energy released as the crack length is extended. From this figure it is clear that the shaded area increases proportional to the crack opening displacement $\Delta$. Therefore, the J-integral value in

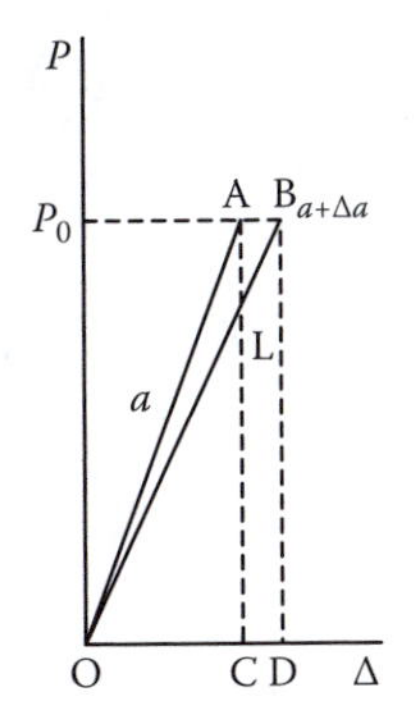


**FIGURE 5.12**
Force ($P$)–displacement ($\Delta$) relation in the linear range for the double cantilever specimen shown in Figure 5.9 for two different crack lengths.

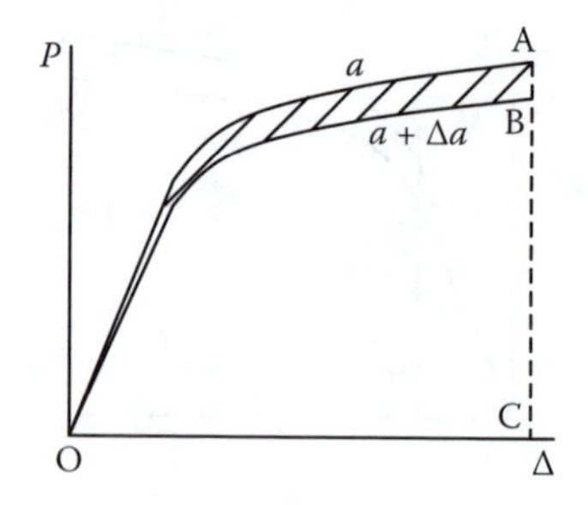


**FIGURE 5.13**
Force ($P$)–displacement ($\Delta$) relation for the double cantilever specimen shown in Figure 5.9 for two different crack lengths.

the nonlinear range should be proportional to $\Delta$, or

$$J = G = -\frac{dU}{dA} = \frac{1}{t\Delta a} \times (\text{Shaded area of Figure 5.13}) = k\Delta \tag{5.23}$$

where $k$ of equation (5.23) is the proportionality constant.

From equations (5.21), (5.22), and (5.23) one can see that $J$ increases nonlinearly for small values of the crack opening displacement ($\Delta$) and it increases linearly for large values of $\Delta$. Therefore, the $J$–$\Delta$ curve obtained experimentally should have the shape shown in Figure 5.14.

From double cantilever specimens of different crack lengths (or, alternately, taking one specimen and gradually increasing its crack length), $P$–$\Delta$ curves are generated for identical specimens with various crack lengths, as shown in Figure 5.15.

To obtain the critical J-integral value ($J_c$), one needs to take a number of double cantilever specimens with different crack lengths and break the specimens while recording the crack opening displacement ($\Delta$, see Figure 5.9) at failure or as the crack starts to propagate. From Figure 5.15 the $J$ values at

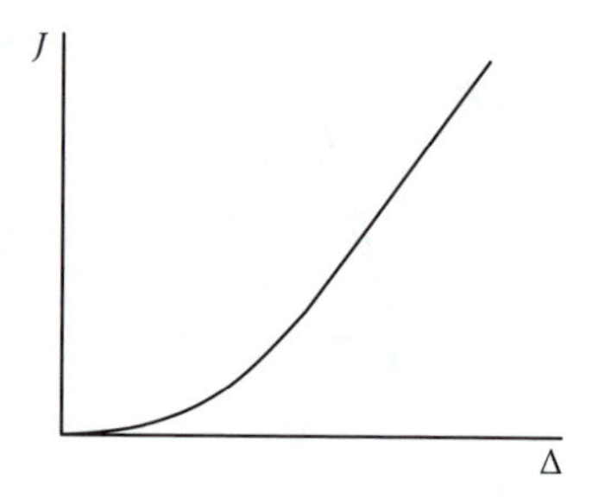


**FIGURE 5.14**
$J$–$\Delta$ variation for the double cantilever specimen shown in Figure 5.9.

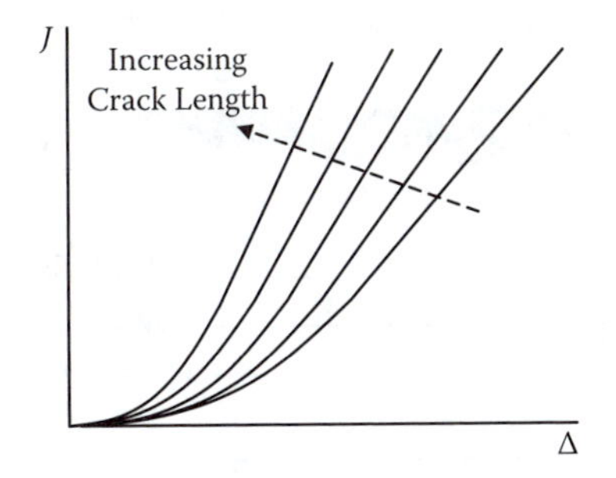


**FIGURE 5.15**
$J$–$\Delta$ variations for the double cantilever specimens shown in Figure 5.9 for different crack lengths.

failure can be obtained. By plotting these $J$ values on $J$–$\Delta$ curves Figure 5.16 is obtained. Square markers of Figure 5.16 give $J$ and $\Delta$ values at failure. Ideally, $J$ values at failure should be independent of crack lengths, as shown in Figure 5.16. The average of these $J$ values at failure is the critical J-integral value or $J_c$ of the material.

## 5.7 Concluding Remarks

The basic theory of J-integral and the prediction of crack propagation from the critical J-integral value are presented in this chapter. The failure prediction from the critical J-integral value as discussed in this chapter works well when the plastic zone size is small compared to the problem dimensions and when the crack is stationary. When a loaded crack starts to propagate, the loading–unloading path of the material just ahead of the crack tip may follow different stress–strain paths during loading and unloading. Limitations

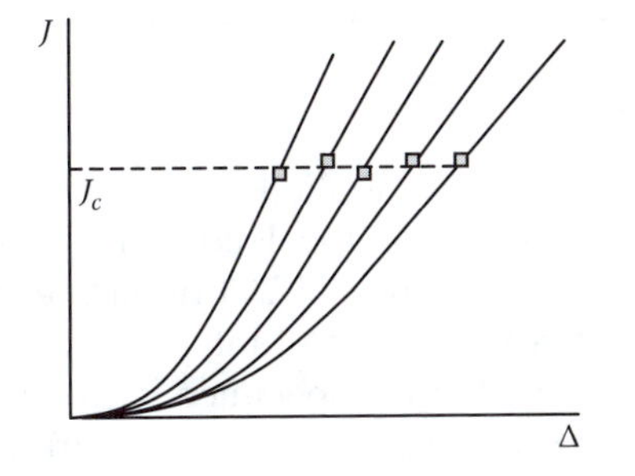


**FIGURE 5.16**
Critical $J$ values are marked by square markers in the $J$–$\Delta$ curves for the double cantilever specimens shown in Figure 5.9.

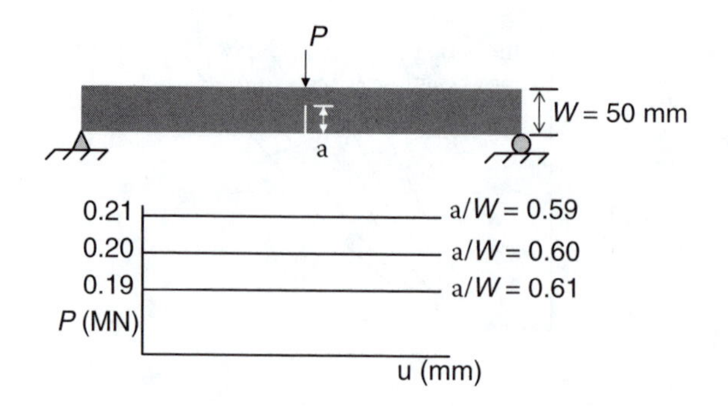


**FIGURE 5.17**

on the applicability of the J-integral technique in such complex situations are not discussed here, but can be found elsewhere (Broek 1997).

## References

ASTM- STP (special technical publication) 514, Fracture toughness, 1972.

ASTM-STP 536, Progress in flaw growth and fracture toughness testing, 1973.

Begely, J. A. and Landis, J. D. The J-integral as a fracture criterion. ASTM STP 514, 1–20, 1972.

Broek, D. *Elementary engineering fracture mechanics*, 4th rev. ed. Dordrecht, the Netherlands: Kluwer Academic Publishers, 1997.

Rice, J. R. A path independent integral and the approximate analysis of strain concentration by notches and cracks. *ASME Journal of Applied Mechanics*, 35, 379–386, 1968.

## Exercise Problems

Problem 5.1: Loading curves (applied load value as a function of the downward displacement of the load) for a three-point bending specimen are shown in Figure 5.17. Three curves are given for three different crack lengths. Give the J-integral value for $a/W = 0.6$ and $u = 1$ mm. Thickness ($t$) of the specimen is 20 mm. Give proper unit of J. (Hint: J-integral value is equal to the strain energy release rate.)

Problem 5.2: In the book, the J-integral expression has been derived for the case when body forces are absent. If body forces are present, what will be the appropriate form of the J-integral?

# 6

# *Fatigue Crack Growth*

## 6.1 Introduction

Under cyclic loadings, pre-existing cracks inside a material may become bigger and cause catastrophic failure of the structure. Structural failure under cyclic loading is also known as fatigue failure. In this chapter crack propagation behavior under cyclic loading or fatigue is studied. Since under fatigue a crack can propagate at a stress level well below the critical stress value, this crack growth phenomenon is also known as subcritical crack growth.

## 6.2 Fatigue Analysis—Mechanics of Materials Approach

If a body is subjected to an oscillating stress between $\sigma_{max}$ and $\sigma_{min}$ as shown in Figure 6.1, then the body might fail even when the applied maximum stress ($\sigma_{max}$) is well below the ultimate stress or failure stress. Number of cycles ($N_f$) required for failure is a function of the stress difference ($S$) between the maximum and minimum stress levels ($S = \Delta\sigma = \sigma_{max} - \sigma_{min}$) and the average stress level $\sigma_{av} = \frac{\sigma_{max}+\sigma_{min}}{2}$, as shown in Figure 6.2. Note that there is a threshold value of $\Delta\sigma$ below which the fatigue phenomenon is not observed. In other words, when $\Delta\sigma$ value is below the threshold value, then the structure does not fail even when it is subjected to a large number of cycles. This threshold value for a given material depends on the surface roughness of the body, surrounding environment, and other parameters.

## 6.3 Fatigue Analysis—Fracture Mechanics Approach

The theory of fatigue crack growth presented in this section is based on the work by Paris and Erdogan (1960). It should be noted here that when a cracked body is subjected to an oscillatory stress field as shown in Figure 6.1, the stress intensity factor also oscillates between $K_{max}$ and $K_{min}$ with an

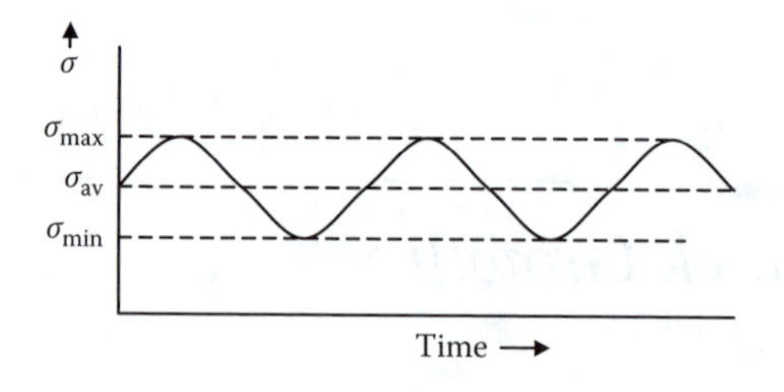


**FIGURE 6.1**
Oscillating stress applied to a body.

average value of $K_m = K_{mean} = \frac{K_{max}+K_{min}}{2}$, as long as the crack length remains constant. However, as the crack starts to propagate, $K_{max}$, $K_{min}$, and $K_m$ values start to vary with time, even when $\sigma_{max}$ and $\sigma_{min}$ values remain unchanged.

In classical fatigue analyses the number of cycles to failure ($N_f$) is considered to be a function of the stress difference $\Delta\sigma$ and the average stress level $\sigma_{av}$. Similarly, in fatigue crack growth analysis, the crack growth rate (crack growth per unit cycle of loading) is assumed to be a function of $\Delta K = (K_{max} - K_{min})$ and $K_m = \frac{K_{max}+K_{min}}{2}$ :

$$\frac{da}{dN} = f(\Delta K, K_m) \tag{6.1}$$

Typical experimental results on fatigue crack growth, as shown in Figure 6.3, justify this assumption. Note that in the log–log scale the crack growth rate is linearly dependent on $\Delta K$. Therefore, in the log–log scale, equation (6.1) should take the following form:

$$\log\left(\frac{da}{dN}\right) = \log C + n\log(\Delta K) \tag{6.2}$$

In equation (6.2) log $C$ is the $y$-intercept of the straight line variation of $\log(\frac{da}{dN})$ against $\log(\Delta K)$, shown in Figure 6.3, and $n$ is the slope of this

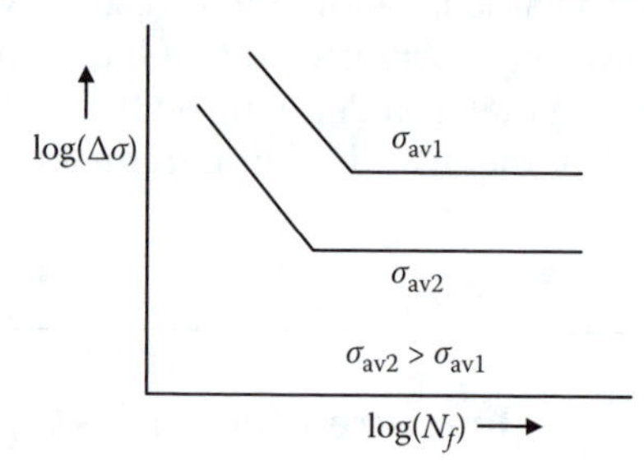


**FIGURE 6.2**
Classical $S$–$N$ curve ($S = \Delta\sigma = \sigma_{max} - \sigma_{min}$, $N = N_f$). Note that the number of cycles required for failure depends on $\Delta\sigma$ and $\sigma_{av}$.

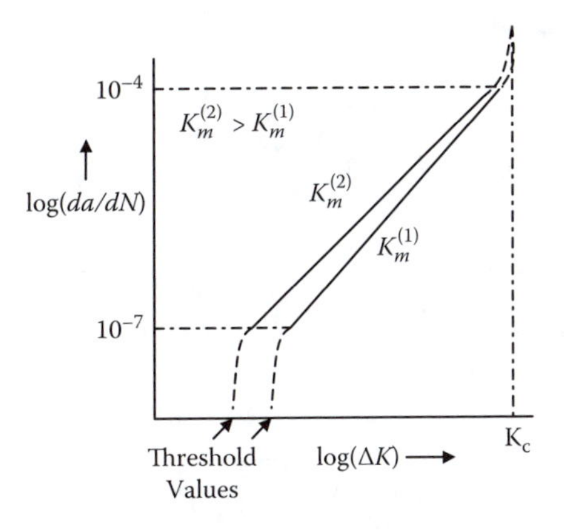


**FIGURE 6.3**
Typical experimental results for fatigue crack growth experiments.

straight line. It should also be noted here that the slope and the *y*-intercept values depend on the mean stress intensity factor $K_m$.

From equation (6.2),

$$\log\left(\frac{da}{dN}\right) = \log C + n\log(\Delta K) = \log C + \log(\Delta K)^n = \log[C(\Delta K)^n] \tag{6.3}$$

$$\therefore \frac{da}{dN} = C(\Delta K)^n$$

Note that $C$ and $n$ are material properties. For bridge steels, $C = 3.4 \times 10^{-10}$ and $n = 3$ when the crack length $a$ is measured in inches and the unit of $\Delta K$ is $\text{ksi}\sqrt{\text{in}}$.

Let us now apply the preceding crack growth formula to calculate the number of cycles required for a circular crack of initial radius $a_0$ to reach a final radius $a$. The stress intensity factor for a circular crack is given by

$$K = \frac{2}{\pi}\sigma\sqrt{\pi a} \tag{6.4}$$

Therefore, if this crack is subjected to an oscillating stress field with a stress difference of $\Delta\sigma$ between the maximum and minimum stress values, then the variation in the stress intensity factor under the fatigue loading is given by

$$\Delta K = \frac{2}{\pi}\Delta\sigma\sqrt{\pi a} \tag{6.5}$$

Substituting equation (6.5) into equation (6.3),

$$\frac{da}{dN} = C(\Delta K)^n = C\left(\frac{2}{\pi}\Delta\sigma\sqrt{\pi a}\right)^n = C\left(\frac{2}{\pi}\Delta\sigma\sqrt{\pi}\right)^n a^{n/2} \tag{6.6}$$

Therefore,

$$\int_{a_0}^{a} \frac{da}{a^{n/2}} = \int_{0}^{N} C\left(\frac{2}{\pi}\Delta\sigma\sqrt{\pi}\right)^n dN$$

$$\therefore \left[\frac{a^{1-\frac{n}{2}}}{1-\frac{n}{2}}\right]_{a_0}^{a} = C\left(\frac{2}{\sqrt{\pi}}\Delta\sigma\right)^n [N]_0^N = C\left(\frac{2}{\sqrt{\pi}}\Delta\sigma\right)^n N \tag{6.7}$$

$$\therefore N = \frac{1}{C\left(\frac{2}{\sqrt{\pi}}\Delta\sigma\right)^n\left(1-\frac{n}{2}\right)}\left[a^{1-\frac{n}{2}} - a_0^{1-\frac{n}{2}}\right]$$

If the final crack radius is equal to the critical crack radius ($a_c$) required for the crack to propagate, then $N$ is equal to the number of cycles to failure and is denoted as $N_f$. Therefore,

$$N_f = \frac{1}{C\left(\frac{2}{\sqrt{\pi}}\Delta\sigma\right)^n\left(1-\frac{n}{2}\right)}\left[a_c^{1-\frac{n}{2}} - a_0^{1-\frac{n}{2}}\right] = \frac{a_0^{1-\frac{n}{2}}}{C\left(\frac{2}{\sqrt{\pi}}\Delta\sigma\right)^n\left(1-\frac{n}{2}\right)}\left[\left(\frac{a_c}{a_0}\right)^{1-\frac{n}{2}} - 1\right]$$

$$= \frac{a_0}{C\left(\frac{2}{\sqrt{\pi}}\Delta\sigma\right)^n a_0^{\frac{n}{2}}\left(1-\frac{n}{2}\right)}\left[\left(\frac{a_0}{a_c}\right)^{\frac{n}{2}-1} - 1\right] = \frac{a_0}{\left(\frac{da}{dN}\right)_0\left(1-\frac{n}{2}\right)}\left[\left(\frac{a_0}{a_c}\right)^{\frac{n}{2}-1} - 1\right] \tag{6.8}$$

$$= \frac{a_0}{\left(\frac{da}{dN}\right)_0\left(\frac{n}{2}-1\right)}\left[1-\left(\frac{a_0}{a_c}\right)^{\frac{n}{2}-1}\right]$$

In the preceding equation, $(\frac{da}{dN})_0 = C(\frac{2}{\pi}\Delta\sigma\sqrt{\pi a_0})^n$ is the initial crack growth rate. Equation (6.7) can also be written in the form similar to the one given in equation (6.8):

$$N = \frac{a_0}{\left(\frac{da}{dN}\right)_0\left(\frac{n}{2}-1\right)}\left[1-\left(\frac{a_0}{a}\right)^{\frac{n}{2}-1}\right] = \frac{a_0}{\left[C\left(\frac{2}{\sqrt{\pi}}\Delta\sigma\right)^n a_0^{n/2}\right]\left(\frac{n}{2}-1\right)}\left[1-\left(\frac{a_0}{a}\right)^{\frac{n}{2}-1}\right] \tag{6.9}$$

Note that from equations (6.8) and (6.9) it is possible to evaluate:

$N_f$ after knowing $a_0$ and $a_c$

$N$ needed to increase the crack radius to $a$ from an initial crack radius of $a_0$

$\Delta\sigma$ needed for given values of $a_0$ and $N_f$

### 6.3.1 Numerical Example

Let a material be subjected to an oscillating stress field between 0 and $\Delta\sigma$. If the material has a circular crack of initial radius $a_0 = 0.02$ in., calculate the number of cycles to failure for six different values of $\Delta\sigma$ (1000, 2000, 3000,…6000 psi) if the fracture toughness of the material is 1000 psi-in.$^{1/2}$ and the constitutive relation for crack propagation for this material is given by $\frac{da}{dN} = 2.4 \times 10^{-24} (\Delta K)^{6.64}$, where unit of $\Delta K$ is psi-in.$^{1/2}$.

Note that, in this case,

$$\sigma_{\min} = 0$$

$$\sigma_{\max} = \Delta\sigma$$

Therefore, for this material the critical crack radius can be obtained in the following manner:

$$K_c = \frac{2}{\pi}\sigma_{\max}\sqrt{\pi a_c} = \frac{2}{\pi}\Delta\sigma\sqrt{\pi a_c} = \frac{2}{\sqrt{\pi}}\Delta\sigma\sqrt{a_c}$$
$$\therefore a_c = \frac{\pi}{4}\left(\frac{K_c}{\Delta\sigma}\right)^2 = \frac{\pi}{4}\left(\frac{1000}{\Delta\sigma}\right)^2 \tag{6.10}$$

Substituting material parameters into equation (6.8),

$$N_f = \frac{a_0}{\left(\frac{da}{dN}\right)_0\left(\frac{n}{2}-1\right)}\left[1-\left(\frac{a_0}{a_c}\right)^{\frac{n}{2}-1}\right] = \frac{a_0^{1-\frac{n}{2}}}{C\left(\frac{2}{\sqrt{\pi}}\Delta\sigma\right)^n\left(\frac{n}{2}-1\right)}\left[1-\left(\frac{a_0}{a_c}\right)^{\frac{n}{2}-1}\right]$$
$$= \frac{a_0^{-2.32}}{2.4\times10^{-24}\times\left(\frac{2}{\sqrt{\pi}}\Delta\sigma\right)^{6.64}\times 2.32}\left[1-\left(\frac{a_0}{a_c}\right)^{2.32}\right] \tag{6.11}$$
$$= 8.04\times10^{22}(\Delta\sigma)^{-6.64}a_0^{-2.32}\left[1-\left(\frac{a_0}{a_c}\right)^{2.32}\right]$$

For a given value of $\Delta\sigma$, equation (6.10) is first used to calculate the critical value of the crack radius and then equation (6.11) is used to calculate the number of cycles to failure. Computed values are given in Table 6.1.

## 6.4 Fatigue Analysis for Materials Containing Microcracks

If the material contains very small cracks or microcracks, then the initial radius ($a_0$) of the circular cracks present in the body is much smaller than the

**TABLE 6.1**

Values of the Critical Crack Radius and the Number of Cycles to Failure for Different Values of $\Delta\sigma$ for the Material Described in Section 6.3.1

| $\Delta\sigma$ (psi) | $a_c$ (inch) | $N_f$ (Cycles to Failure) |
|---|---|---|
| 1000 | 0.785 | $8.46 \times 10^6$ |
| 2000 | 0.196 | $8.44 \times 10^4$ |
| 3000 | 0.0873 | $5.56 \times 10^3$ |
| 4000 | 0.0491 | 745 |
| 5000 | 0.0314 | 126 |
| 6000 | 0.0218 | 11 |

critical crack radius $a_c$. Then equation (6.8) can be simplified to

$$N_f = \frac{a_0^{1-\frac{n}{2}}}{C\left(\frac{2}{\sqrt{\pi}}\Delta\sigma\right)^n\left(\frac{n}{2}-1\right)}\left[1-\left(\frac{a_0}{a_c}\right)^{\frac{n}{2}-1}\right] = \frac{a_0^{1-\frac{n}{2}}}{C\left(\frac{2}{\sqrt{\pi}}\Delta\sigma\right)^n\left(\frac{n}{2}-1\right)} = B_0(\Delta\sigma)^{-n} \quad (6.12)$$

where constant $B_0$ is given by

$$B_0 = \frac{a_0^{1-\frac{n}{2}}}{C\left(\frac{2}{\sqrt{\pi}}\right)^n\left(\frac{n}{2}-1\right)} \quad (6.13)$$

Equation (6.12) can be rewritten in the following form:

$$\frac{1}{B_0}N_f = (\Delta\sigma)^{-n}$$
$$\therefore \Delta\sigma = \left(\frac{1}{B_0}N_f\right)^{-\frac{1}{n}} = D_0(N_f)^{-\frac{1}{n}} \quad (6.14)$$

In the preceding equation, $D_0 = B_0^{1/n}$. Taking log on both sides of equation (6.14), one gets

$$\log(\Delta\sigma) = \log D_0 - \frac{1}{n}\log N_f \quad (6.15)$$

Equation (6.15) is plotted in Figure 6.4. Note that Figure 6.4 is identical to the slant portion of the $S$–$N$ curve shown in Figure 6.2. Thus, the $S$–$N$ curve of the classical fatigue analysis can be explained from the fracture mechanics analysis simply by taking very small values for the initial crack dimensions.

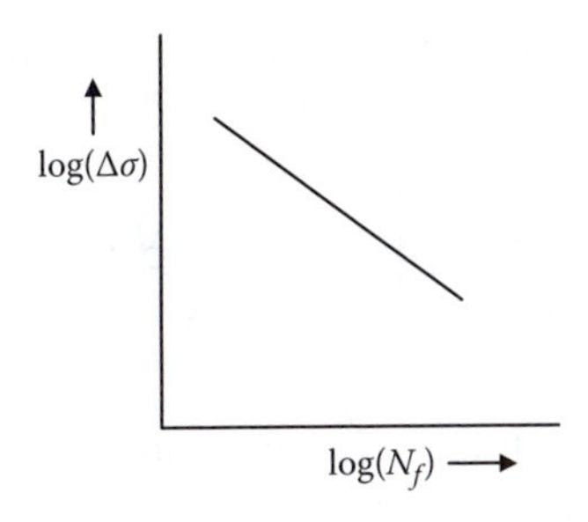


**FIGURE 6.4**
Plot of equation (6.15).

## 6.5 Concluding Remarks

The number of cycles required for failure of a cracked body subjected to a certain level of cyclic loading has been derived in this chapter based on a relatively simple and popular constitutive law relating the crack growth rate and the level of loading. For a more comprehensive discussion on fatigue crack growth and the failure phenomenon or for studying other mathematical models for fatigue prediction, readers are referred to the publications given in the reference list (Bolotin, 1999; Liu and Iinno, 1969; McClintock, 1963; Schijve, 1967; Weertman, 1984).

## References

Bolotin, V. V. *Mechanics of fatigue*. Boca Raton, FL: CRC Press, 1999.

Liu, H. W. and Iinno, N. A mechanical model for fatigue crack propagation. *Proceedings of the 2nd International Conference on Fracture*, 812–824, 1969.

McClintock, F. A. On the plasticity of the growth of fatigue cracks. In *Fracture of solids*. New York: John Wiley & Sons, pp. 65–102, 1963.

Paris, P. C. and Erdogan, F. A critical analysis of crack propagation laws. *Journal of Basic Engineering*, 85, 528–534, 1960.

Schijve, J. Significance of fatigue cracks in micro-range and macro-range. *ASTM STP* 415, 415–459, 1967.

Weertman, J. Rate of growth of fatigue cracks calculated from the theory of infinitesimal dislocations distributed on a plane. *International Journal of Fracture* 26, 308–315, 1984.

## Exercise Problems

Problem 6.1: (a) A large plate has a small internal crack through the thickness. The crack has the initial lengths $2a_0$ and is normal to a remote tension $\sigma_0$, which pulsates between the two levels $\sigma_{max}$

and $\sigma_{min}$. The fatigue crack growth equation for this material is given by

$$\frac{da}{dN} = C\left[\left(\frac{\Delta K}{E}\right)^4 - \left(\frac{\Delta K_t}{E}\right)^4\right]$$

where $\Delta K_t$ is the threshold stress intensity factor. Prove that the number of cycles required for failure is given by

$$N_c = A.\ln\left[\frac{(a_c - B)(a_0 + B)}{(a_c + B)(a_0 - B)}\right]$$

where $A = E^4/[2c\pi(\Delta\sigma.\Delta K_t)^2]$, $B = (1/\pi)(\Delta K_t/\Delta\sigma)^2$, $2a_c$ is the critical crack length.

(b) Express $a_c$ in terms of $K_c$ and $\sigma_{max}$, $\sigma_{min}$, or $\Delta\sigma$, whichever is appropriate.

Problem 6.2: For a pressure vessel steel $da/dN = 10^{-6}$ in./cycle, when $\Delta K = 12$ ksi.in.$^{1/2}$, and $da/dN = 10^{-4}$ in./cycle, when $\Delta K = 96$ ksi.in.$^{1/2}$. Assuming that $da/dN = C(\Delta K)^n$,

(a) Find the values of $C$ and $n$.

(b) If fatigue cycling with constant stress range starts at $\Delta K = 12$ ksi.in.$^{1/2}$ and $a = a_0 = 0.01$ in., how many cycles are required to reach a crack growth rate of $10^{-4}$ in./cycle for a Griffith crack of length $2a$ in an infinite medium?

(c) If $K_c = 1000$ ksi.in.$^{1/2}$, after how many cycles will the material fail? Consider the minimum stress to be zero.

# 7

# *Stress Intensity Factors for Some Practical Crack Geometries*

## 7.1 Introduction

In chapter 3, section 3.6, stress intensity factors (SIFs) for different problem geometries have been discussed. For most of these problem geometries, the crack is assumed to be present in an infinite solid. In other words, stress-free boundaries were not present near the crack tip except for the two crack surfaces. In this chapter we introduce some realistic problem geometries in which a stress-free boundary in addition to the two crack surfaces may be present near the crack tip. Stress intensity factors for some two-dimensional and some three-dimensional cracks, such as circular cracks, elliptical cracks, and part-through semi-elliptical cracks, are discussed in this chapter.

## 7.2 Slit Crack in a Strip

Figure 7.1 shows a slit crack of length $2a$ in a strip of width $2b$ subjected to uniaxial tension $\sigma$. Note that the problem geometry of Figure 7.1 has some similarities with collinear cracks in an infinite medium, as shown in Figure 7.2. In this Figure the region bounded by two consecutive dashed lines forms a strip of width $2b$, subjected to uniaxial tension $\sigma$ and containing a crack of length $2a$ at the center of the strip. The region bounded by two dashed lines in Figure 7.2 and the strip geometry of Figure 7.1 are almost identical. The only difference is that the boundaries of Figure 7.1 are traction free, $\sigma_{xx} = \sigma_{xy} = 0$, and the horizontal displacement $u_x \neq 0$; these two stress components and the horizontal displacement along the dashed lines of Figure 7.2 are given by $\sigma_{xy} = 0$, $\sigma_{xx} \neq 0$, $u_x = 0$. Since the problem geometry is symmetric about any dashed line of Figure 7.2, the horizontal displacement and the shear stress along these lines must be zero, but the normal stress component is not necessarily zero.

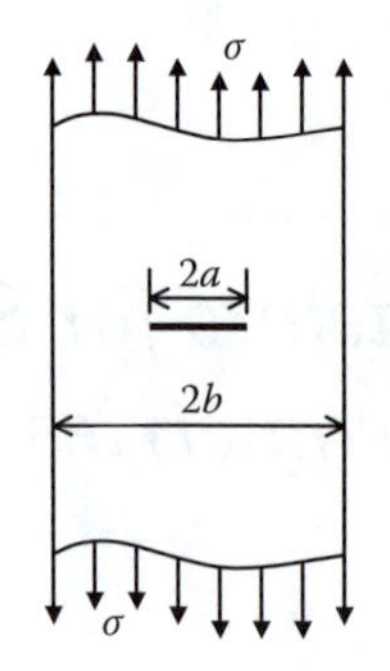


**FIGURE 7.1**
Slit crack in a strip.

The SIF for the problem geometry of Figure 7.2 is given by

$$K = \left\{ \frac{2b}{\pi a} \tan\left( \frac{\pi a}{2b} \right) \right\}^{1/2} \sigma\sqrt{\pi a} \tag{7.1}$$

Equation (7.1) can also be used as the stress intensity factor of the problem geometry of Figure 7.1 when the crack tips are not too close to the boundary surfaces, when $\frac{a}{b} < 0.5$. Isida (1955) gave a power series solution for the SIF of the problem geometry of Figure 7.1. This power series solution can be used for $\frac{a}{b} \le 0.9$. Feddersen (1967) proposed the following solution for the SIF of the slit crack in a strip of Figure 7.1:

$$K = \left\{ \sec\left( \frac{\pi a}{2b} \right) \right\}^{1/2} \sigma\sqrt{\pi a} \tag{7.2}$$

Feddersen's expression has been found to give excellent result, producing an error less than 5% when compared with Isida's power series solution. Plots of equations (7.1) and (7.2) are given in Figure 7.3.

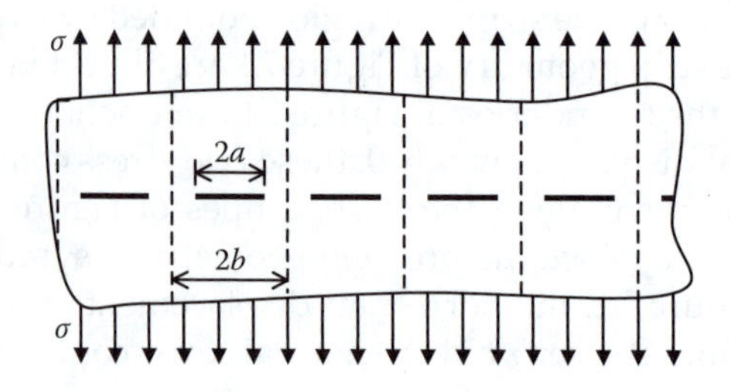


**FIGURE 7.2**
Infinite number of collinear cracks in an infinite medium.

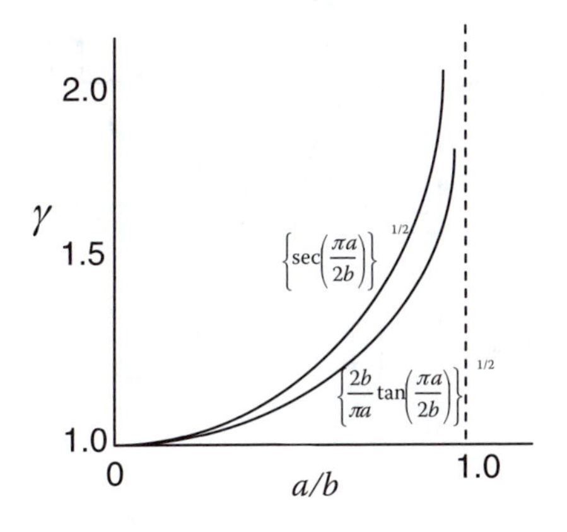


**FIGURE 7.3**
Normalized stress intensity factors (normalized with respect to the SIF of Griffith crack) for problem geometries shown in Figures 7.1 (the higher curve) and 7.2 (the lower curve).

## 7.3 Crack Intersecting a Free Surface

To estimate the SIF of a crack of length $a$ intersecting a free surface and subjected to the opening mode loads, as shown in Figure 7.4, one can proceed in the following manner. Without knowing the true solution, if one wants to estimate the SIF of this problem geometry knowing only the SIF of Griffith crack, the first thing one needs to decide is how long a Griffith crack should give the same SIF. Should it be $a$ or something different? To answer this question, it is necessary to compare the problem geometry of Figure 7.4 with the two problem geometries given in Figure 7.5.

Note that the right sides of the dashed lines of Figures 7.5(a) and 7.5(b) look identical to the problem geometry of Figure 7.4. However, on the vertical free surface of Figure 7.4 the normal and shear stress components are zero,

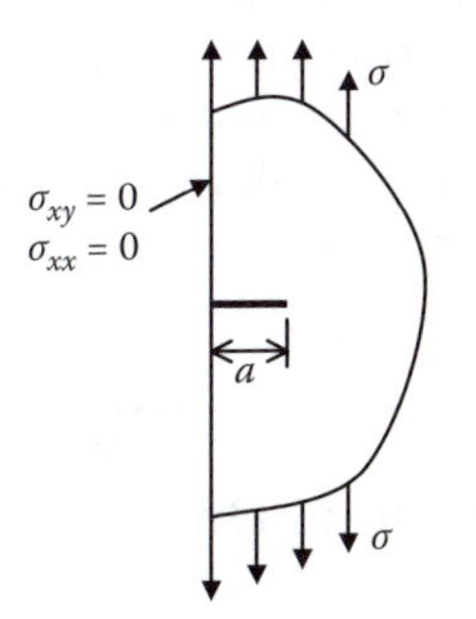


**FIGURE 7.4**
Crack intersecting a free surface.

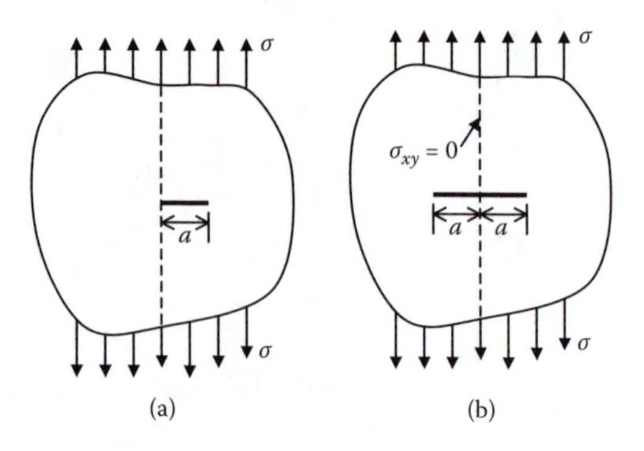


**FIGURE 7.5**
Griffith cracks of length $a$ (left figure) and $2a$ (right figure) subjected to the opening mode loadings.

$\sigma_{xx} = \sigma_{xy} = 0$, while neither of these two stress components is zero along the dashed line of Figure 7.5a. In fact, these stresses become infinity on the dashed line when it goes through the crack tip. Therefore, problem geometries shown in Figures 7.4 and 7.5a are completely different. Figure 7.5b, on the other hand, is symmetric about the dashed line; therefore, the shear stress component must be zero along this line, but this is not true for the normal stress component. Therefore, on the dashed line of Figure 7.5b, $\sigma_{xy} = 0$, $\sigma_{xx} \neq 0$. However, the nonzero normal stress component is not infinity anywhere on the dashed line since this line is not going through the crack tip. Therefore, the problem geometries of Figures 7.5b and 7.4 are very similar (although not identical) since the stress fields along the vertical boundary of Figure 7.4 are not identical to those along the dashed line of Figure 7.5b. Then, one can conclude that the SIF of Figure 7.4 should be close to that of Figure 7.5b, but not identical. It can be shown that the SIF for the problem geometry of Figure 7.4 is given by (Paris and Sih, 1965)

$$K = 1.122\sigma\sqrt{\pi a} \tag{7.3}$$

Note that the difference between the SIF of Figures 7.4 and 7.5b is 12.2%. Thus, one can say that a free surface increases the SIF by about 12%. This is the reason why a brittle material containing a large number of microcracks when loaded shows first signs of failure on the surface. This phenomenon is commonly known as the surface effect.

## 7.4 Strip with a Crack on Its One Boundary

For the problem geometry shown in Figure 7.6, the SIF is given by

$$K = \gamma\sigma\sqrt{\pi a} \tag{7.4}$$

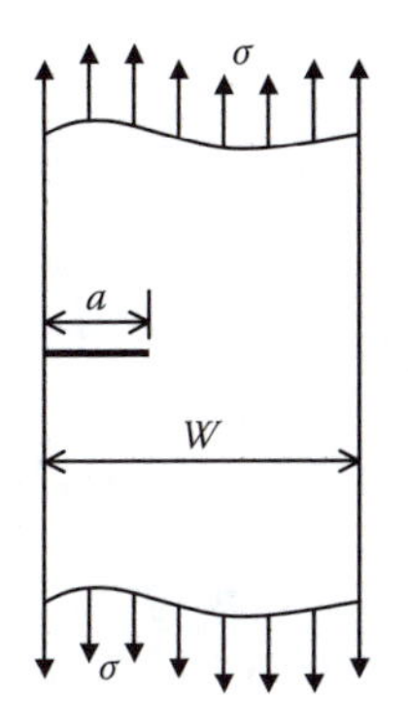


**FIGURE 7.6**
Crack on one side of a strip.

where

$$\gamma = \frac{1}{\sqrt{\pi}}\left\{1.99 - 0.41\left(\frac{a}{W}\right) + 18.7\left(\frac{a}{W}\right)^2 - 38.48\left(\frac{a}{W}\right)^3 + 53.85\left(\frac{a}{W}\right)^4\right\} \tag{7.5}$$

Gamma of equations (7.4) and (7.5) is sometimes called the normalizing factor or normalized stress intensity factor, where it is implied that the SIF is normalized with respect to the SIF of a Griffith crack of crack length $2a$.

## 7.5 Strip with Two Collinear Identical Cracks on Its Two Boundaries

The SIF for the problem geometry of Figure 7.7 can be expressed as given in equation (7.4), where the normalized stress intensity factor $\gamma$ is given

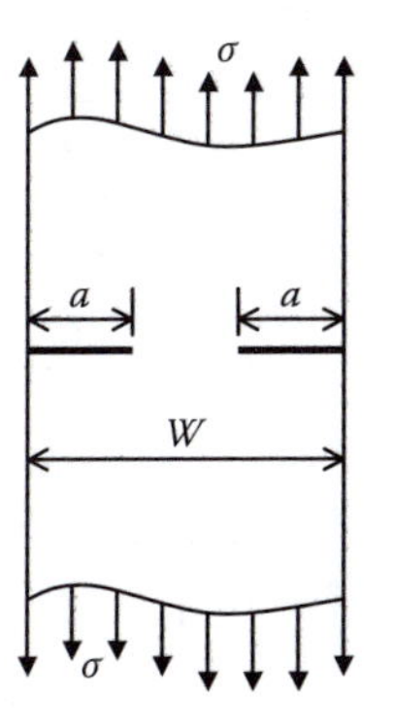


**FIGURE 7.7**
Two identical collinear cracks on two boundaries of a strip.

by

$$\gamma = \frac{1}{\sqrt{\pi}}\left\{1.99 + 0.76\left(\frac{a}{W}\right) - 8.48\left(\frac{a}{W}\right)^2 + 27.36\left(\frac{a}{W}\right)^3\right\} \tag{7.6}$$

## 7.6 Two Half Planes Connected over a Finite Region Forming Two Semi-infinite Cracks in a Full Space

The SIF for the problem geometry of Figure 7.8 is given by

$$K = \frac{P}{\sqrt{\pi b}} \tag{7.7}$$

Equation (7.7) can be written in a different form as

$$K = \frac{P}{\sqrt{\pi b}} = \frac{2}{\pi}\bar{\sigma}\sqrt{\pi b} = \gamma\bar{\sigma}\sqrt{\pi b} \tag{7.8}$$

where

$$\gamma = \frac{2}{\pi}, \quad \bar{\sigma} = \frac{P}{2b} \tag{7.9}$$

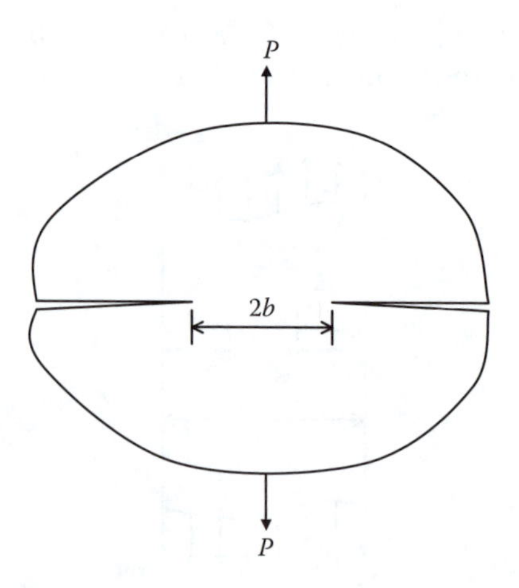


**FIGURE 7.8**
Two identical collinear semi-infinite cracks in a full space.

## 7.7 Two Cracks Radiating Out from a Circular Hole

The problem geometry is shown in Figure 7.9. The exact solution of this problem has been given by Bowie (1956). In this section the engineering solution to this problem is presented without using the exact solution.

Note that for crack length much greater than the hole radius, $a >> R$, two cracks with the circular hole behaves like a Griffith crack of length $2(a + R)$, since the hole at the center of the crack is far away from the crack tip. Therefore, the stress intensity factor in this case is given by

$$K = \sigma\sqrt{\pi(a+R)} = \sigma\sqrt{\pi a\left(1+\frac{R}{a}\right)} = \gamma\sigma\sqrt{\pi a} \tag{7.10}$$

where

$$\gamma = \sqrt{\left(1+\frac{R}{a}\right)} = \frac{\sqrt{\left(1+\frac{a}{R}\right)}}{\sqrt{\frac{a}{R}}} \tag{7.11}$$

When the crack length is much smaller than the hole radius, the crack should behave almost like a crack of length $a$ intersecting a free surface, as discussed in section 7.3. However, due to the stress concentration around the circular hole, the circumferential stress $\sigma_{\theta\theta}$ around the circular hole should be $2\sigma$ under biaxial state of stress, as shown in Figure 7.9. Considering additional 12.2% increase of the stress intensity factor due to the presence of the free surface, we get

$$K = 1.122 \times 2\sigma\sqrt{\pi a} = 2.244\sigma\sqrt{\pi a} = \gamma\sigma\sqrt{\pi a} \tag{7.12}$$

Equation (7.12) can be improved further by considering the radial variation of the circumferential stress $\sigma_{\theta\theta}$. From the theory of elasticity (see chapter 1,

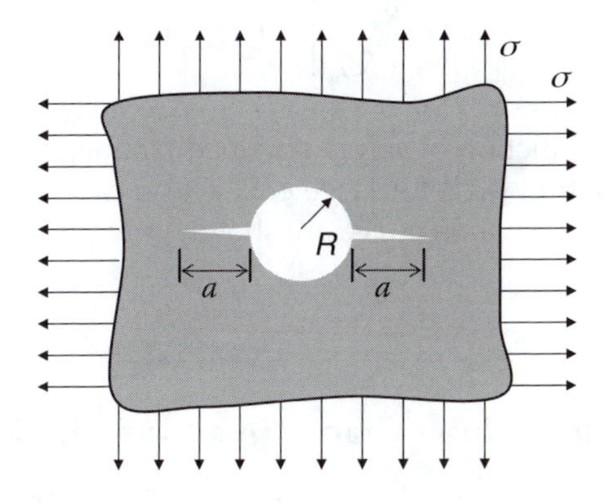


**FIGURE 7.9**
Circular hole with two radial cracks under biaxial state of stress.

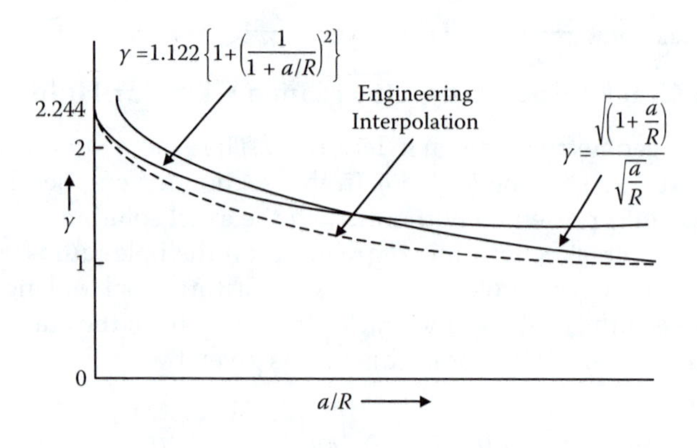


**FIGURE 7.10**
Variation of the normalized stress intensity factor with $a/R$ ratio for the problem geometry shown in Figure 7.9.

example 1.13), one can compute the circumferential stress $\sigma_{\theta\theta}$ at the crack tip position in absence of the crack:

$$\sigma_{\theta\theta} = \sigma\left\{1+\left(\frac{R}{a+R}\right)^2\right\} \tag{7.13}$$

Therefore, instead of $2\sigma$, if one uses equation (7.13) a more accurate estimate of $K$ is obtained:

$$K = 1.122\sigma\left\{1+\left(\frac{R}{a+R}\right)^2\right\}\sqrt{\pi a} = \gamma\sigma\sqrt{\pi a} \tag{7.14}$$

where

$$\gamma = 1.122\left\{1+\left(\frac{R}{a+R}\right)^2\right\} = 1.122\left\{1+\left(\frac{1}{1+a/R}\right)^2\right\} \tag{7.15}$$

Variation of $\gamma$ as a function of $a/R$ is obtained from equation (7.15) for small $a/R$ and from equation (7.11) for large $a/R$. These two variations along with the engineering interpolation curve are shown in Figure 7.10.

## 7.8 Two Collinear Finite Cracks in an Infinite Plate

The problem of two collinear finite cracks in an infinite plate as shown in Figure 7.11 was solved by Willmore (1949). Cracks of length $2a$ are separated by a center-to-center distance of $2b$. From the symmetry of the problem, one

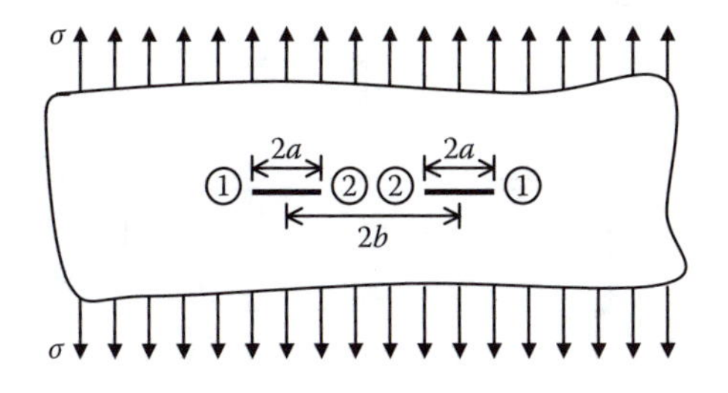


**FIGURE 7.11**
Two collinear finite cracks in an infinite plate.

can conclude that the SIF of crack ends marked as 1 should have one value and crack ends marked as 2 should have another value.

If the problem geometries of Figures 7.11 and 7.2 are compared, then one can see that Figure 7.11 is obtained simply by removing all cracks but only leaving the last two. Therefore, the SIF given in equation (7.1) and shown in Figure 7.3 can approximately show the variation of SIF for Figure 7.11. However, for the exact variation of SIF, the true solution will have to be looked at. The SIF variation for this problem geometry is shown in Figure 7.12.

From Figure 7.12 it is clear that $\gamma_2$ is greater than $\gamma_1$; therefore, as the applied load increases, the crack end 2 propagates, first bringing the two cracks closer; then they eventually join to form one Griffith crack whose $\gamma$ value will be 1.414 because when the two cracks join, the new crack length becomes $4a$ instead of $2a$. Note that in absence of the true solution, if one uses the solution of the problem geometry shown in Figure 7.2, then it should be a more conservative estimate and therefore acceptable.

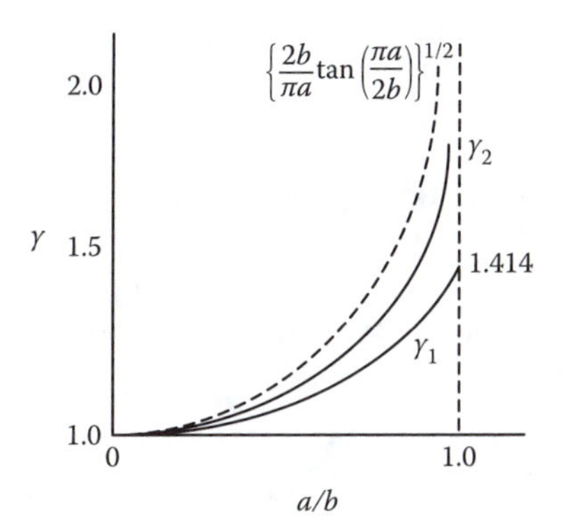


**FIGURE 7.12**
Variations of the normalized stress intensity factor with $a/b$ ratio for the problem geometry shown in Figure 7.11 (continuous lines) and Figure 7.2 (dashed line). Note that $\gamma_1$ and $\gamma_2$ are SIF for crack ends 1 and 2, respectively.

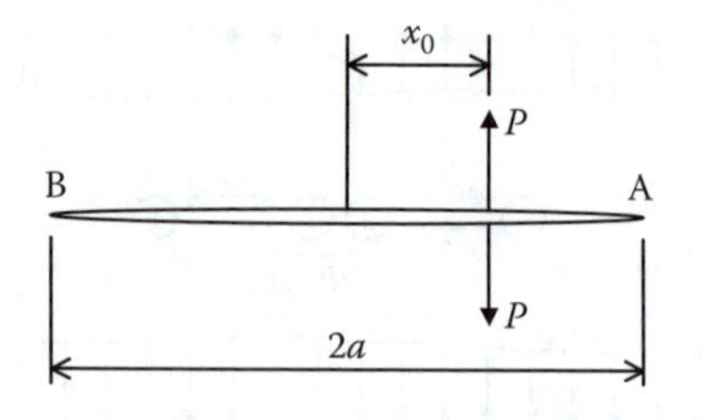


**FIGURE 7.13**
Griffith crack subjected to two opening loads on its surface.

## 7.9 Cracks with Two Opposing Concentrated Forces on the Surface

For a Griffith crack subjected to two opposing concentrated loads $P$ at a distance $x_0$ from the center of the crack, as shown in Figure 7.13, the stress intensity factors at ends A and B are given by

$$K_A = \frac{P}{\sqrt{\pi a}}\left(\frac{a+x_0}{a-x_0}\right)^{1/2}$$
$$K_B = \frac{P}{\sqrt{\pi a}}\left(\frac{a-x_0}{a+x_0}\right)^{1/2} \tag{7.16}$$

Equation (7.16) has been derived in section 9.5. Clearly, $K_A > K_B$; therefore, end A will propagate first. It can be justified intuitively also; since loads $P$ are applied closer to end A, it is easier for the loads to open end A. Note that for $x_0 = 0,\ K_A = K_B = \frac{P}{\sqrt{\pi a}}$.

## 7.10 Pressurized Crack

From the fundamental solution given in section 7.9, the pressurized crack problem can be solved. If a Griffith crack on its top and bottom surfaces is subjected to a pressure field $p(x)$ trying to open the crack, as shown in Figure 7.14, then the SIFs at crack ends A and B are obtained by integrating equation (7.16) as shown:

$$K_A = \frac{1}{\sqrt{\pi a}}\int_{-a}^{a} p(x)\left(\frac{a+x}{a-x}\right)^{1/2} dx$$
$$K_B = \frac{1}{\sqrt{\pi a}}\int_{-a}^{a} p(x)\left(\frac{a-x}{a+x}\right)^{1/2} dx \tag{7.17}$$

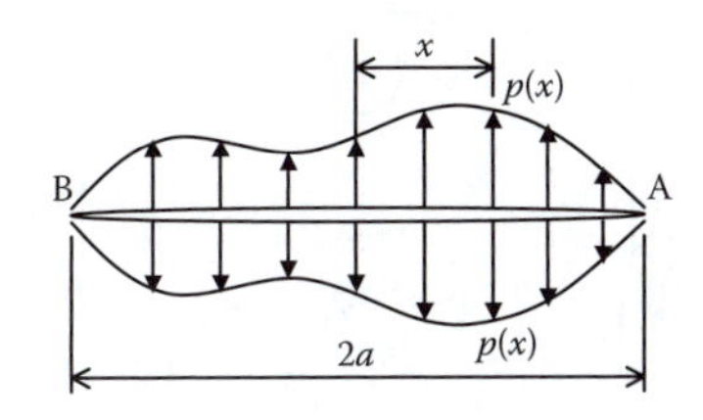


**FIGURE 7.14**
Pressurized Griffith crack.

Note that for uniform pressure $p(x) = p_0$,

$$K_A = K_B = \frac{p_0}{\sqrt{\pi a}} \int_{-a}^{a} \left( \frac{a+x}{a-x} \right)^{1/2} dx = \frac{p_0}{\sqrt{\pi a}} \pi a = p_0 \sqrt{\pi a} \tag{7.18}$$

## 7.11 Crack in a Wide Strip with a Concentrated Force at Its Midpoint and a Far Field Stress Balancing the Concentrated Force

Problem geometry is shown in Figure 7.15. From the vertical force equilibrium the relation between the concentrated force $P$ and the applied stress $\sigma$ are obtained, $P = \sigma W t$, where $t$ is the thickness of the strip. Assuming $t = 1$ or $\sigma$ equal to the force per unit length, one obtains

$$\sigma = \frac{P}{W} \tag{7.19}$$

Note that the problem geometry of Figure 7.15 can be obtained from the linear combination of the three problems shown in Figure 7.16. If problems

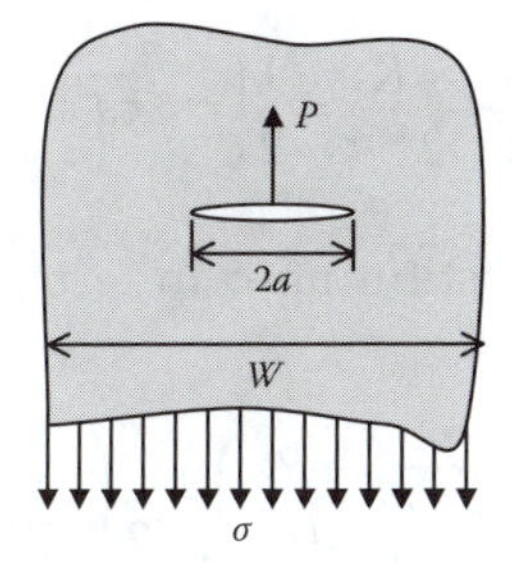


**FIGURE 7.15**
Wide cracked strip subjected to a downward stress over a width of $W$ and an upward concentrated force $P$ at the center of the crack.

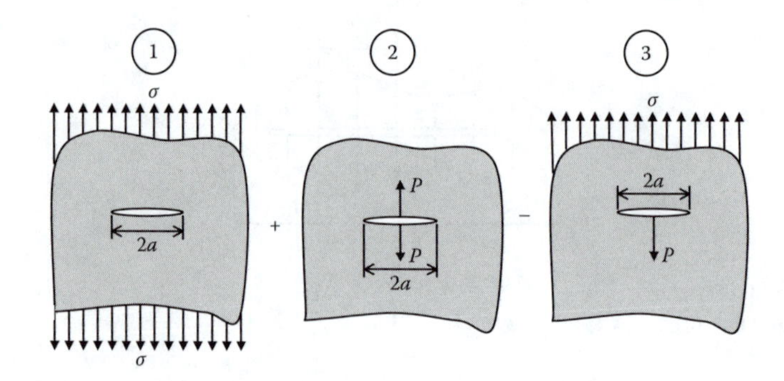


**FIGURE 7.16**
Subtraction of problem 3 from the superposition of problems 1 and 2 gives the problem geometry of Figure 7.15.

1 and 2 are superimposed and problem 3 is subtracted, then the problem geometry of Figure 7.15 is obtained. Therefore, if the SIF of problems 1, 2, and 3 of Figure 7.16 are denoted as $K_1$, $K_2$, and $K_3$, respectively, then the SIF $K$ of Figure 7.15 can be written as

$$K = K_1 + K_2 - K_3 \tag{7.20}$$

The problem geometry of Figure 7.15 and the third problem of Figure 7.16 are identical; one problem is simply rotated by 180° with respect to the other. Therefore, $K$ and $K_3$ of equation (7.20) should have the same value. Substituting $K_3 = K$ in equation (7.20), one gets

$$K = \frac{K_1 + K_2}{2} \tag{7.21}$$

Note that for $W >> a$, the SIF of problem 1 of Figure 7.16 is simply equal to that of a Griffith crack and the SIF of problem 2 can be obtained from equation (7.16) by substituting $x_0 = 0$. Therefore, for $W >> a$, equation (7.21) can be written as

$$K = \frac{K_1 + K_2}{2} = \frac{1}{2}\left(\sigma\sqrt{\pi a} + \frac{P}{\sqrt{\pi a}}\right) \tag{7.22}$$

From equation (7.19), substituting $\sigma$ in terms of $P$ into equation (7.22), one obtains

$$\begin{aligned} K &= \frac{1}{2}\left(\frac{P}{W}\sqrt{\pi a} + \frac{P}{\sqrt{\pi a}}\right) = \frac{P}{2}\left(\frac{\sqrt{\pi a}}{W} + \frac{1}{\sqrt{\pi a}}\right) \\ &= \frac{P}{2\sqrt{W}}\left(\sqrt{\frac{\pi a}{W}} + \sqrt{\frac{W}{\pi a}}\right) = \frac{P}{2\sqrt{W}}\left(x + \frac{1}{x}\right) \end{aligned} \tag{7.23}$$

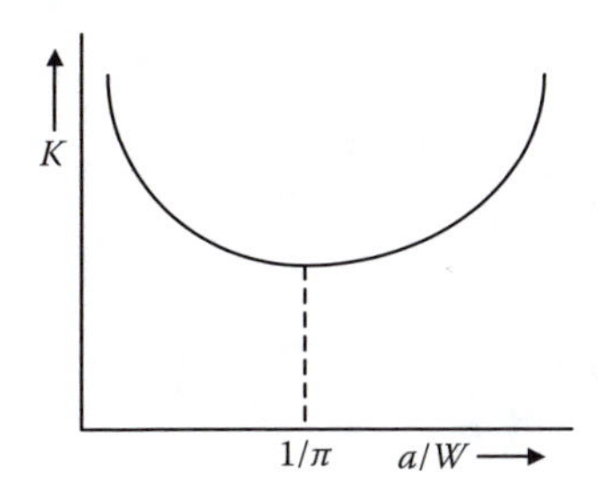


**FIGURE 7.17**
Schematic of the variation of the SIF of the problem geometry shown in Figure 7.15 as a function of $a/W$.

where

$$x = \sqrt{\frac{\pi a}{W}} \tag{7.24}$$

Therefore, for a stationary (maximum or minimum) value of $K$, its derivative with respect to $x$ must vanish:

$$\frac{dK}{dx} = \frac{P}{2\sqrt{W}}\left(1 - \frac{1}{x^2}\right) = 0 \tag{7.25}$$

From equations (7.24) and (7.25) one can see that $K$ will be a maximum or a minimum for

$$\begin{aligned} x &= \sqrt{\frac{\pi a}{W}} = 1 \\ \therefore \frac{a}{W} &= \frac{1}{\pi} \end{aligned} \tag{7.26}$$

For this value of $\frac{a}{W}$ the double derivative of $K$ with respect to $x$ is a positive value, as shown:

$$\frac{d^2K}{dx^2} = \frac{P}{2\sqrt{W}}\frac{2}{x^3} = \frac{P}{\sqrt{W}}\left(\frac{\pi a}{W}\right)^{\frac{3}{2}} \tag{7.27}$$

Therefore, at $\frac{a}{W} = \frac{1}{\pi}$, $K$ should reach its minimum value as shown in Figure 7.17.

## 7.12 Circular or Penny-Shaped Crack in a Full Space

The stress field surrounding a circular crack, also known as a penny-shaped crack, is available in the literature (Sneddon 1946). Figure 7.18 shows a circular crack in an infinite medium. The crack plane is $z = 0$. When this solid is

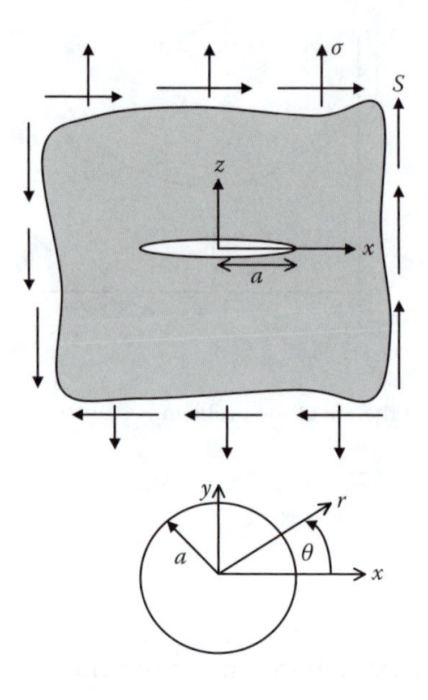


**FIGURE 7.18**
Circular crack (top figure—elevation view; bottom figure—plan view) in an infinite solid medium subjected to a normal stress $\sigma$ and a shear stress $S$.

subjected to both normal stress $\sigma$ and shear stress $S$, as shown in Figure 7.18, then the stress field on plane $z = 0$ is given by

$$\sigma_{zz}(r,\theta,0) = \frac{2\sigma}{\pi}\left\{\frac{1}{\sqrt{\left(\frac{r}{a}\right)^2 - 1}} - \sin^{-1}\left(\frac{a}{r}\right)\right\}H(r-a) + \sigma H(r-a)$$

$$\sigma_{\theta z}(r,\theta,0) = \frac{2S}{\pi}\left\{\frac{-1}{\sqrt{\left(\frac{r}{a}\right)^2 - 1}} + \sin^{-1}\left(\frac{a}{r}\right) + \frac{\nu}{(2-\nu)}\frac{1}{\left(\frac{r}{a}\right)^2\sqrt{\left(\frac{r}{a}\right)^2 - 1}} - \frac{\pi}{2}\right\} H(r-a)\sin\theta \tag{7.28}$$

$$\sigma_{rz}(r,\theta,0) = -\frac{2S}{\pi}\left\{\frac{-1}{\sqrt{\left(\frac{r}{a}\right)^2 - 1}} + \sin^{-1}\left(\frac{a}{r}\right) - \frac{\nu}{(2-\nu)}\frac{1}{\left(\frac{r}{a}\right)^2\sqrt{\left(\frac{r}{a}\right)^2 - 1}} - \frac{\pi}{2}\right\} H(r-a)\cos\theta$$

To obtain the SIF from the given stress field (equation 7.28), one can take a point $P$ very close to the crack tip. The coordinate of this point is given by

$r = a + \rho$, where $\rho << a$. Therefore, for point $P$ one can write

$$\frac{r}{a} = \frac{a+\rho}{a} = 1 + \frac{\rho}{a} \approx 1$$

$$\therefore \left(\frac{r}{a}\right)^2 = \left(1+\frac{\rho}{a}\right)^2 = 1 + 2\frac{\rho}{a} + \left(\frac{\rho}{a}\right)^2 \approx 1 + 2\frac{\rho}{a} \tag{7.29}$$

$$\therefore \left(\frac{r}{a}\right)^2 - 1 \approx 2\frac{\rho}{a}$$

Substituting equation (7.29) into equation (7.28), one obtains for $\frac{\rho}{a} << 1$:

$$\sigma_{zz}(r,\theta,0) = \frac{2\sigma}{\pi}\left\{\frac{1}{\sqrt{\frac{2\rho}{a}}} - \sin^{-1}(1)\right\} + \sigma = \frac{2\sigma}{\pi}\left\{\sqrt{\frac{a}{2\rho}} - \frac{\pi}{2}\right\} + \sigma = \frac{2\sigma}{\pi}\sqrt{\frac{a}{2\rho}} = \frac{2}{\pi}\frac{\sigma\sqrt{\pi a}}{\sqrt{2\pi\rho}} \tag{7.30}$$

$$\begin{aligned}
\sigma_{\theta z}(r,\theta,0) &= \frac{2S}{\pi}\left\{\frac{-1}{\sqrt{\frac{2\rho}{a}}} + \sin^{-1}(1) + \frac{\nu}{(2-\nu)}\frac{1}{\sqrt{\frac{2\rho}{a}}} - \frac{\pi}{2}\right\}\sin\theta \\
&= \frac{2S}{\pi}\left\{-\sqrt{\frac{a}{2\rho}} + \frac{\pi}{2} + \frac{\nu}{(2-\nu)}\sqrt{\frac{a}{2\rho}} - \frac{\pi}{2}\right\}\sin\theta \\
&= \frac{2S}{\pi}\left\{-1 + \frac{\nu}{(2-\nu)}\right\}\sqrt{\frac{a}{2\rho}}\sin\theta = \frac{2S}{\pi}\frac{(-2+2\nu)}{(2-\nu)}\sqrt{\frac{a}{2\rho}}\sin\theta \\
&= \frac{2}{\pi}\frac{S\sqrt{\pi a}}{\sqrt{2\pi\rho}}\left(\frac{2\nu-2}{2-\nu}\right)\sin\theta
\end{aligned} \tag{7.31}$$

$$\begin{aligned}
\sigma_{rz}(r,\theta,0) &= -\frac{2S}{\pi}\left\{\frac{-1}{\sqrt{\frac{2\rho}{a}}} + \sin^{-1}(1) - \frac{\nu}{(2-\nu)}\frac{1}{1\sqrt{\frac{2\rho}{a}}} - \frac{\pi}{2}\right\}\cos\theta \\
&= -\frac{2S}{\pi}\left\{-\sqrt{\frac{a}{2\rho}} + \frac{\pi}{2} - \frac{\nu}{(2-\nu)}\sqrt{\frac{a}{2\rho}} - \frac{\pi}{2}\right\}\cos\theta \\
&= \frac{2S}{\pi}\left\{1 + \frac{\nu}{(2-\nu)}\right\}\sqrt{\frac{a}{2\rho}}\cos\theta \\
&= \frac{2S}{\pi}\frac{2}{(2-\nu)}\sqrt{\frac{a}{2\rho}}\cos\theta = \frac{4}{\pi}\frac{S\sqrt{\pi a}}{\sqrt{2\pi\rho}}\frac{\cos\theta}{(2-\nu)}
\end{aligned} \tag{7.32}$$

Since the stress field very close to the crack tip on the crack plane is given by $\sigma = \frac{K}{\sqrt{2\pi\rho}}$, the stress intensity factors for modes I, II, and III for the circular crack can be obtained from equations (7.30), (7.31), and (7.32) as

$$
\begin{aligned}
K_I &= \frac{2}{\pi}\sigma\sqrt{\pi a} \\
K_{II} &= \frac{4}{\pi}\frac{S\sqrt{\pi a}}{(2-\nu)}\cos\theta \\
K_{III} &= \frac{4}{\pi}\left(\frac{\nu-1}{2-\nu}\right)S\sqrt{\pi a}\sin\theta
\end{aligned}
\tag{7.33}
$$

## 7.13 Elliptical Crack in a Full Space

The stress and displacement fields near an elliptical crack were given by Green and Sneddon (1950). This solution was used by Irwin (1962) to obtain the SIF expression for elliptical cracks. Consider an elliptical crack with semimajor and semiminor axes equal to $a$ and $b$, respectively, as shown in Figure 7.19. This crack is present in an infinite solid that is subjected to a normal stress $\sigma$ in the direction perpendicular to the crack plane. Then the crack is subjected to mode I or opening mode loading.

The crack surface displacement in the z direction and the normal stress field on $z = 0$ plane near the crack front are given by

$$
\begin{aligned}
u_z &= \frac{2(1-\nu^2)}{\Phi E} b\sigma \left\{1 - \left(\frac{x}{a}\right)^2 - \left(\frac{y}{b}\right)^2\right\}^{\frac{1}{2}} \\
\sigma_{zz} &= \frac{\sigma}{\Phi}\sqrt{\frac{b}{a}}\{a^2\sin^2\phi + b^2\cos^2\phi\}^{\frac{1}{4}}\frac{1}{\sqrt{2\rho}}
\end{aligned}
\tag{7.34}
$$

Note that in equation (7.34) the displacement field is valid on the crack surface only and the stress field is valid in front of the crack tip at a distance $\rho$ from the crack tip when the point of interest is very close to the crack tip,

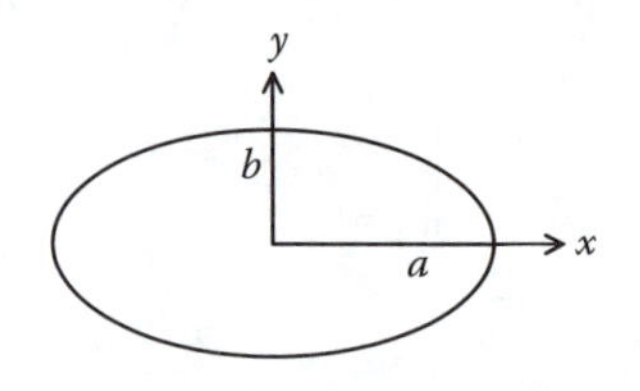


**FIGURE 7.19**
Elliptical crack in an infinite solid.

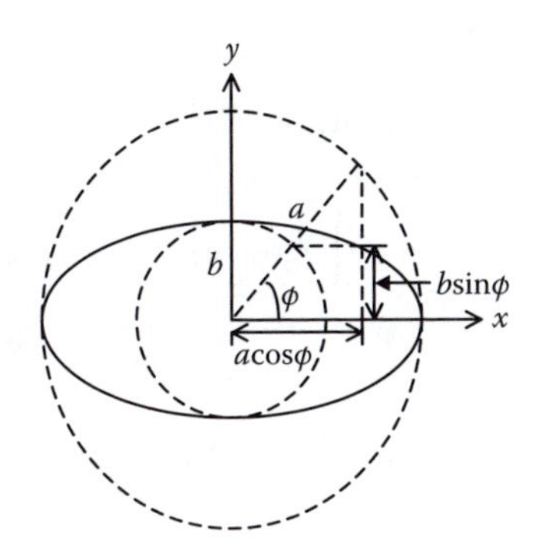


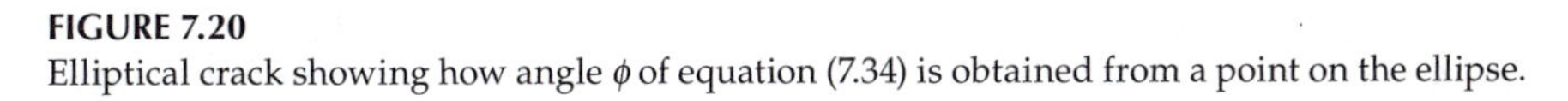

**FIGURE 7.20**
Elliptical crack showing how angle $\phi$ of equation (7.34) is obtained from a point on the ellipse.

$\frac{\rho}{b} \ll 1$. $\rho$ is the perpendicular distance of the point from the crack front. In equation (7.34) $\Phi$ in the denominator is the elliptic integral,

$$\Phi = \int_0^{\pi/2} \left( \sin^2\theta + \frac{b^2}{a^2}\cos^2\theta \right)^{1/2} d\theta \tag{7.35}$$

Angle $\phi$ in equation (7.34) is obtained from the $x$ and $y$ coordinates of the crack tip point as shown in Figure 7.20. One can clearly see from Figure 7.20 that

$$\begin{aligned} x &= a\cos\phi \\ y &= b\sin\phi \end{aligned} \tag{7.36}$$

From equation (7.34) the SIF for an elliptical crack is easily obtained:

$$K = \frac{\sigma}{\Phi}\sqrt{\frac{\pi b}{a}}\{a^2\sin^2\phi + b^2\cos^2\phi\}^{\frac{1}{4}} \tag{7.37}$$

### 7.13.1 Special Case 1—Circular Crack

For $a = b$, equation (7.35) gives

$$\Phi = \int_0^{\pi/2} (\sin^2\theta + \cos^2\theta)^{1/2} d\theta = \int_0^{\pi/2} d\theta = \frac{\pi}{2}$$

Therefore, from equation (7.37),

$$K = \frac{\sigma}{\Phi}\sqrt{\frac{\pi b}{a}}\{a^2\sin^2\phi + b^2\cos^2\phi\}^{\frac{1}{4}} = \frac{2\sigma}{\pi}\sqrt{\pi}\sqrt{a}\{\sin^2\phi + \cos^2\phi\}^{\frac{1}{4}} = \frac{2}{\pi}\sigma\sqrt{\pi a} \tag{7.38}$$

### 7.13.2 Special Case 2—Elliptical Crack with Very Large Major Axis

For $a \gg b$, $\frac{b}{a} \to 0$; then, from equation (7.35),

$$\Phi = \int_0^{\pi/2} \left( \sin^2\theta + \frac{b^2}{a^2}\cos^2\theta \right)^{1/2} d\theta = \int_0^{\pi/2} (\sin^2\theta)^{1/2} d\theta = \int_0^{\pi/2} \sin\theta \, d\theta = [-\cos\theta]_0^{\pi/2} = 1$$

$$\therefore K = \frac{\sigma}{\Phi}\sqrt{\frac{\pi b}{a}}\{a^2\sin^2\phi + b^2\cos^2\phi\}^{\frac{1}{4}} = \sigma\sqrt{\pi b}\left\{\sin^2\phi + \frac{b^2}{a^2}\cos^2\phi\right\}^{\frac{1}{4}}$$

$$= \sigma\sqrt{\pi b}\sqrt{\sin\phi} \tag{7.39}$$

From equation (7.20) one can clearly see that if $a$ is large, then $\phi$ is close to 90° for all crack tip points whose $x$ coordinates are much smaller than $a$. Therefore, equation (7.39) becomes $K = \sigma\sqrt{\pi b}$. Note that this is the stress intensity factor of a Griffith crack of length $2b$, as it should be.

### 7.13.3 SIF at the End of Major and Minor Axes of Elliptical Cracks

For a point at the end of the major axis $\phi = 0°$,

$$K = \frac{\sigma}{\Phi}\sqrt{\frac{\pi b}{a}}\{a^2\sin^2\phi + b^2\cos^2\phi\}^{\frac{1}{4}} = \frac{\sigma}{\Phi}\sqrt{\frac{\pi b}{a}}\{b^2\}^{\frac{1}{4}} = \frac{\sigma}{\Phi}\sqrt{\frac{\pi b}{a}}\sqrt{b} = \frac{\sigma}{\Phi}\sqrt{\pi b}\sqrt{\frac{b}{a}} \tag{7.40}$$

and for a point at the end of the minor axis $\phi = 90°$,

$$K = \frac{\sigma}{\Phi}\sqrt{\frac{\pi b}{a}}\{a^2\sin^2\phi + b^2\cos^2\phi\}^{\frac{1}{4}} = \frac{\sigma}{\Phi}\sqrt{\frac{\pi b}{a}}\{a^2\}^{\frac{1}{4}} = \frac{\sigma}{\Phi}\sqrt{\frac{\pi b}{a}}\sqrt{a} = \frac{\sigma}{\Phi}\sqrt{\pi b} \tag{7.41}$$

From equations (7.40) and (7.41) it is clear that the SIF at the end of the minor axis is greater than that at the end of the major axis. In fact, among all points on the elliptical crack the maximum SIF is observed at the end of the minor axis and the minimum SIF is observed at the end of the major axis. Therefore, the crack starts to propagate from the tip of the minor axis and the elliptical crack becomes closer to a circular crack.

## 7.14 Part-through Surface Crack

Consider a surface crack in a plate of thickness $B$ as shown in Figure 7.21. Crack depth is $b$ and its length (or surface width) is $2a$. We would like to obtain the SIF for this problem geometry when the plate is subjected to a

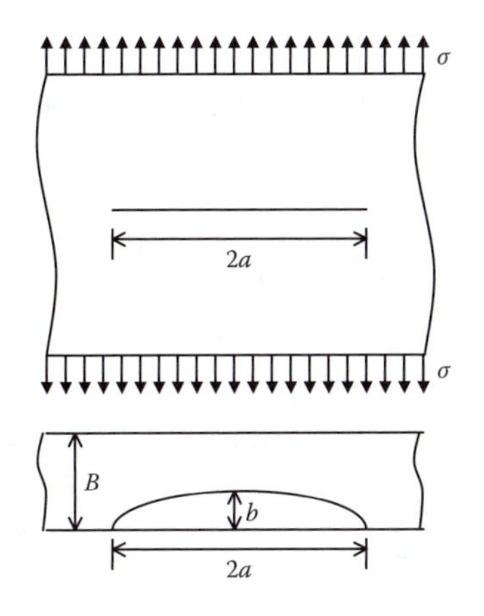


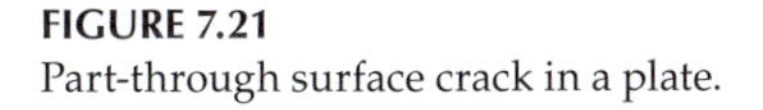
**FIGURE 7.21**
Part-through surface crack in a plate.

tensile stress $\sigma$ normal to the crack surface. The following steps are taken to come up with the final stress intensity factor expression.

### 7.14.1 First Approximation

As a first approximation, the crack can be assumed to be a half ellipse and SIF for the elliptical crack can be assumed to be its SIF. Then, from equation (7.41),

$$K = \frac{\sigma}{\Phi}\sqrt{\frac{\pi b}{a}}\left[a^2 \sin^2\phi + b^2 \cos^2\phi\right]^{\frac{1}{4}}\Bigg|_{\phi=\frac{\pi}{2}} = \frac{\sigma}{\Phi}\sqrt{\pi b}$$

### 7.14.2 Front Face Correction Factor

The preceding equation is valid for an elliptical crack in a full space in absence of any free surface near the crack. Since the crack intersects a stress-free surface (we will call it front face of the plate), the SIF should be increased by a factor of 1.12, as discussed in section 7.3. This factor is called the front face correction factor. With this correction the SIF becomes

$$K = 1.12\frac{\sigma}{\Phi}\sqrt{\pi b} \tag{7.42}$$

### 7.14.3 Plasticity Correction

To take into account the effect of plasticity, the crack depth $b$ is replaced by $b + r_p$ to obtain

$$K = 1.12\frac{\sigma}{\Phi}\sqrt{\pi(b + r_p)} = 1.12\frac{\sigma}{\Phi}\sqrt{\pi\left[b + \frac{1}{2\pi}\left(\frac{K}{\sigma_{YS}}\right)^2\right]}$$

$$\therefore K^2 = \left(1.12\frac{\sigma}{\Phi}\right)^2\left[\pi b + \frac{1}{2}\left(\frac{K}{\sigma_{YS}}\right)^2\right] \tag{7.43}$$

$$\therefore K^2\left[1 - \left(1.12\frac{\sigma}{\Phi}\right)^2\frac{1}{2\sigma_{YS}^2}\right] = \pi b\left(1.12\frac{\sigma}{\Phi}\right)^2$$

$$\therefore K = \frac{\left(1.12\frac{\sigma}{\Phi}\right)\sqrt{\pi b}}{\sqrt{\left[1 - \frac{1}{2}\left(1.12\frac{\sigma}{\Phi\sigma_{YS}}\right)^2\right]}}$$

### 7.14.4 Back Face Correction Factor

If the back face is far away from the crack tip, then no correction is necessary for the back face. When it is at a moderate distance (not too close to the crack tip), then also no correction is necessary because the front face correction factor 1.12 is a bit too high for the part-through surface crack since, unlike the problem of section 7.3, the crack length along the front face is restricted to $2a$. Note that in the problem geometry of section 7.3 the crack extends to infinity along the front face. However, for the crack geometry of Figure 7.21, the plate material beyond length $2a$ restricts the opening of the crack under external loads to some extent, reducing its stress intensity factor. Therefore, one may argue that the effect of the back face is indirectly taken into account by taking a relatively high front face correction factor. However, when the back face comes very close to the crack, its effect on the SIF is not compensated by only the front face correction factor; a separate back face correction factor ($M_k$) needs to be introduced, as shown below:

$$K = \frac{M_k\left(1.12\frac{\sigma}{\Phi}\right)\sqrt{\pi b}}{\sqrt{\left[1 - \frac{1}{2}\left(1.12\frac{\sigma}{\Phi\sigma_{YS}}\right)^2\right]}} \tag{7.44}$$

Variation of $M_k$ is shown in Figure 7.22.

For more discussions on part-through surface cracks, please refer to Irwin (1962), Thresher and Smith (1972), and ASTM STP 410.

## 7.15 Corner Cracks

Stress intensity factors for three different types of corner cracks are given next.

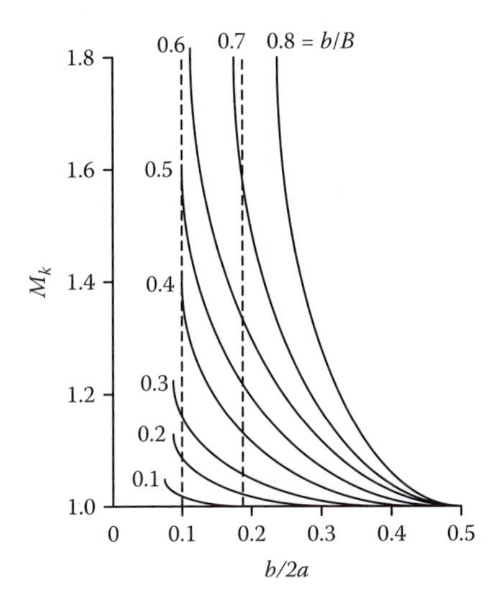


**FIGURE 7.22**
Back face correction factor for part-through surface cracks as a function of crack depth/crack length ($b/2a$) ratio for different values of crack depth/plate width ($b/B$) ratio. $B$, $b$, and $a$ are shown in Figure 7.21.

### 7.15.1 Corner Cracks with Almost Equal Dimensions

For a penny-shaped crack subjected to a far-field normal stress $\sigma$, the SIF $K = \frac{2}{\pi}\sigma\sqrt{\pi a}$. Since the corner crack shown in Figure 7.23 has two free surfaces intersecting the crack and SIF is increased by 12.2% for each intersecting surface; therefore, for two free surfaces the multiplying factor should be $1.12 \times 1.12 = 1.25$. The SIF for this corner crack is then

$$K = 1.25 \times \frac{2}{\pi}\sigma\sqrt{\pi a} = \frac{2.5}{\pi}\sigma\sqrt{\pi a} \tag{7.45}$$

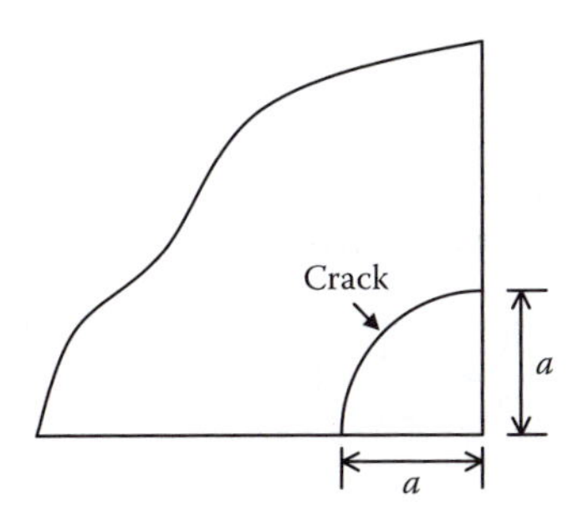


**FIGURE 7.23**
Corner crack—quarter of a circle.

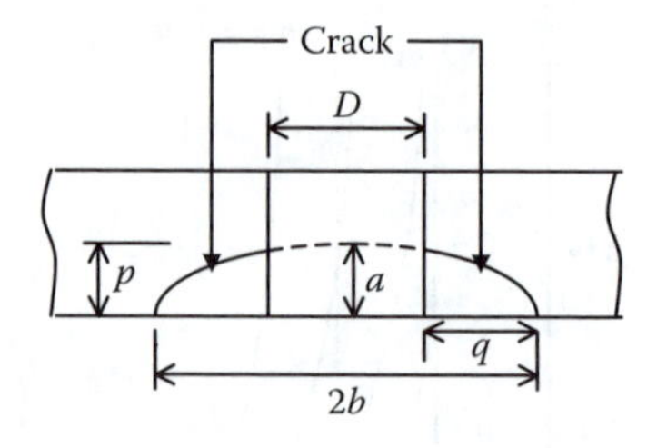


**FIGURE 7.24**
Cracks at two edges of a circular hole.

### 7.15.2 Corner Cracks at Two Edges of a Circular Hole

For the crack geometry shown in Figure 7.24, the stress intensity factor is given by

$$K = \gamma \sigma_c \sqrt{\pi a} \tag{7.46}$$

where

$$\gamma = \frac{1.2}{\Phi} \left\{ \frac{p^2(D+q)^2(D-q)^2(Dq)^{-1} + 4p^2(D+q)^2}{4Dq[4p^2 + (D-q)^2]} \right\}^{\frac{1}{4}} \tag{7.47}$$

In the preceding equations $\Phi$ is the elliptic integral defined in section 7.13 and $\sigma_c$ is the normal stress near the periphery of the circular hole. $\sigma_c$ should account for the stress concentration effect due to the presence of the circular hole in a thin plate or cylindrical hole in a thick plate. Note that the far field applied stress $\sigma$ is different from the stress $\sigma_c$ around the circular hole.

### 7.15.3 Corner Crack at One Edge of a Circular Hole

For a single edge crack at the corner of a circular hole, the equation given in section 7.15.2 is valid with the new definition of dimension $2b$ as shown in Figure 7.25.

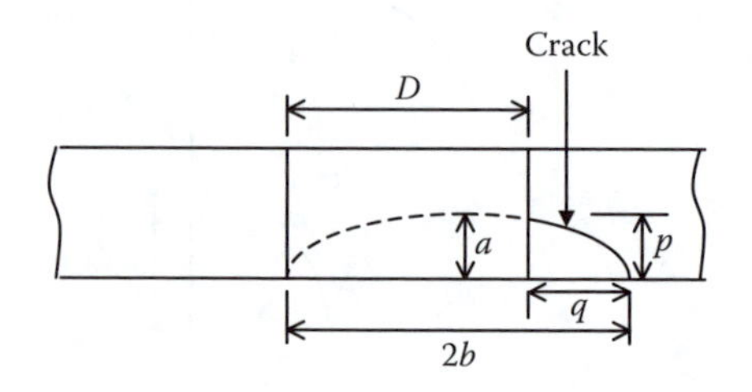


**FIGURE 7.25**
Corner crack at one edge of a circular hole.

## 7.16 Concluding Remarks

In this chapter the SIFs for a number of problem geometries of practical interest are presented. Stress intensity factor expressions for other problem geometries that are not covered here or in chapter 3 can be found in Tada, Paris, and Irwin (1973) and Broek (1986).

## References

American Society of Testing Methods. Special technical publication #410 (ASTM STP 410).

Bowie, O. Analysis of an infinite plate containing radial cracks originating from the boundry of an internal circular hole. *Journal of Mathematics and Physics,* 35, 60–71, 1956.

Broek, D. *Elementary engineering fracture mechanics,* Dordrecht, the Netherlands: Kluwer, 1986.

Feddersen, C. E. Discussion, ASTM STP 410, 77–79, 1967.

Green, A. E. and Sneddon, I. N. The stress distribution in the neighborhood of a flat elliptical crack in an elastic solid. *Proceedings of the Cambridge Philosophical Society,* 46, 159–164, 1950.

Irwin, G. R. The crack extension force for a part-through crack in a plate. *Transactions of the ASME Journal of Applied Mechanics,* 33, 651–654, 1962.

Isida, M. On the tension of a strip with a central elliptical hole. *Transactions of the Japan Society of Mechanical Engineering,* 21, 507–518, 1955.

Paris, P. C. and Sih, G. C., Stress analysis of cracks, ASTM STP 391, 30–81, 1965.

Sneddon, I. N. The distribution of stress in the neighborhood of a crack in an elastic solid. *Proceedings of the Royal Society of London,* A187, 229–260, 1946.

Tada, H., Paris, P. C., and Irwin, G. R. *The stress analysis of cracks handbook.* Hellertown, PA: Del Research Corp., 1973.

Thresher, R.W. and Smith, F. W. Stress intensity factors for a surface crack in a finite solid. *Journal of Applied Mechanics,* ASME, 39, 195–200, March 1972.

Willmore, T. J. The distribution of stress in the neighborhood of a crack. *Journal of Mechanics and Applied Mathematics,* 2, 53–67, 1949.

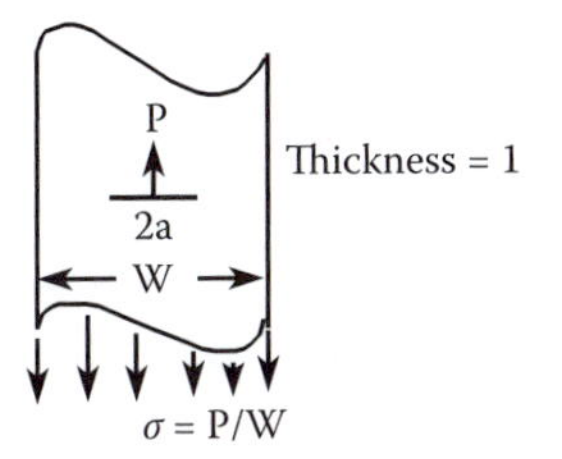


**FIGURE 7.26**

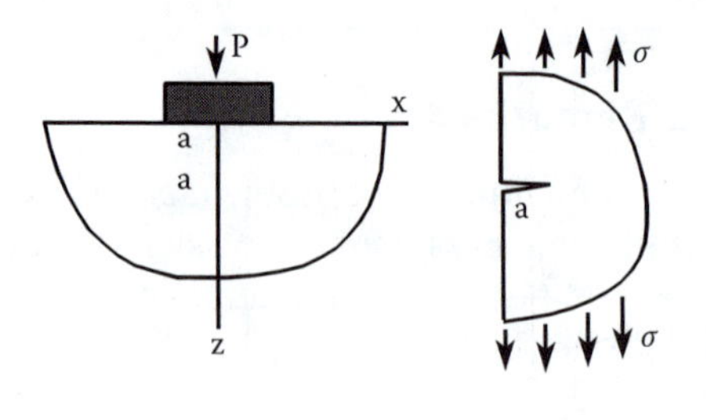


FIGURE 7.27

## Exercise Problems

Problem 7.1: If $W$ is constant and $a$ varies, then for what value of $a/W$ should the SIF of the problem geometry shown in Figure 7.26 be an extremum (maximum or minimum)? Identify if your answer of $a/W$ ratio corresponds to a maximum or a minimum value of $K$. For simplicity use the $K$ expression that is derived in the book for $W$ much greater than $a$ (then the Griffith crack approximation is justified).

Problem 7.2: Solution of a two-dimensional punch problem (left diagram of Figure 7.27) is given by $z = 0$, $\sigma_{zz} = 0$ for $x > a$ or $x < -a$, and $\sigma_{zz} = -\frac{P}{\pi\sqrt{a^2-x^2}}$ for $-a < x < a$; $P$ is the $z$ direction force per unit length in the $y$ direction. For a surface crack (right diagram of Figure 7.27), the stress intensity factor is given by $K = 1.122\sigma(\pi a)^{1/2}$. Knowing the solution of the preceding two problems, apply your engineering judgment to obtain the stress intensity factor for the problem geometry shown in Figure 7.28 as the ratio $a/b$ varies.

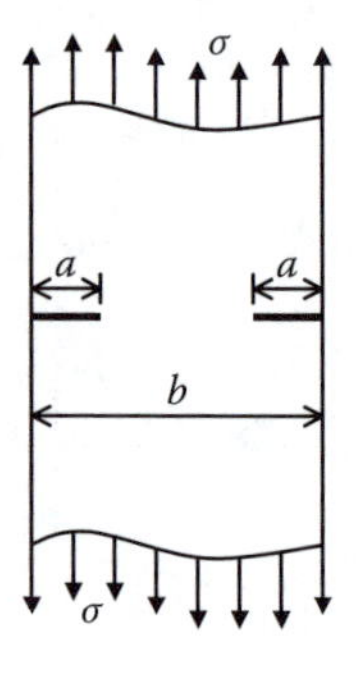

FIGURE 7.28

# 8

# *Numerical Analysis*

## 8.1 Introduction

Only a small number of problems having simple crack geometries (such as Griffith cracks, circular cracks, and elliptical cracks in an unbounded medium) can be solved analytically using the knowledge of the theory of elasticity. Analytical solutions of a few such simple problems are presented in chapter 9. One can use these fundamental solutions and one's engineering judgment to obtain approximate solutions for a variety of other problem geometries with cracks, as discussed in the previous chapters. However, a large number of problems that still remain unsolved can be solved only numerically. Different numerical techniques that can be used to obtain the stress intensity factors (SIFs) for problem geometries with various loading conditions are discussed in this chapter.

## 8.2 Boundary Collocation Technique

From equation (2.29) the following relations are obtained:

$$\begin{aligned}\sigma_{\theta\theta} = &\sum_{n=1,3,5} \frac{n}{2}\left(\frac{n}{2}+1\right) r^{\frac{n}{2}-1}\left[c_{1n}\left(\cos\left\{\left(\frac{n}{2}-1\right)\theta\right\} - \frac{n-2}{n+2}\cos\left\{\left(\frac{n}{2}+1\right)\theta\right\}\right)\right.\\ &\left.+c_{2n}\left(\sin\left\{\left(\frac{n}{2}-1\right)\theta\right\} - \sin\left\{\left(\frac{n}{2}+1\right)\theta\right\}\right)\right]\\ &+\sum_{n=2,4,6} \frac{n}{2}\left(\frac{n}{2}+1\right) r^{\frac{n}{2}-1}\left[c_{1n}\left(\cos\left\{\left(\frac{n}{2}-1\right)\theta\right\} - \cos\left\{\left(\frac{n}{2}+1\right)\theta\right\}\right)\right.\\ &\left.+c_{2n}\left(\sin\left\{\left(\frac{n}{2}-1\right)\theta\right\} - \frac{n-2}{n+2}\sin\left\{\left(\frac{n}{2}+1\right)\theta\right\}\right)\right]\end{aligned} \tag{8.1a}$$

$$\sigma_{r\theta} = \sum_{n=1,3,5} \frac{n}{2} r^{\frac{n}{2}-1} \left[ c_{1n} \left( \left(\frac{n}{2}-1\right) \sin\left\{\left(\frac{n}{2}-1\right)\theta\right\} - \frac{n-2}{n+2}\left(\frac{n}{2}+1\right) \sin\left\{\left(\frac{n}{2}+1\right)\theta\right\} \right) \right.$$
$$\left. + c_{2n} \left( -\left(\frac{n}{2}-1\right) \cos\left\{\left(\frac{n}{2}-1\right)\theta\right\} + \left(\frac{n}{2}+1\right) \cos\left\{\left(\frac{n}{2}+1\right)\theta\right\} \right) \right] \tag{8.1b}$$
$$+ \sum_{n=2,4,6} \frac{n}{2} r^{\frac{n}{2}-1} \left[ c_{1n} \left( \left(\frac{n}{2}-1\right) \sin\left\{\left(\frac{n}{2}-1\right)\theta\right\} - \left(\frac{n}{2}+1\right) \sin\left\{\left(\frac{n}{2}+1\right)\theta\right\} \right) \right.$$
$$\left. + c_{2n} \left( -\left(\frac{n}{2}-1\right) \cos\left\{\left(\frac{n}{2}-1\right)\theta\right\} + \frac{n-2}{n+2}\left(\frac{n}{2}+1\right) \cos\left\{\left(\frac{n}{2}+1\right)\theta\right\} \right) \right]$$

Similarly, from equations (2.35) and (2.27),

$$\sigma_{rr} = \sum_{n} r^{\left(\frac{n}{2}-1\right)} \left[ \left\{ \left(\frac{n}{2}+1\right) - \left(\frac{n}{2}-1\right)^2 \right\} \left\{ c_{1n} \cos\left(\frac{n}{2}-1\right)\theta + c_{2n} \sin\left(\frac{n}{2}-1\right)\theta \right\} \right.$$
$$\left. + \left\{ \left(\frac{n}{2}+1\right) - \left(\frac{n}{2}+1\right)^2 \right\} \left\{ c_{3n} \cos\left(\frac{n}{2}+1\right)\theta + c_{4n} \sin\left(\frac{n}{2}+1\right)\theta \right\} \right]$$
$$= \sum_{n=1,3,5} r^{\left(\frac{n}{2}-1\right)} \left[ \left\{ \left(\frac{n}{2}+1\right) - \left(\frac{n}{2}-1\right)^2 \right\} \left\{ c_{1n} \cos\left(\frac{n}{2}-1\right)\theta + c_{2n} \sin\left(\frac{n}{2}-1\right)\theta \right\} \right.$$
$$\left. - \left\{ \left(\frac{n}{2}+1\right) - \left(\frac{n}{2}+1\right)^2 \right\} \left\{ \frac{n-2}{n+2} c_{1n} \cos\left(\frac{n}{2}+1\right)\theta + c_{2n} \sin\left(\frac{n}{2}+1\right)\theta \right\} \right]$$
$$+ \sum_{n=2,4,6} r^{\left(\frac{n}{2}-1\right)} \left[ \left\{ \left(\frac{n}{2}+1\right) - \left(\frac{n}{2}-1\right)^2 \right\} \left\{ c_{1n} \cos\left(\frac{n}{2}-1\right)\theta + c_{2n} \sin\left(\frac{n}{2}-1\right)\theta \right\} \right.$$
$$\left. - \left\{ \left(\frac{n}{2}+1\right) - \left(\frac{n}{2}+1\right)^2 \right\} \left\{ c_{1n} \cos\left(\frac{n}{2}+1\right)\theta + \frac{n-2}{n+2} c_{2n} \sin\left(\frac{n}{2}+1\right)\theta \right\} \right] \tag{8.2}$$

If computing the stress field near the crack tip is of interest, then one needs to keep the dominant term (corresponding to $n = 1$) only. However, for stress field computation at a point away from the crack tip, several terms of the preceding series expressions (equations 8.1 and 8.2) should be kept.

If a total of $n$ terms are kept ($n = 1, 2, 3,\ldots n$) in the series expressions given in equations (8.1) and (8.2), then there are a total of $2n$ unknown constants in the series. These unknown constants are $c_{11}$, $c_{21}$, $c_{12}$, $c_{22}$, $c_{13}$, $c_{23}$, $c_{14}$, $c_{24}$,...

$c_{1n}$, $c_{2n}$. If these constants are somehow obtained, then the SIFs $K_I$ and $K_{II}$ can be evaluated from the relations given in equations (2.31) and (2.32):

$$K_I = c_{11}\sqrt{2\pi}$$
$$K_{II} = c_{21}\sqrt{2\pi} \tag{8.3}$$

Now the question is how to obtain these unknown constants. The answer to this question is that these can be obtained from the known boundary conditions of the problem geometry as illustrated in the following examples.

### 8.2.1 Circular Plate with a Radial Crack

The problem geometry is shown in Figure 8.1. The crack tip is at its center. Applied stresses $\sigma_{rr}$ and $\sigma_{r\theta}$ along its boundary are functions of $\theta$. These stress fields keep the cracked plate in equilibrium. Note that the applied boundary tractions or stresses $\sigma_{rr}$ and $\sigma_{r\theta}$ are known at all points on the boundary. We also know the general expressions of these two components of stress as shown in equations (8.1) and (8.2). If $n$ points on the boundary are considered and the two given stress components at every boundary point are equated to the stress expressions of $\sigma_{r\theta}$ and $\sigma_{rr}$ (given in equations 8.1 and 8.2), then a total of $2n$ equations are obtained to solve for $2n$ unknowns $c_{11}$, $c_{21}$, $c_{12}$, $c_{22}$, $c_{13}$, $c_{23}$, $c_{14}$, $c_{24}$,...$c_{1n}$, $c_{2n}$. Equation (8.3) is then used to obtain the stress intensity factors. A closer observation of the stress expressions of equations (8.1) and (8.3) reveals that the coefficient of $c_{22}$ is zero for both $\sigma_{r\theta}$ and $\sigma_{rr}$. Therefore, from $n$ boundary points, only $(2n - 1)$ equations need to be satisfied.

### 8.2.2 Rectangular Cracked Plate

For a rectangular cracked plate (shown in Figure 8.2), the applied boundary tractions give $\sigma_{xx}$ and $\sigma_{xy}$ values on the vertical boundaries and $\sigma_{yy}$ and $\sigma_{xy}$ values on the horizontal boundaries. Using stress transformation laws, these stress components can be expressed in terms of $\sigma_{r\theta}$, $\sigma_{\theta\theta}$, and $\sigma_{rr}$. Thus,

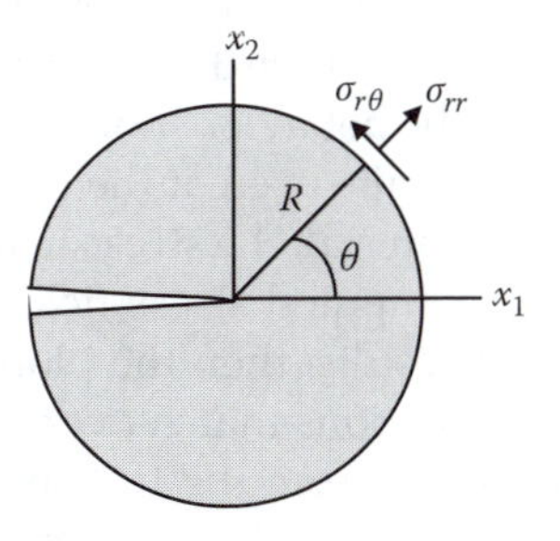


**FIGURE 8.1**
A cracked circular plate of radius $R$ is subjected to normal and shear stresses at the boundary.

$\sigma_{xx}$, $\sigma_{yy}$, and $\sigma_{xy}$ are expressed in terms of a combination of series expressions given in equations (8.1) and (8.2). Then, satisfying boundary conditions at $n$ points, $2n$ unknown constants are evaluated from a system of simultaneous equations, and $K_I$ and $K_{II}$ are obtained from equation (8.3).

If the plate is neither circular nor rectangular, but rather has a more complex geometry, then a similar approach can also be followed to solve the unknown coefficients. This is possible because normal and shear stresses at any point on the boundary can be expressed in terms of $\sigma_{r\theta}$, $\sigma_{\theta\theta}$, and $\sigma_{rr}$ using the stress transformation laws.

Note that during the boundary collocation technique no point on the crack surface is considered because the stress-free boundary conditions on the crack surface are automatically satisfied by the series expressions given in equations (8.1) and (8.2).

## 8.3 Conventional Finite Element Methods

A number of investigators (see the reference list) have followed the conventional finite element method (FEM) to obtain the SIF for cracked plate geometries. The SIF can be obtained:

by matching stresses or displacements at selected points
by matching local strain energy
from the strain energy release rate
by evaluating the J-integral

These different techniques are described next.

### 8.3.1 Stress and Displacement Matching

If the plate geometry shown in Figure 8.2 is discretized into a finite element mesh by putting finer mesh or smaller elements near the crack tip (where a high stress gradient is expected) and coarser mesh or larger elements away from the crack tip and analyzed, then the stress field computed near the crack tip is expected to show the behavior shown in Figure 8.3. Note that the FEM predicted stress value is finite even at the crack tip. However, it is well known that for linear elastic material, the stress field should be infinite at the crack tip. If the analytically computed stress field (equation 2.47) is plotted on top of the FEM predicted results, then the plots shown in Figure 8.4 are obtained. Note that the two solutions match over a region that is neither too close to the crack tip nor too far from the tip. Since the analytical solution (equation 2.47) is obtained considering the singular term only ($n$ = 1), it is good for the points very close to the crack tip. Therefore, this analytical solution is not reliable at large distances. In short, analytical solution is good near

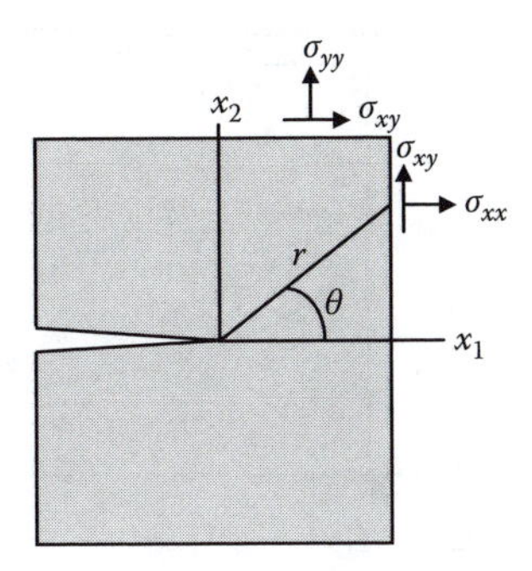


**FIGURE 8.2**
Rectangular cracked plate subjected to boundary stresses.

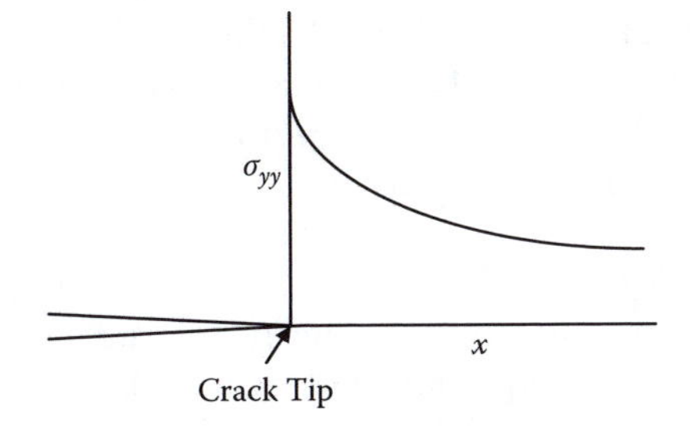

**FIGURE 8.3**
Finite element method computed stress field in front of a crack tip.

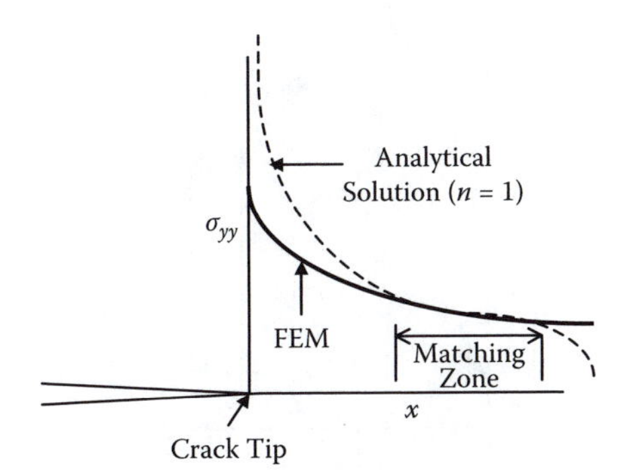


**FIGURE 8.4**
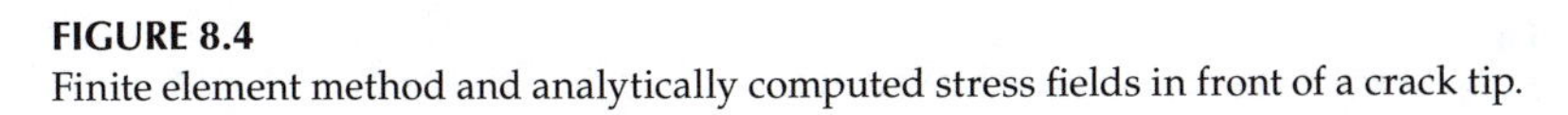
Finite element method and analytically computed stress fields in front of a crack tip.

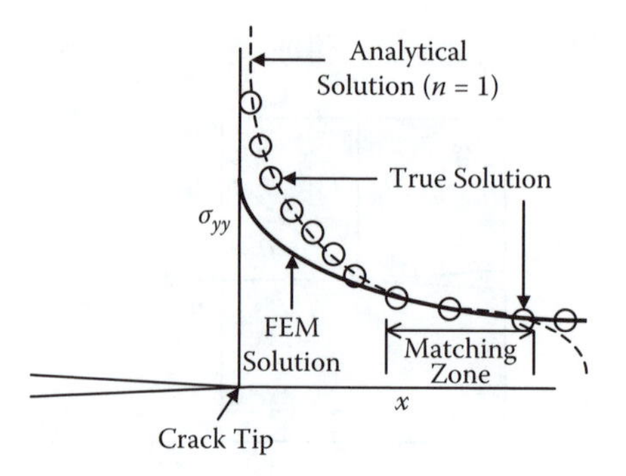


**FIGURE 8.5**
True stress field (circles) matches with the analytical solution (dashed line) near the crack tip and with the FEM solution (continuous line) away from the crack tip.

the crack tip and FEM solution is good away from the crack tip. Fortunately, there is a region, marked in Figure 8.4 as the matching zone, where both analytical and FEM solutions overlap. The SIF can be obtained equating the analytical expression to the FEM solution in this region.

True solutions are shown by circles in Figure 8.5. Note that they match with the analytical solution near the crack tip and with the FEM solution away from the crack tip. Over a small region close (but not too close) to the crack tip, marked in the Figure as the matching zone, all three solutions match. True solution can be obtained by considering a large number of terms in the series expressions given in equations (8.1) and (8.2).

In one of the early works of FEM applied to fracture mechanics problems Watwood (1969) considered a cracked plate subjected to uniaxial tension as shown in Figure 8.6 ($a = 10$, $b = 40$, and $L = 85$). Since the problem is symmetric

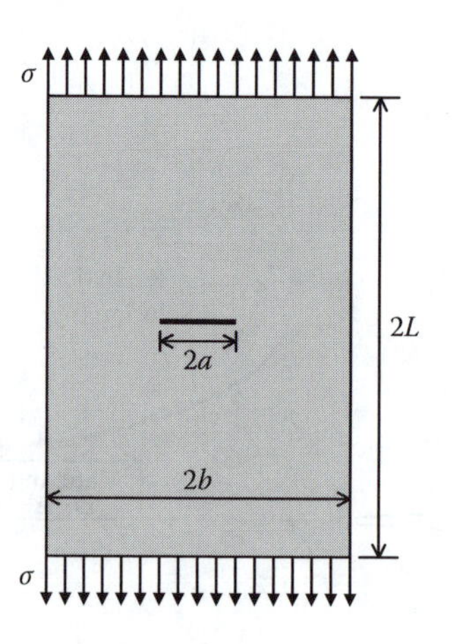


**FIGURE 8.6**
Cracked plate under uniaxial tension.

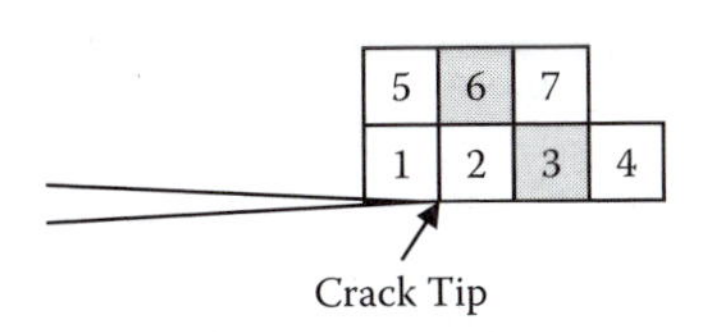


**FIGURE 8.7**
Seven elements near the crack tip from which SIF values given in Table 8.1 are obtained. Best results are obtained from elements 3 and 6.

about both its horizontal and vertical central axes of symmetry, it is possible to consider only a quarter of the plate and discretize the quarter plate into a large number of finite elements. Watwood considered 470 elements and 478 nodes in the quarter plate.

The SIF is obtained by equating the stress computed in seven elements near the crack tip to the analytical stress expression as given:

$$\sigma_y\Big|_{FEM} = \frac{K_I}{\sqrt{2\pi r}}\left[\cos\left(\frac{\theta}{2}\right) + \frac{1}{2}\sin\theta\sin\left(\frac{3\theta}{2}\right)\right] \tag{8.4}$$

Seven elements near the crack tip from which the SIF values are computed using equation (8.4) are shown in Figure 8.7. Stress intensity factor values are presented in Table 8.1. True value of $K_I$ for this problem geometry is 5.83. Clearly, best results are obtained for elements 3 and 6. These two elements are close to the crack tip, but not the closest elements. Figures 8.4 and 8.5 explain why elements 3 and 6 give best results. Average SIF value obtained from elements 2, 3, 6, and 7 is 5.87, which is close to the true value of 5.83.

Instead of stress matching the FEM, computed displacements can also be matched with the analytical expressions (see equation 2.48) and SIF values can be obtained. For best results from the displacement matching consider the points on the crack surface.

For this specific mesh, elements 3 and 6 had the ideal distance to give most accurate SIF prediction. However, if the mesh grid is changed—say, made more refined—then instead of the second layer of elements (from the crack tip), maybe a third or fourth layer of elements will give most accurate results. Chan, Tubo, and Wilson (1970) (also see Wilson, 1973) suggested a method to avoid this problem of mesh dependence, element size, and position dependence on SIF calculation. They took extremely refined mesh near the crack

**TABLE 8.1**

$K_I$ Values Obtained from Seven Elements Shown in Figure 8.7

| Element No. | 1 | 2 | 3 | 4 | 5 | 6 | 7 |
|---|---|---|---|---|---|---|---|
| $K_I$ | 12.5 | 5.55 | 5.88 | 6.54 | 7.38 | 5.88 | 6.19 |

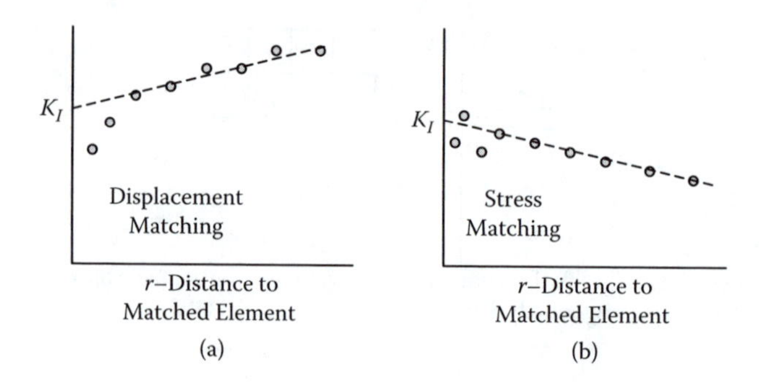


**FIGURE 8.8**
Estimating $K_I$ by extrapolating predicted $K_I$ values from different elements at various distances from the crack tip: (a) displacement matching and (b) stress matching.

tip (100 to 500 times smaller element size in comparison to Watwood's element size). Then they estimated $K_I$ from different elements near the crack tip and plotted the computed values as a function of the radial distance of the element from the crack tip. Figures 8.8a and 8.8b show the numerical values (shown by small circles) obtained by matching FEM computed displacement and stress, respectively, to the analytical expressions of displacement and stress given in equations (2.47) and (2.48). Note that from both displacement and stress matching the $K_I$ value is obtained as a function of the element distance ($r$) from the crack tip. The dependence on $r$ is approximately linear when the points are not too close to the crack tip. If one extrapolates the straight line, then the intersection of this line with the vertical axis (or $K_I$ axis) gives an accurate estimate (error less than 5%) of the SIF. However, this method is expensive due to the refined mesh requirement and the accuracy is not that good considering the expense.

### 8.3.2 Local Strain Energy Matching

In an effort to average the numerical results over a region and match the average property over that region with an expression involving $K_I$ this method has been developed. Instead of trying to match stress or displacement at specific points, local strain energy near the crack tip computed from the finite element (FE) analysis is matched with the analytical expression for the strain energy over the same region.

Note that, for two-dimensional problems, the strain energy density expressions for plane stress and plane strain problems are given by

$$U_0 = \frac{1}{2E}\left[\sigma_{xx}^2 + \sigma_{yy}^2 + 2(1+\nu)\sigma_{xy}^2 - 2\nu\sigma_{xx}\sigma_{yy}\right] \quad \text{for plane stress}$$

$$U_0 = \frac{1+\nu}{2E}\left[(1-\nu)\left(\sigma_{xx}^2 + \sigma_{yy}^2\right) + 2\left(\sigma_{xy}^2 - \nu\sigma_{xx}\sigma_{yy}\right)\right] \quad \text{for plane strain} \tag{8.5}$$

Substituting the analytical expressions (equation 2.47) of the three stress components near a crack tip in equation (8.5) and integrating this over a circular region of radius $R$ around the crack tip (shown in Figure 8.9a), the local energy in this circular region surrounding the crack tip can be obtained as

$$U = \left(\frac{5-3\nu}{8}\right)\frac{K_I^2 R}{E} \quad \text{for plane stress}$$
$$= \frac{(5-8\nu)(1+\nu)}{8}\frac{K_I^2 R}{E} \quad \text{for plane strain} \tag{8.6}$$

Since equation (8.6) is obtained with only one term ($n = 1$) analytical expression for stresses, equation (8.6) should give good results only if the radius of the circular region $R$ is very small or $R \ll a$ (the crack length).

Since FEM does not give good results very close to the crack tip (because of the singular stress behavior), the local strain energy is computed over an annular region (Figure 8.9b), instead of a circular region. The FEM predicted strain energy over the annular region is matched with the analytical expressions given in equation (8.7) to obtain $K_I$:

$$U = \left(\frac{5-3\nu}{8}\right)\frac{K_I^2 (R_1 - R_2)}{E} \quad \text{for plane stress}$$
$$= \frac{(5-8\nu)(1+\nu)}{8}\frac{K_I^2 (R_1 - R_2)}{E} \quad \text{for plane strain} \tag{8.7}$$

This method has many of the disadvantages of the method of stress or displacement matching at specific points, except that it takes into account all values of $\theta$ as well as both stresses and displacements.

### 8.3.3 Strain Energy Release Rate

Relation between the strain energy release rate and SIF can be used to evaluate SIF. If the strain energy of a plate for two crack lengths $a_0$ and $a_1$ is

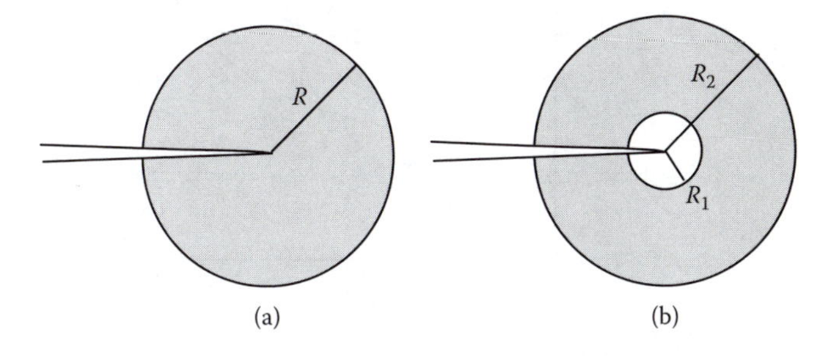


**FIGURE 8.9**
(a) Circular region and (b) annular region surrounding the crack tip where local strain energy is computed by FEM.

computed from FE analysis as $U(a_0)$ and $U(a_1)$, respectively, then the strain energy release rate predicted by FEM is

$$\frac{U(a_1)-U(a_0)}{t(a_1-a_0)} = \frac{dU}{dA} = \alpha\frac{K_I^2}{E} \tag{8.8}$$

In this equation $t$ is the plate thickness, $\alpha = 1$ for plane stress, and $\alpha = 1 - \nu^2$ for plane strain. Then, from equation (8.8),

$$K_I^2 = \frac{E}{\alpha}\left\{\frac{U(a_1)-U(a_0)}{t(a_1-a_0)}\right\} = E\left\{\frac{U(a_1)-U(a_0)}{t(a_1-a_0)}\right\} \quad \text{for thin plate} \tag{8.9}$$

While computing $U(a_0)$ and $U(a_1)$ it is not necessary to change the finite element mesh for the two runs; simply advance the crack length by one or a few elements for the second run. Note that FEM computed $U(a_0)$ and $U(a_1)$ may be different from true $U(a_0)$ and $U(a_1)$ values as shown in Figure 8.10. However, since in equation (8.9) only the difference value between the two energy values is needed, the FEM predicted value may not be much different from the true value of $U(a_1) - U(a_0)$.

For plate thickness = 1, equation (8.9) can be written as

$$\begin{aligned} K_I^2 &= E\left\{\frac{U(a_1)-U(a_0)}{(a_1-a_0)}\right\} = E\frac{\Delta U}{\Delta a} \\ \therefore \frac{K_I^2}{\sigma^2} &= \frac{E\cdot\Delta U}{\sigma^2\Delta a} \\ \therefore \frac{K_I}{\sigma} &= \sqrt{\frac{E\cdot\Delta U}{\sigma^2\Delta a}} \end{aligned} \tag{8.10}$$

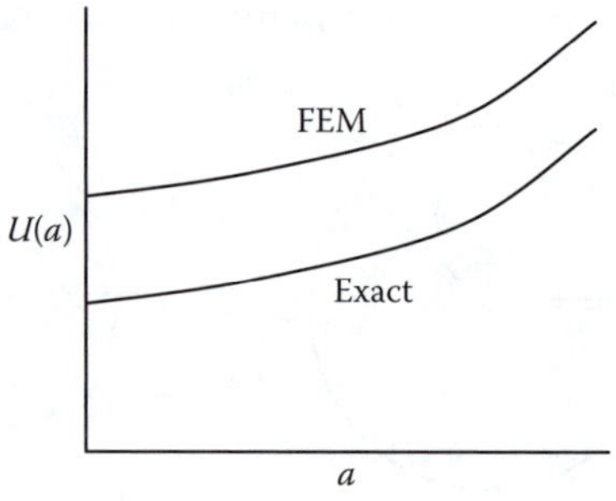

**FIGURE 8.10**
Variation of strain energy in the plate as a function of crack length. Two curves correspond to the exact variation and the FEM prediction.

**TABLE 8.2**

Computation of $K_I$ from Strain Energy Release Rate from Different Values of the Crack Length[a]

| $a$ | $\frac{EU}{\sigma^2}$ | $\frac{E \cdot \Delta U}{\sigma^2}$ | $\frac{K_I}{\sigma} = \frac{E \cdot \Delta U}{\sigma^2}$ | Exact $\frac{K_I}{\sigma}$ | $a$ | % Difference |
|---|---|---|---|---|---|---|
| 7 | 3160.04 | | | | | |
| | | 21.50 | 4.86 | 4.96 | 7.5 | –2.0% |
| 8 | 3181.54 | | | | | |
| | | 24.76 | 5.22 | 5.32 | 8.5 | –1.9% |
| 9 | 3206.30 | | | | | |
| | | 28.14 | 5.56 | 5.66 | 9.5 | –1.8% |
| 10 | 3234.44 | | | | | |
| | | 31.66 | 5.90 | 6.00 | 10.5 | –1.7% |
| 11 | 3266.10 | | | | | |
| | | 35.38 | 6.23 | 6.34 | 11.5 | –1.8% |
| 12 | 3301.10 | | | | | |
| | | 39.26 | 6.57 | 6.67 | 12.5 | –1.5% |
| 13 | 3340.74 | | | | | |
| | | 43.36 | 6.90 | 7.01 | 13.5 | –1.6% |
| 14 | 3384.10 | | | | | |

[a] Problem geometry is shown in Figure 8.6.

In equation (8.10) $\sigma$ is applied stress. If one applies this technique to Watwood's problem (shown in Figure 8.6) for different values of $a = 7, 8, 9, \ldots 14$, then $\Delta a = 1$. Equation (8.10) is thus simplified to

$$\frac{K_I}{\sigma} = \sqrt{\frac{E \cdot \Delta U}{\sigma^2}} \tag{8.11}$$

Table 8.2 shows the computed results.

Note that instead of the crack increment length $\Delta a = 1$, as shown in Table 8.2, if we take $\Delta a = 4$ ($a_0 = 8$ and $a_1 = 12$), then, from equation (8.10),

$$\frac{K_I}{\sigma} = \sqrt{\frac{E \cdot \Delta U}{\Delta a \cdot \sigma^2}} = \sqrt{\frac{3301.48 - 3181.54}{4}} = \sqrt{\frac{119.94}{4}} = 5.48$$

while the true value is 5.83. Thus, it gives an error of –6%.

The advantage of this method is that with coarser mesh accurate results can be obtained. Its disadvantage is that the FE analysis will have to be carried out twice for two different crack lengths even when we are interested in computing the stress intensity factor for a single value of the crack length.

Watwood suggested that if $K_I$ increases with the crack length, then this method gives a lower bound of SIF, while if $K_I$ decreases with the crack length, then it gives an upper bound.

### 8.3.4 J-Integral Method

Recall from chapter 5 (equation 5.8) that the strain energy release rate is equal to the J-integral value

$$G = J = \int_S \left( U dx_2 - T_i \frac{\partial u_i}{\partial x_1} dS \right) \tag{8.12}$$

Therefore, instead of evaluating the strain energy release rate by running the finite element program twice, as done in section 8.3.3, one can evaluate it from one run of the finite element mesh by choosing an appropriate path $S$ for the J-integral enclosing the crack tip. This path should be selected such that the line integral can be evaluated relatively easily. For example, for the problem geometry shown in Figure 8.6, the J-integral path should be taken along ABCDEFG, as shown in Figure 8.11(a). Since the problem is symmetric about the horizontal central axis, the J-integral value over path ABCDEFG is simply twice the J-integral value over path DEFG.

On path $DE$ the traction (**T**) is zero; therefore, the J-integral value on this path is given by

$$J_{DE} = \int_D^E U dx_2 \tag{8.13}$$

Since from the finite element analysis $U$ is known for all elements along line $DE$, equation (8.13) can be evaluated by adding the strain energy density values for all elements along line $DE$.

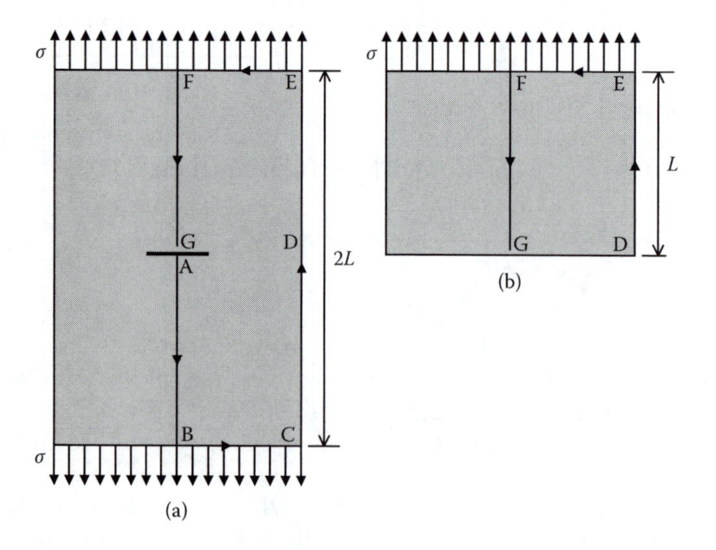


**FIGURE 8.11**
(a) J-integral path ABCDEFG for the complete plate and (b) J-integral path DEFG for the half plate.

On path *EF* $dx_2 = 0$ and $T_1 = 0$, $T_2 = \sigma$; therefore,

$$J_{EF} = \int_{EF}\left(-T_i\frac{\partial u_i}{\partial x_1}dS\right) = -\int_{EF} T_2\frac{\partial u_2}{\partial x_1}(-dx_1) = \int_E^F T_2\frac{\partial u_2}{\partial x_1}dx_1$$
$$= \int_E^F \sigma\frac{\partial u_2}{\partial x_1}dx_1 = \sigma\int_E^F du_2 = \sigma\left[u_2(F) - u_2(E)\right] \tag{8.14}$$

On path *FG*,

$$J_{FG} = \int_{FG}\left(Udx_2 - T_i\frac{\partial u_i}{\partial x_1}dS\right) = \int_F^G\left(Udx_2 - T_i\frac{\partial u_i}{\partial x_1}(-dx_2)\right) = \int_F^G\left(Udx_2 + T_i\frac{\partial u_i}{\partial x_1}dx_2\right) \tag{8.15}$$

From the symmetry condition on path *FG*, the shear stress is zero; therefore, $T_2 = 0$ and $T_1 = -\sigma_{11}$

$$\therefore T_1\frac{\partial u_1}{\partial x_1} = -\sigma_{11}\frac{\partial u_1}{\partial x_1} = -\sigma_{11}\varepsilon_{11} = -\frac{\sigma_{11}}{E}(\sigma_{11} - \nu\sigma_{22}) \tag{8.16}$$

From equations (8.15) and (8.16),

$$J_{FG} = \int_F^G\left(U + T_i\frac{\partial u_i}{\partial x_1}\right)dx_2 = \int_F^G\left(U - \frac{\sigma_{11}}{E}(\sigma_{11} - \nu\sigma_{22})\right)dx_2 \tag{8.17}$$

Equation (8.17) can be evaluated from the FEM generated results. Then the stress intensity factor is obtained from the following relation:

$$\alpha\frac{K_I^2}{E} = J_{ABCDEFG} = 2\times(J_{DE} + J_{EF} + J_{FG}) \tag{8.18}$$

## 8.4 Special Crack Tip Finite Elements

The fundamental limitation of the conventional finite element method for solving fracture mechanics problems is trying to model a singular stress field near the crack tip by nonsingular expressions of stresses in conventional finite elements. For simple three-node triangular elements (constant strain triangles), stress is constant inside the element. In triangular and quadrilateral elements, the stress can be varied inside the element in linear, quadratic, or in higher order polynomial forms by changing the number of nodes of the element; however, the field will not have any singularity. Some investigators (Byskov 1970; Walsh 1971; Tracey 1971; Tong and Pian 1973; Wilson 1973) have

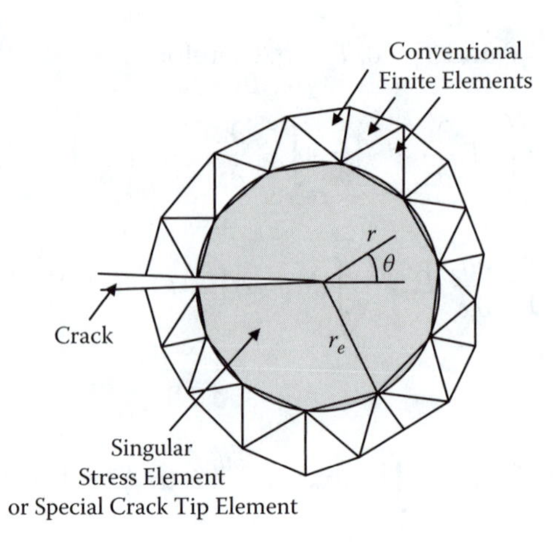


**FIGURE 8.12**
Special crack tip element surrounded by conventional triangular finite elements.

suggested using special crack tip elements with unconventional interpolation functions to produce singular stress fields at the crack tip. A circular crack element surrounding the crack tip, as shown in Figure 8.12, can have a singular stress field at the crack tip if the displacement variation in the element is expressed in the following form:

$$
\begin{aligned}
u_1 &= u_{10} + \sum_{n=1}^{N} \delta_n \left(\frac{r}{r_e}\right)^{\frac{n}{2}} f_n(\theta) \\
u_2 &= \sum_{n=1}^{N} \delta_n \left(\frac{r}{r_e}\right)^{\frac{n}{2}} g_n(\theta)
\end{aligned}
\tag{8.19}
$$

If the crack is located on a horizontal line of symmetry, as shown in Figure 8.1, 8.2, or 8.6, and the load is applied symmetrically, then from the symmetry condition it is easy to see that the crack tip element located on the axis of symmetry should not have any rigid body motion in the vertical direction, although it can have it in the horizontal direction. That is why $u_{10}$, the rigid body motion in the horizontal direction, is introduced in the $u_1$ expression only in equation (8.19). Variables $r$ and $\theta$ are the polar coordinates measured from the crack tip, as shown in Figure 8.12. Radius of the crack tip element is $r_e$; equation (8.19) is valid only inside the element, $r \le r_e$ and $-\pi \le \theta \le \pi$. Note that, for the displacement variation of equation (8.19), strain and stress should have square root singularity $(\frac{1}{\sqrt{r}})$ at the crack tip as predicted by the stress analysis near the crack tip. A layer of conventional finite elements (triangular or quadrilateral) is to be placed around the special crack tip element as shown in Figure 8.12.

In equation (8.19) $u_{10}$ and $\delta_n$ are unknown constants, and $N$ is an integer greater than or equal to 1. Functions $f_n(\theta)$ and $g_n(\theta)$ are known functions of $\theta$, obtained from the infinite series expressions of the displacement field near a crack tip. The infinite series expressions for the stress field are given in equations (8.1) and (8.2). Similarly, the series expressions for the displacement field can be obtained. It is possible to show that, for plane strain problems,

$$f_1(\theta) = -\left(\frac{5}{2} - 4v\right)\cos\left(\frac{\theta}{2}\right)\cos\theta + \frac{1}{2}\cos\left(\frac{3\theta}{2}\right)\cos\theta$$
$$-\left(\frac{7}{2} - 4v\right)\sin\left(\frac{\theta}{2}\right)\sin\theta + \frac{1}{2}\sin\left(\frac{3\theta}{2}\right)\sin\theta \quad (8.20)$$

For plane stress problems, $v$ should be replaced by $\frac{v}{1+v}$. The constant $\delta_1$ is proportional to $K_I$:

$$K_I = -\frac{E}{1+v}\sqrt{\frac{2\pi}{r_e}}\delta_1 \quad (8.21)$$

Note that the interelement compatibility (or displacement continuity) conditions across the singular stress element and conventional finite element boundaries is violated because the displacement interpolation functions are different types for these two types of elements. In spite of this incompatibility, the results show good convergence, especially when four-term expansion ($N = 4$ in equation 8.19) is considered. Error in the computed value of $K_I$ as a function of the radius of the circular element is shown in Figure 8.13 for SSC-1 and SSC-4 elements. Note that SSC-N stands for singular stress, circular element with $N$ term expansion.

Error also depends on the number of conventional finite elements, such as constant stress triangles (CST), surrounding the singular stress element as shown in Figure 8.14.

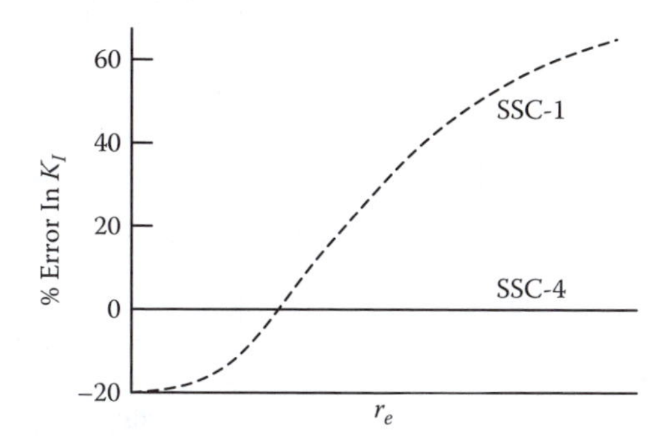


**FIGURE 8.13**
Computational error as a function of the radius of SSC-1 and SSC-4 elements.

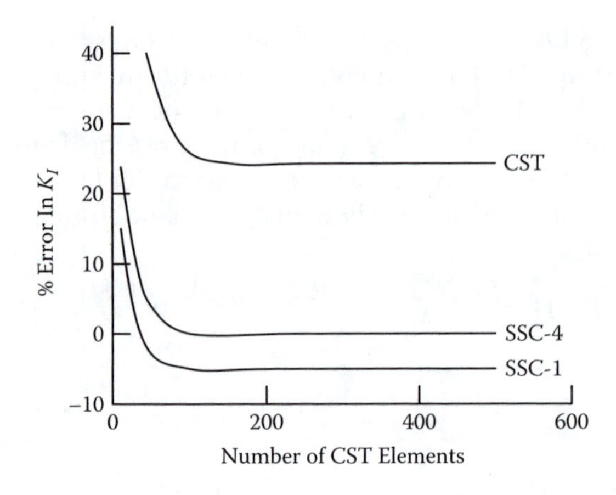


**FIGURE 8.14**
Error variation as a function of the number of CST elements surrounding the crack tip elements SSC-1 and SSC-4 (see Figure 8.12) and when only CST elements are used in the absence of SSC.

## 8.5 Quarter Point Quadrilateral Finite Element

Barsoum (1976) and Henshell and Shaw (1975) showed that in eight-node quadrilateral elements it is possible to model the singular stress behavior at the crack tip simply by moving two middle nodes (node numbers 2 and 8 in Figure 8.15b) in the derived element at quarter point positions toward the crack tip as shown in Figure 8.15b.

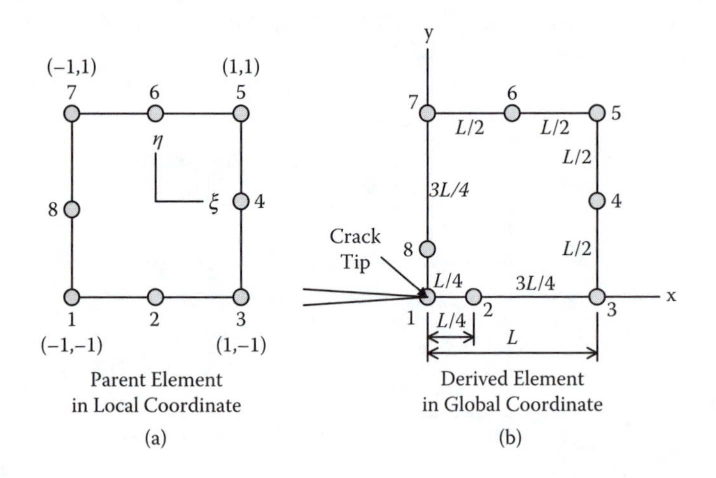


**FIGURE 8.15**
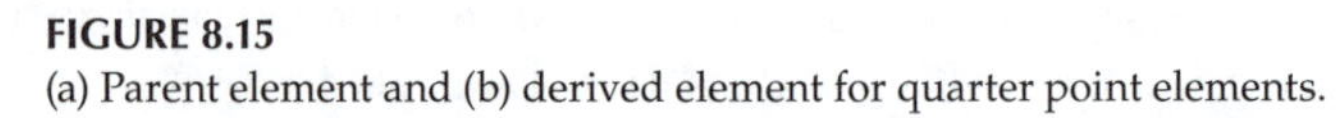
(a) Parent element and (b) derived element for quarter point elements.

Interpolation functions $N_1$, $N_2$, and $N_3$ for the eight-node quadrilateral elements are given by

$$
\begin{aligned}
N_1 &= -\frac{1}{4}(1-\xi)(1-\eta)(1+\xi+\eta) \\
N_2 &= \frac{1}{2}(1-\xi^2)(1-\eta) \\
N_3 &= \frac{1}{4}(1+\xi)(1-\eta)(\xi-\eta-1)
\end{aligned}
\tag{8.22}
$$

Variations of these three interpolation functions along line 123 ($\eta = -1$) are given by

$$
\begin{aligned}
N_1 &= -\frac{\xi}{2}(1-\xi) = \frac{\xi}{2}(\xi-1) \\
N_2 &= 1-\xi^2 \\
N_3 &= \frac{\xi}{2}(\xi+1)
\end{aligned}
\tag{8.23}
$$

Substituting $x_1 = 0$, $x_2 = L/4$, $x_3 = L$, and equation (8.23) in the local coordinate-global coordinate relation,

$$
\begin{aligned}
x &= N_1x_1 + N_2x_2 + N_3x_3 = N_1 \times 0 + N_2 \times \frac{L}{4} + N_3 \times L \\
&= (1-\xi^2)\times\frac{L}{4} + \frac{\xi}{2}(\xi+1)\times L = \frac{L}{4}\xi^2 + \frac{L}{2}\xi + \frac{L}{4}
\end{aligned}
\tag{8.24}
$$

Equation (8.24) can be written as

$$
\begin{aligned}
&\frac{L}{4}\xi^2 + \frac{L}{2}\xi + \left(\frac{L}{4} - x\right) = 0 \\
&\therefore \xi = \frac{-\frac{L}{2} \pm \sqrt{\frac{L^2}{4} - 4\times\frac{L}{4}\left(\frac{L}{4}-x\right)}}{2\times\frac{L}{4}} = -1 \pm \frac{2}{L}\sqrt{\frac{L^2}{4} - \frac{L^2}{4} + Lx} = -1 \pm 2\sqrt{\frac{x}{L}}
\end{aligned}
\tag{8.25}
$$

In Figure 8.15 it is easy to see that node 3 corresponds to $x = L$ and $\xi = 1$. Therefore,

$$
\xi = -1 + 2\sqrt{\frac{x}{L}} \tag{8.26}
$$

Then the displacement field is given by

$$u = N_1u_1 + N_2u_2 + N_3u_3 = \frac{\xi}{2}(\xi-1)u_1 + (1-\xi^2)u_2 + \frac{\xi}{2}(\xi+1)u_3$$

$$= \frac{1}{2}\left(2\sqrt{\frac{x}{L}}-1\right)\left(2\sqrt{\frac{x}{L}}-2\right)u_1 + \left\{1-\left(2\sqrt{\frac{x}{L}}-1\right)^2\right\}u_2 + \frac{1}{2}\left(2\sqrt{\frac{x}{L}}-1\right)2\sqrt{\frac{x}{L}}u_3$$

$$= u_1\left(2\sqrt{\frac{x}{L}}-1\right)\left(\sqrt{\frac{x}{L}}-1\right) + u_2\left\{1-\left(2\sqrt{\frac{x}{L}}-1\right)^2\right\} + u_3\left(2\sqrt{\frac{x}{L}}-1\right)\sqrt{\frac{x}{L}} \tag{8.27}$$

Therefore,

$$\varepsilon_{xx} = \frac{\partial u}{\partial x} = -\frac{u_1}{2}\left[\frac{3}{\sqrt{xL}}-\frac{4}{L}\right] + u_2\left[\frac{2}{\sqrt{xL}}-\frac{4}{L}\right] + \frac{u_3}{2}\left[-\frac{1}{\sqrt{xL}}+\frac{4}{L}\right] \tag{8.28}$$

Clearly, the strain shows square root singularity at the crack tip ($x = 0$). The stress field should also show the same square root singularity.

The main advantage of the quarter point finite element is that it can produce accurate results without requiring a special crack tip element. Therefore, conventional codes that can handle eight-point quadrilateral elements or triangular elements (collapsed quadrilateral element) can be used to solve fracture mechanics problems simply by shifting the middle nodes on the element boundaries adjacent to the crack tip to quarter point positions toward the crack tip, as shown in Figure 8.15b. Barsoum (1977) showed that triangular quarter point elements are even better than quadrilateral quarter point elements for modeling cracked geometry. Barsoum thought that in the case of quadrilateral elements the stress was singular only along the two edges of the finite element, whereas in the case of triangular elements the stress is singular along any ray from the crack tip.

Later, Banks-Sills and Bortman (1984) re-examined the quarter point quadrilateral element and concluded that for quadrilateral elements the behavior of stresses is also square-root singular on all rays emanating from the crack tip in a small region of the element. There have been some suggestions to use one or two layers of transition elements (Lynn and Ingraffea 1978) between quarter point elements and conventional finite elements. In transition elements the middle nodes are placed in between the quarter point and the midpoint, as shown in Figure 8.16. Transition elements may produce slightly better results for some problems, but often the effect of transition elements has been found to be insignificant. Banks-Sills and Sherman (1986) solved a fracture mechanics problem by several techniques and concluded that the quarter point element technique is overall the most efficient technique.

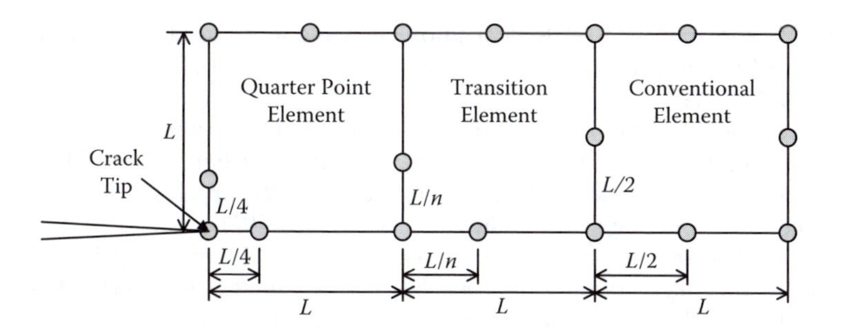


**FIGURE 8.16**
Transition element placed between quarter point element and conventional finite element. For the transition element, the shorter length $L/n$ can be any value between $L/2$ and $L/4$.

## 8.6 Concluding Remarks

Among various numerical techniques discussed in this chapter the most popular technique for solving crack problems has been the use of quarter point elements at the crack tips in a finite element mesh made of eight-node quadrilateral isoparametric elements. Although use of special crack tip elements (discussed in section 8.4) became popular in the 1970s, this approach lost its charm after the invention of the quarter point elements.

## References

Banks-Sills, L. and Bortman, Y. Reappraisal of the quarter-point quadrilateral element in linear elastic fracture mechanics. *International Journal of Fracture,* 25, 169–180, 1984.

Banks-Sills, L. and Sherman, D. Comparison of methods for calculating stress intensity factors with quarter-point elements. *International Journal of Fracture,* 32, 127–140, 1986.

Barsoum, R. S. On the use of isoparametric finite elements in linear fracture mechanics. *International Journal of Numerical Methods in Engineering,* 10, 25–37, 1976.

Barsoum, R. S. Triangular quarter point elements as elastic and perfectly plastic crack tip elements. *International Journal of Numerical Methods in Engineering,* 11, 85–98, 1977.

Byskov, E. The calculation of stress intensity factors using the finite element method with cracked elements. *International Journal of Fracture Mechanics,* 6, 159–167, 1970.

Chan, S. K., Tubo, I. S., and Wilson, W. K. On the finite element method in linear fracture mechanics. *Engineering Fracture Mechanics,* 2, 1–17, 1970.

Henshell, R. D. and Shaw, K. G. Crack tip finite elements are unnecessary. *International Journal of Numerical Methods in Engineering,* 9, 495–507, 1975.

Hilton, P. D. and Sih, G. C. Applications of the finite element method to the calculation of stress intensity factors. In *Methods of analysis and solution of crack problems, Mechanics of fracture I,* ed. G. C. Sih. Groningen, the Netherlands: P. Noordhoff, pp. 426–483, 1973.

Lynn, P. P. and Ingraffea, A. R. Transition elements to be used with quarter-point crack tip elements. *International Journal of Numerical Methods in Engineering,* 12, 1031–1036, 1978.

Tong, P. and Pian, T. H. H. On the convergence of the finite element method for problems with singularity. *International Journal of Solids and Structures,* 9, 313–321, 1973.

Tracey, D. M. Finite elements for determination of crack tip elastic stress intensity factors. *Engineering Fracture Mechanics,* 3, 255–265, 1971.

Walsh, P. F. The computation of stress intensity factors by a special finite element technique. *International Journal of Solids and Structures,* 7, 1333–1342, 1971.

Watwood, V. B. The finite element method for prediction of crack behavior. *Nuclear Engineering Design,* 11, 323–332, 1969.

Wilson, W. K. Finite element methods for elastic bodies containing cracks. In *Methods of analysis and solution of crack problems, Mechanics of fracture I,* ed. G. C. Sih. Groningen, the Netherlands: Noordhoff, pp. 484–515, 1973.

# 9

# *Westergaard Stress Function*

## 9.1 Introduction

For solving crack problems analytically, several complex forms of Airy stress function have been proposed (Westergaard 1939; Muskhelishvili 1953; Paris and Sih 1965; Sih 1966, 1973; Rice 1969; Goodier 1969; Eftis and Liebowitz 1972). Among these different stress functions the Westergaard stress function has become most popular because it can solve a wide range of crack problems. A few sample problems are solved analytically in this chapter following the Westergaard stress function technique.

## 9.2 Background Knowledge

As described in chapter 1, if $\Phi(x, y)$ is the Airy stress function, then stresses in a two-dimensional problem geometry are defined in terms of the Airy stress function in the following manner (equation 1.101):

$$\begin{aligned} \sigma_{xx} &= \Phi_{,yy} \\ \sigma_{yy} &= \Phi_{,xx} \\ \sigma_{xy} &= -\Phi_{,xy} \end{aligned} \tag{9.1}$$

Note that the preceding three equations can be written in index notation:

$$\sigma_{\alpha\beta} = \Phi_{,\gamma\gamma}\,\delta_{\alpha\beta} - \Phi_{,\alpha\beta} \tag{9.2}$$

where $\alpha$, $\beta$ take values $x$ and $y$. Repeated index $\gamma$ means summation over both $x$ and $y$ $\Phi_{,\gamma\gamma} = \Phi_{,xx} + \Phi_{,yy}$.

Substituting these expressions in the compatibility equation [($\varepsilon_{xx,yy} + \varepsilon_{yy,xx} - 2\varepsilon_{xy,xy} = 0$) in terms of strain or ($\sigma_{\alpha\alpha,\beta\beta} = \sigma_{xx,xx} + \sigma_{xx,yy} + \sigma_{yy,xx} + \sigma_{yy,yy} = 0$) in terms of stress], one gets (see equation 1.102)

$$\nabla^4\Phi = 0 \tag{9.3}$$

Therefore, $\Phi$ must be biharmonic. One can show that if

$$\Phi = \phi_1 + x\phi_2 + y\phi_3 + (x^2 + y^2)\phi_4 \tag{9.4}$$

then $\Phi$ is a biharmonic function when $\phi_1$, $\phi_2$, $\phi_3$, $\phi_4$ are harmonic:

$$\nabla^2\phi_1 = \nabla^2\phi_2 = \nabla^2\phi_3 = \nabla^2\phi_4 = 0 \tag{9.5}$$

Therefore, any biharmonic function can be represented in the form given in equation (9.4).

From the special properties of an analytic function $\phi(z)$, where $z$ is a complex variable $z = x + iy$ and $\phi(z) = U(x, y) + iV(x, y)$, one can write

$$\nabla^2 U = \nabla^2 V = 0 \tag{9.6}$$

In other words, when $\phi(z)$ has a derivative with respect to $z$ (definition of an analytic function is that its derivative exists), its real and imaginary parts $[U = \mathrm{Re}(\phi), V = \mathrm{Im}(\phi)]$ must be harmonic. Note that $U(x, y)$ and $V(x, y)$ are real functions of $x$ and $y$.

## 9.3 Griffith Crack in Biaxial State of Stress

Let us consider the problem geometry shown in Figure 9.1. A Griffith crack of length $2a$ is subjected to a biaxial state of stress. From the symmetry condition one can see that the shear stress ($\sigma_{xy}$) along two lines of symmetry $y = 0$ and $x = 0$ must be zero.

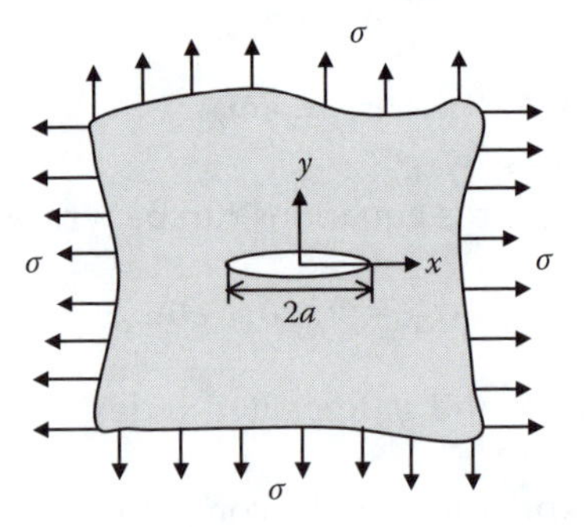


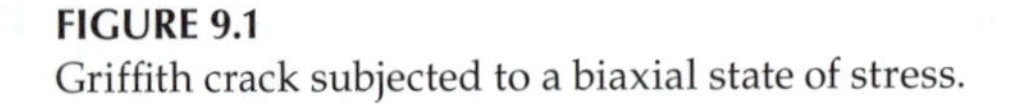
**FIGURE 9.1**
Griffith crack subjected to a biaxial state of stress.

In terms of the Westergaard stress function $Z(z)$, where $z$ is a complex variable ($z = x + iy$), the harmonic functions $\phi_i$ of equation (9.4) are defined as

$$\begin{aligned} \phi_1 &= \mathrm{Re}\left(\iint Z(z)dz\right) \\ \phi_3 &= \mathrm{Im}\left(\int Z(z)dz\right) \\ \phi_2 &= \phi_4 = 0 \end{aligned} \tag{9.7}$$

Note that single and double integrals of a function $Z(z)$ must be analytic because the derivative of the integral of a function is the function itself and the derivative of the double integral of a function is the single integral of the function. Thus, derivatives of both single and double integrals of any complex function must exist. Therefore, from equation (9.6) both real and imaginary parts of $\iint Z(z)dz$ and $\int Z(z)dz$ must be harmonic.

### 9.3.1 Stress and Displacement Fields in Terms of Westergaard Stress Function

Since $z = x + iy$,

$$\begin{aligned} \frac{\partial z}{\partial x} &= 1 \\ \frac{\partial z}{\partial y} &= i \end{aligned} \tag{9.8}$$

Therefore,

$$\begin{aligned} \frac{\partial \phi(z)}{\partial x} &= \frac{\partial \phi}{\partial z}\frac{\partial z}{\partial x} = \frac{\partial \phi}{\partial z} = \phi'(z) \\ \frac{\partial \phi(z)}{\partial y} &= \frac{\partial \phi}{\partial z}\frac{\partial z}{\partial y} = i\frac{\partial \phi}{\partial z} = i\phi'(z) \end{aligned} \tag{9.9}$$

From equations (9.4) and (9.7),

$$\Phi = \phi_1 + x\phi_2 + y\phi_3 + (x^2 + y^2)\phi_4 = \mathrm{Re}\left(\iint Z(z)dz\right) + y\,\mathrm{Im}\left(\int Z(z)dz\right) \tag{9.10}$$

Therefore,

$$\begin{aligned}
\Phi_{,x} &= \mathrm{Re}\left(\int Z(z)dz\right) + y\,\mathrm{Im}(Z(z)) \\
\Phi_{,xx} &= \mathrm{Re}(Z) + y\,\mathrm{Im}(Z') \\
\Phi_{,y} &= \mathrm{Re}\left(i\int Z(z)dz\right) + \mathrm{Im}\left(\int Z(z)dz\right) + y\,\mathrm{Im}(iZ(z)) \\
&= -\mathrm{Im}\left(\int Z(z)dz\right) + \mathrm{Im}\left(\int Z(z)dz\right) + y\,\mathrm{Im}(iZ) = y\,\mathrm{Re}(Z) \\
\Phi_{,yy} &= \mathrm{Re}(Z) + y\,\mathrm{Re}(iZ') = \mathrm{Re}(Z) - y\,\mathrm{Im}(Z') \\
\Phi_{,xy} &= y\,\mathrm{Re}(Z')
\end{aligned} \tag{9.11}$$

From equations (9.1) and (9.11), the stress expressions are obtained:

$$\begin{aligned}
\sigma_{xx} &= \Phi_{,yy} = \mathrm{Re}(Z) - y\,\mathrm{Im}(Z') \\
\sigma_{yy} &= \Phi_{,xx} = \mathrm{Re}(Z) + y\,\mathrm{Im}(Z') \\
\sigma_{xy} &= -\Phi_{,xy} = -y\,\mathrm{Re}(Z')
\end{aligned} \tag{9.12}$$

From the preceding equation set, one can see that as $y$ approaches zero the shear stress $\sigma_{xy}$ becomes zero if $\mathrm{Re}(Z')$ is finite.

From equation (9.12) and stress–strain relations, the strain components can be expressed in terms of the Westergaard stress function. Then, integrating the strain expressions, the displacement field is obtained:

$$\begin{aligned}
\frac{E}{1+\nu}u &= \frac{\kappa-1}{2}\mathrm{Re}\left(\int Z(z)dz\right) - y\,\mathrm{Im}(Z) \\
\frac{E}{1+\nu}v &= \frac{\kappa+1}{2}\mathrm{Im}\left(\int Z(z)dz\right) - y\,\mathrm{Re}(Z)
\end{aligned} \tag{9.13}$$

$$\begin{aligned}
\kappa &= 3 - 4\nu \quad \text{for plane strain} \\
\kappa &= \frac{3-\nu}{1+\nu} \quad \text{for plane stress}
\end{aligned} \tag{9.14}$$

### 9.3.2 Westergaard Stress Function for the Griffith Crack under Biaxial Stress Field

Let us investigate the following stress function:

$$Z(z) = \frac{\sigma z}{\sqrt{z^2 - a^2}} \tag{9.15}$$

As $z$ approaches infinity, the stress function $Z(z)$ simplifies to

$$\lim_{z \to \infty} Z(z) = \lim_{z \to \infty} \frac{\sigma z}{\sqrt{z^2 - a^2}} = \frac{\sigma z}{\sqrt{z^2}} = \frac{\sigma z}{z} = \sigma \tag{9.16}$$

Therefore, for the large $x$ and $y$ values in Figure 9.1, the complex variable $z$ is large, since $z = x + iy$, and $Z(z) = \sigma$. From equation (9.12),

$$\begin{aligned} \sigma_{xx} &= \mathrm{Re}(Z) - y\,\mathrm{Im}(Z') = \sigma \\ \sigma_{yy} &= \mathrm{Re}(Z) + y\,\mathrm{Im}(Z') = \sigma \\ \sigma_{xy} &= -y\,\mathrm{Re}(Z') = 0 \end{aligned} \tag{9.17}$$

Clearly, equation (9.17) shows that the stress function of equation (9.15) satisfies the regularity condition or conditions at infinity.

From equations (9.12) and (9.13), the stress and displacement fields closer to the crack can be obtained. In these equations derivative and integral of $Z(z)$ appear:

$$\begin{aligned} \int Z(z)dz &= \int \frac{\sigma z}{\sqrt{z^2 - a^2}} dz = \frac{\sigma}{2} \int \frac{2z}{\sqrt{z^2 - a^2}} dz = \sigma\sqrt{z^2 - a^2}\,w \\ Z'(z) &= \frac{d}{dz}\left(\frac{\sigma z}{\sqrt{z^2 - a^2}}\right) = \sigma\left(\frac{\sqrt{z^2 - a^2} - z\frac{1}{2}(z^2 - a^2)^{-1/2}2z}{z^2 - a^2}\right) \\ &= \sigma\left(\frac{z^2 - a^2 - z^2}{(z^2 - a^2)^{3/2}}\right) = -\frac{\sigma a^2}{(z^2 - a^2)^{3/2}} = -\frac{1}{\sqrt{z^2 - a^2}}\frac{\sigma a^2}{(z^2 - a^2)} \end{aligned} \tag{9.18}$$

The function $\sqrt{z^2 - a^2}$ is a multivalued function explained as follows:

$$\sqrt{z^2 - a^2} = \sqrt{z - a}\sqrt{z + a} = \sqrt{r_1 e^{i\theta_1}}\sqrt{r_2 e^{i\theta_2}} = \sqrt{r_1 r_2}\, e^{\frac{i}{2}(\theta_1 + \theta_2)} \tag{9.19}$$

In equation (9.19),

$$\begin{aligned} r_1 &= |z - a| = AP \\ r_2 &= |z + a| = BP \end{aligned} \tag{9.20}$$

$AP$ and $BP$ are distances of the point of interest ($P$) from the two crack tips A and B. Angles $\theta_1$ and $\theta_2$ are shown in Figure 9.2. Depending on how angles $\theta_1$ and $\theta_2$ are defined, different values of the function $\sqrt{z^2 - a^2}$ are possible at the same point. Let us assume that both $\theta_1$ and $\theta_2$ vary from 0 to $2\pi$. Then,

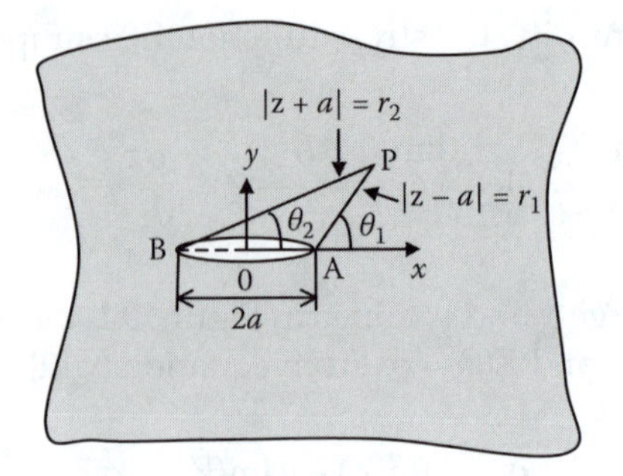


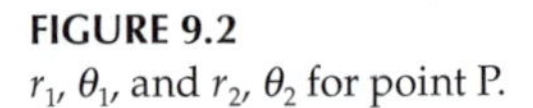

**FIGURE 9.2**
$r_1$, $\theta_1$, and $r_2$, $\theta_2$ for point P.

for different points on the $x$- and $y$-axes, shown in Figure 9.3, the function $\sqrt{z^2 - a^2}$ should have the following values:

For point P of Figure 9.3, $\theta_1 = \theta_2 = 0$, $r_1 = |z - a| = x - a$, $r_2 = |z + a| = x + a$; therefore, at P,

$$\sqrt{z^2 - a^2} = \sqrt{r_1 r_2 e^{\frac{i}{2}(\theta_1 + \theta_2)}} = \sqrt{(x-a)(x+a)e^{\frac{i}{2}0}} = \sqrt{x^2 - a^2}$$

Similarly, for point Q, $\theta_1 = \pi/2 + \alpha$, $\theta_2 = \pi/2 - \alpha$, where $\alpha$ is the angle between AQ and OR:

$$r_1 = |z - a| = \sqrt{y^2 + a^2}, \quad r_2 = |z + a| = \sqrt{y^2 + a^2}$$

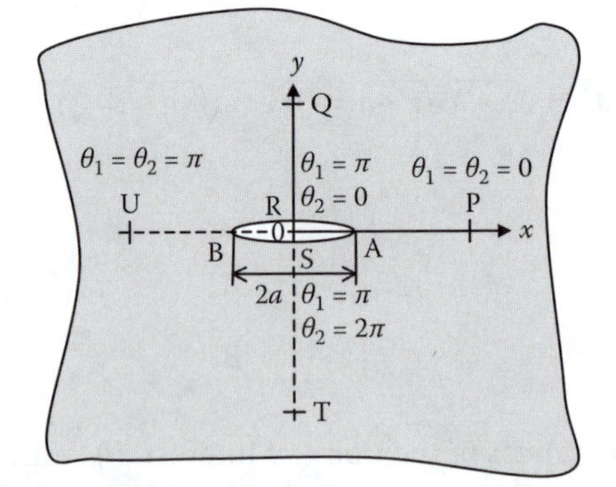


**FIGURE 9.3**
$\theta_1$ and $\theta_2$ values for different points P, Q, R, S, T on the cracked plate.

Therefore, at Q,

$$\sqrt{z^2-a^2}=\sqrt{r_1r_2}e^{\frac{i}{2}(\theta_1+\theta_2)}=\sqrt{(y^2+a^2)^{\frac{1}{2}}(y^2+a^2)^{\frac{1}{2}}e^{\frac{i}{2}\pi}}=i\sqrt{y^2+a^2}$$

For point R, $\theta_1=\pi$, $\theta_2=0$, $r_1=|z-a|=a-x$, $r_2=|z+a|=a+x$; therefore, at R,

$$\sqrt{z^2-a^2}=\sqrt{r_1r_2}e^{\frac{i}{2}(\theta_1+\theta_2)}=\sqrt{(a-x)(a+x)e^{\frac{i}{2}\pi}}=i\sqrt{a^2-x^2}$$

For point S, $\theta_1=\pi$, $\theta_2=2\pi$, $r_1=|z-a|=a-x$, $r_2=|z+a|=a+x$; therefore, at S,

$$\sqrt{z^2-a^2}=\sqrt{r_1r_2}e^{\frac{i}{2}(\theta_1+\theta_2)}=\sqrt{(a-x)(a+x)e^{\frac{i}{2}3\pi}}=-i\sqrt{a^2-x^2}$$

For point T, $\theta_1=3\pi/2-\alpha$, $\theta_2=3\pi/2+\alpha$, where $\alpha$ is the angle between AS and OS:

$$r_1=|z-a|=\sqrt{y^2+a^2},\quad r_2=|z+a|=\sqrt{y^2+a^2}$$

Therefore, at T,

$$\sqrt{z^2-a^2}=\sqrt{r_1r_2}e^{\frac{i}{2}(\theta_1+\theta_2)}=\sqrt{(y^2+a^2)^{\frac{1}{2}}(y^2+a^2)^{\frac{1}{2}}e^{\frac{i}{2}3\pi}}=-i\sqrt{y^2+a^2}$$

For point U, $\theta_1=\theta_2=\pi$, $r_1=|z-a|=-x-a$, $r_2=|z+a|=-x+a$; therefore, at U,

$$\sqrt{z^2-a^2}=\sqrt{r_1r_2}e^{\frac{i}{2}(\theta_1+\theta_2)}=\sqrt{(-x-a)(-x+a)e^{\frac{i}{2}2\pi}}=-\sqrt{x^2-a^2}$$

Note that at points R and S, the $\theta_1$ value is the same ($\pi$), but the values of $\theta_2$ are different (0 for point R, and $2\pi$ for point S); this results in different values of the function $\sqrt{z^2-a^2}$ between these two points. However, for points just above and below point U, both $\theta_1$ and $\theta_2$ values are the same ($\pi$) and hence there is no jump in the value of the function $\sqrt{z^2-a^2}$ between two points just above and below point U.

However, $\theta_1$ and $\theta_2$ values above and below point P are different. For a point slightly above point P ($y=0^+$), $\theta_1=\theta_2=0$ and for a point slightly below point P ($y=0^-$), $\theta_1=\theta_2=2\pi$. Therefore, for a point just below point P, the function $\sqrt{z^2-a^2}$ value is given by $\sqrt{z^2-a^2}=\sqrt{r_1r_2}e^{\frac{i}{2}(\theta_1+\theta_2)}=\sqrt{(x-a)(x+a)e^{\frac{i}{2}4\pi}}$ $=\sqrt{x^2-a^2}$. Note that this value is the same as that of a point just above point P. The only line across which the function value changes from one side of the line to the other side is line AB. This line is called the branch cut.

Values of the multivalued function $\sqrt{z^2 - a^2}$ at different points of Figure 9.3 are summarized below:

$$
\begin{aligned}
\sqrt{z^2 - a^2} &= \sqrt{x^2 - a^2} && \text{at point P} \\
&= i\sqrt{y^2 + a^2} && \text{at point Q} \\
&= i\sqrt{a^2 - x^2} && \text{at point R} \\
&= -i\sqrt{a^2 - x^2} && \text{at point S} \\
&= -i\sqrt{y^2 + a^2} && \text{at point T} \\
&= -\sqrt{x^2 - a^2} && \text{at point U}
\end{aligned}
\tag{9.21}
$$

From equations (9.12), (9.18), and (9.21), on line AP of Figure 9.3, since $y = 0$,

$$
\begin{aligned}
\sigma_{xx} &= \mathrm{Re}(Z) - y\,\mathrm{Im}(Z') = \mathrm{Re}\left(\frac{\sigma z}{\sqrt{z^2 - a^2}}\right) = \frac{\sigma x}{\sqrt{x^2 - a^2}} \\
\sigma_{yy} &= \mathrm{Re}(Z) + y\,\mathrm{Im}(Z') = \mathrm{Re}\left(\frac{\sigma z}{\sqrt{z^2 - a^2}}\right) = \frac{\sigma x}{\sqrt{x^2 - a^2}} \\
\sigma_{xy} &= -y\,\mathrm{Re}(Z') = 0
\end{aligned}
\tag{9.22}
$$

and the displacement field from equation (9.13),

$$
\begin{aligned}
\frac{E}{1+\nu}u &= \frac{\kappa - 1}{2}\mathrm{Re}\left(\int Z(z)dz\right) - y\,\mathrm{Im}(Z) = \frac{\kappa - 1}{2}\mathrm{Re}\left(\sigma\sqrt{z^2 - a^2}\right) \\
&= \frac{\kappa - 1}{2}\mathrm{Re}\left(\sigma\sqrt{x^2 - a^2}\right) = \frac{\kappa - 1}{2}\sigma\sqrt{x^2 - a^2} \\
\frac{E}{1+\nu}v &= \frac{\kappa + 1}{2}\mathrm{Im}\left(\int Z(z)dz\right) - y\,\mathrm{Re}(Z) = \frac{\kappa + 1}{2}\mathrm{Im}\left(\sigma\sqrt{z^2 - a^2}\right) \\
&= \frac{\kappa + 1}{2}\mathrm{Im}\left(\sigma\sqrt{x^2 - a^2}\right) = 0 \\
\therefore u &= \frac{\sigma(1+\nu)(\kappa - 1)}{2E}\sqrt{x^2 - a^2}
\end{aligned}
\tag{9.23}
$$

On line AB, on the top surface of the crack, $y = 0^+$,

$$
\begin{aligned}
\sigma_{xx} &= \mathrm{Re}(Z) - y\,\mathrm{Im}(Z') = \mathrm{Re}\left(\frac{\sigma z}{\sqrt{z^2 - a^2}}\right) = \mathrm{Re}\left(\frac{\sigma x}{i\sqrt{a^2 - x^2}}\right) = 0 \\
\sigma_{yy} &= \mathrm{Re}(Z) + y\,\mathrm{Im}(Z') = \mathrm{Re}\left(\frac{\sigma z}{\sqrt{z^2 - a^2}}\right) = \mathrm{Re}\left(\frac{\sigma x}{i\sqrt{a^2 - x^2}}\right) = 0 \\
\sigma_{xy} &= -y\,\mathrm{Re}(Z') = 0
\end{aligned}
\tag{9.24}
$$

Therefore, the traction-free boundary conditions on the top crack surface are satisfied. The displacement field on the top surface of the crack (line AB, $y = 0^+$) can be obtained from equation (9.13):

$$\begin{aligned}
\frac{E}{1+\nu}u &= \frac{\kappa-1}{2}\mathrm{Re}\left(\int Z(z)dz\right) - y\,\mathrm{Im}(Z) = \frac{\kappa-1}{2}\mathrm{Re}\left(\int Z(z)dz\right) \\
&= \frac{\kappa-1}{2}\mathrm{Re}\left(i\sigma\sqrt{a^2-x^2}\right) = 0 \\
\frac{E}{1+\nu}v &= \frac{\kappa+1}{2}\mathrm{Im}\left(\int Z(z)dz\right) - y\,\mathrm{Re}(Z) = \frac{\kappa+1}{2}\mathrm{Im}\left(\int Z(z)dz\right) \\
&= \frac{\kappa+1}{2}\mathrm{Im}\left(i\sigma\sqrt{a^2-x^2}\right) = \frac{\sigma(\kappa+1)}{2}\sqrt{a^2-x^2} \\
\therefore v &= \frac{\sigma(1+\nu)(\kappa+1)}{2E}\sqrt{a^2-x^2}
\end{aligned} \tag{9.25}$$

In the same manner on line AB, on the bottom surface of the crack, $y = 0^-$, we obtain $\sigma_{xx} = \sigma_{yy} = \sigma_{xy} = 0$:

$$\begin{aligned}
\frac{E}{1+\nu}u &= \frac{\kappa-1}{2}\mathrm{Re}\left(\int Z(z)dz\right) - y\,\mathrm{Im}(Z) = \frac{\kappa-1}{2}\mathrm{Re}\left(\int Z(z)dz\right) \\
&= \frac{\kappa-1}{2}\mathrm{Re}\left(-i\sigma\sqrt{a^2-x^2}\right) = 0 \\
\frac{E}{1+\nu}v &= \frac{\kappa+1}{2}\mathrm{Im}\left(\int Z(z)dz\right) - y\,\mathrm{Re}(Z) = \frac{\kappa+1}{2}\mathrm{Im}\left(\int Z(z)dz\right) \\
&= \frac{\kappa+1}{2}\mathrm{Im}\left(-i\sigma\sqrt{a^2-x^2}\right) = -\frac{\sigma(\kappa+1)}{2}\sqrt{a^2-x^2} \\
\therefore v &= -\frac{\sigma(1+\nu)(\kappa+1)}{2E}\sqrt{a^2-x^2}
\end{aligned} \tag{9.26}$$

Note that equations (9.25) and (9.26) can be written as

$$\left(\frac{v}{c}\right)^2 + x^2 = a^2$$

or

$$\left(\frac{v}{ca}\right)^2 + \left(\frac{x}{a}\right)^2 = 1 \tag{9.27}$$

Therefore, the open crack forms an ellipse whose semimajor axis is the half-crack length $a$ and the semiminor axis is the half of the maximum crack opening displacement at the center:

$$ca = \frac{\sigma(1+\nu)(\kappa+1)}{2E} \tag{9.28}$$

Similarly, on line RQ, $x = 0$, $y > 0$:

$$\begin{aligned}
\sigma_{xx} &= \mathrm{Re}(Z) - y\,\mathrm{Im}(Z') = \mathrm{Re}\left(\frac{\sigma z}{\sqrt{z^2-a^2}}\right) - y\,\mathrm{Im}\left(-\frac{\sigma a^2}{\sqrt{z^2-a^2}\,(z^2-a^2)}\right)\\
&= \mathrm{Re}\left(\frac{\sigma(x+iy)}{i\sqrt{a^2+y^2}}\right) + y\,\mathrm{Im}\left(\frac{\sigma a^2}{i\sqrt{a^2+y^2}\,\{-(a^2+y^2)\}}\right)\\
&= \mathrm{Re}\left(\frac{\sigma(-ix+y)}{\sqrt{a^2+y^2}}\right) + y\,\mathrm{Im}\left(\frac{i\sigma a^2}{\sqrt{a^2+y^2}\,(a^2+y^2)}\right)\\
&= \sigma\left\{\frac{y}{\sqrt{a^2+y^2}} + \frac{a^2}{(a^2+y^2)^{3/2}}\right\}\\
\sigma_{yy} &= \mathrm{Re}(Z) + y\,\mathrm{Im}(Z') = \sigma\left\{\frac{y}{\sqrt{a^2+y^2}} - \frac{a^2}{(a^2+y^2)^{3/2}}\right\}\\
\sigma_{xy} &= -y\,\mathrm{Re}(Z') = -y\,\mathrm{Re}\left(-\frac{\sigma a^2}{\sqrt{z^2-a^2}\,(z^2-a^2)}\right) = y\,\mathrm{Re}\left(\frac{\sigma a^2}{i\sqrt{a^2+y^2}\,\{-(a^2+y^2)\}}\right) = 0
\end{aligned} \tag{9.29}$$

Note that the stress and displacement fields at any point of interest can be obtained from equations (9.12), (9.13), (9.15), and (9.18).

### 9.3.3 Stress Field Close to a Crack Tip

For a point P close to the crack tip as shown in Figure 9.4,

$$z = a + re^{i\theta} \tag{9.30}$$

For $\frac{r}{a} \ll 1$,

$$z + a = 2a + re^{i\theta} \approx 2a \tag{9.31}$$

$$z - a = a + re^{i\theta} - a = re^{i\theta} \tag{9.32}$$

$$\begin{aligned}
&\therefore z^2 - a^2 = (z-a)(z+a) = 2are^{i\theta}\\
&\therefore \sqrt{z^2-a^2} = \sqrt{2ar}\,e^{i\frac{\theta}{2}}
\end{aligned} \tag{9.33}$$

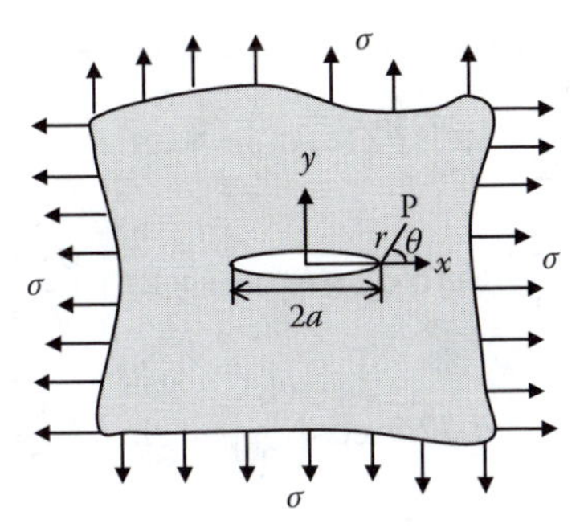


**FIGURE 9.4**
Point P near a crack tip is denoted by the $(r, \theta)$ coordinates.

Therefore,

$$Z(z) = \frac{\sigma z}{\sqrt{z^2 - a^2}} = \frac{\sigma(a + re^{i\theta})}{\sqrt{2are^{i\frac{\theta}{2}}}} \approx \frac{\sigma a}{\sqrt{2are^{i\frac{\theta}{2}}}} = \frac{\sigma\sqrt{\pi a}}{\sqrt{2\pi r}} e^{-i\frac{\theta}{2}} = \frac{K}{\sqrt{2\pi(z - a)}} \tag{9.34}$$

and

$$\int Z(z)dz = \sigma\sqrt{z^2 - a^2} = \sigma\sqrt{2are^{i\theta/2}}$$

$$Z'(z) = -\frac{1}{\sqrt{z^2 - a^2}} \frac{\sigma a^2}{(z^2 - a^2)} = -\frac{\sigma a^2}{(2are^{i\theta})^{3/2}} = -\frac{\sigma a^2 e^{-3i\theta/2}}{(2ar)^{3/2}} \tag{9.35}$$

Therefore,

$$\sigma_{xx} = \text{Re}(Z) - y\,\text{Im}(Z') = \text{Re}\left(\frac{\sigma\sqrt{\pi a}}{\sqrt{2\pi r}} e^{-i\theta/2}\right) - y\,\text{Im}\left(-\frac{\sigma a^2 e^{-3i\theta/2}}{(2ar)^{3/2}}\right)$$

$$= \frac{\sigma\sqrt{\pi a}}{\sqrt{2\pi r}} \cos\left(\frac{\theta}{2}\right) - r\sin\theta \frac{\sigma a^2}{(2ar)^{3/2}} \sin\left(\frac{3\theta}{2}\right)$$

$$= \frac{\sigma\sqrt{\pi a}}{\sqrt{2\pi r}} \left\{\cos\left(\frac{\theta}{2}\right) - \frac{1}{2}\sin\theta\sin\left(\frac{3\theta}{2}\right)\right\} \tag{9.36}$$

$$\sigma_{yy} = \text{Re}(Z) + y\,\text{Im}(Z') = \frac{\sigma\sqrt{\pi a}}{\sqrt{2\pi r}} \left\{\cos\left(\frac{\theta}{2}\right) + \frac{1}{2}\sin\theta\sin\left(\frac{3\theta}{2}\right)\right\}$$

$$\sigma_{xy} = -y\,\text{Re}(Z') = -y\,\text{Re}\left(-\frac{\sigma a^2 e^{-3i\theta/2}}{(2ar)^{3/2}}\right) = r\sin\theta \frac{\sigma a^2}{(2ar)^{3/2}} \cos\left(\frac{3\theta}{2}\right)$$

$$= \frac{\sigma\sqrt{\pi a}}{\sqrt{2\pi r}} \frac{1}{2} \sin\theta\cos\left(\frac{3\theta}{2}\right)$$

Similarly,

$$
\begin{aligned}
\frac{E}{1+\nu}u &= \frac{\kappa-1}{2}\mathrm{Re}\left(\int Z(z)dz\right) - y\,\mathrm{Im}(Z) \\
&= \frac{\kappa-1}{2}\mathrm{Re}(\sigma\sqrt{2ar}e^{i\theta/2}) - y\,\mathrm{Im}\left(\frac{\sigma\sqrt{\pi a}}{\sqrt{2\pi r}}e^{-i\theta/2}\right) \\
&= \frac{\kappa-1}{2}\sigma\sqrt{2ar}\cos\left(\frac{\theta}{2}\right) + r\sin\theta\frac{\sigma\sqrt{\pi a}}{\sqrt{2\pi r}}\sin\left(\frac{\theta}{2}\right) \\
&= \sigma\sqrt{\pi a}\sqrt{\frac{r}{2\pi}}\left\{(\kappa-1)\cos\left(\frac{\theta}{2}\right) + \sin\theta\sin\left(\frac{\theta}{2}\right)\right\} \\
\frac{E}{1+\nu}v &= \frac{\kappa+1}{2}\mathrm{Im}\left(\int Z(z)dz\right) - y\,\mathrm{Re}(Z) \\
&= \frac{\kappa+1}{2}\mathrm{Im}\left(\sigma\sqrt{2ar}e^{i\theta/2}\right) - y\,\mathrm{Re}\left(\frac{\sigma\sqrt{\pi a}}{\sqrt{2\pi r}}e^{-i\theta/2}\right) \\
&= \frac{\kappa+1}{2}\sigma\sqrt{2ar}\sin\left(\frac{\theta}{2}\right) - r\sin\theta\frac{\sigma\sqrt{\pi a}}{\sqrt{2\pi r}}\cos\left(\frac{\theta}{2}\right) \\
&= \sigma\sqrt{\pi a}\sqrt{\frac{r}{2\pi}}\left\{(\kappa+1)\sin\left(\frac{\theta}{2}\right) - \sin\theta\cos\left(\frac{\theta}{2}\right)\right\}
\end{aligned}
\tag{9.37}
$$

It should be noted here that if we were interested in computing the stress and displacement fields near a crack tip that was located at the origin—instead of at point $x = a$ (as shown in Figure 9.4), then the Westergaard stress function expression given in equation (9.34) would have changed to

$$
Z(z) = \frac{K}{\sqrt{2\pi z}} = \frac{K}{\sqrt{2\pi r}}e^{-i\frac{\theta}{2}} \tag{9.38}
$$

Note that, because of the two different locations of the crack tip, although the $Z(z)$ expressions given in equations (9.34) and (9.38) are slightly different when $Z(z)$ is expressed in terms of $z$, these two expressions are identical when $Z$ is expressed in terms of $r$ and $\theta$ measured from the crack tip. Thus, the $Z(r,\theta)$ expression is independent of the crack tip position while $Z(z)$ is not.

## 9.4 Concentrated Load on a Half Space

It is shown below that the Westergaard stress function for the problem geometry of Figure 9.5 is given by

$$
Z(z) = \frac{P}{i\pi z} = \frac{P}{i\pi r}e^{-i\theta} = -\frac{P}{\pi r}(\sin\theta + i\cos\theta) \tag{9.39}
$$

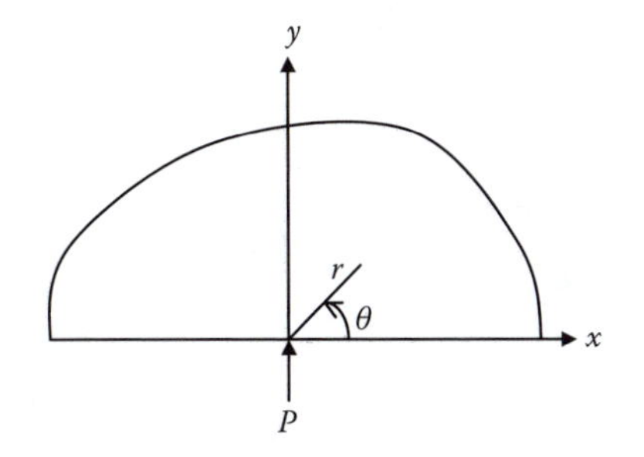


**FIGURE 9.5**
A concentrated load P is acting on the surface of the half space at the origin of the coordinate system.

Note that for the preceding stress function,

$$\mathrm{Re}(Z) = -\frac{P\sin\theta}{\pi r},\ \mathrm{Im}(Z) = -\frac{P\cos\theta}{\pi r}$$

and

$$\begin{aligned} Z'(z) &= -\frac{P}{i\pi z^2} = \frac{iP}{\pi r^2}e^{-2i\theta} = \frac{iP}{\pi r^2}(\cos 2\theta - i\sin 2\theta) \\ \mathrm{Im}(Z') &= \frac{P\cos 2\theta}{\pi r^2} \end{aligned} \tag{9.40}$$

Therefore,

$$\begin{aligned} \sigma_{xx} &= \mathrm{Re}(Z) - y\,\mathrm{Im}(Z') \\ &= -\frac{P}{\pi r}\sin\theta - r\sin\theta\cdot\frac{P}{\pi r^2}\cos 2\theta = -\frac{P}{\pi r}\sin\theta(1+\cos 2\theta) = -\frac{2P}{\pi r}\sin\theta\cos^2\theta \\ \sigma_{yy} &= \mathrm{Re}(Z) + y\,\mathrm{Im}(Z') = -\frac{P}{\pi r}\sin\theta(1-\cos 2\theta) = -\frac{2P}{\pi r}\sin^3\theta \\ \sigma_{xy} &= -y\,\mathrm{Re}(Z') = -r\sin\theta\cdot\frac{P}{\pi r^2}\sin 2\theta = -\frac{2P}{\pi r}\sin^2\theta\cos\theta \end{aligned} \tag{9.41}$$

From equation (9.41) one can clearly see that $\sigma_{yy} = \sigma_{xy} = 0$ on the half space surface at $y = 0$, for $r \neq 0$, $\theta = 0$, and $\pi$. Therefore, the stress-free boundary conditions are satisfied on the free surface.

The resultant internal vertical force generated by the stress field on the semicircle shown in Figure 9.6 is then computed to check if the stress field of equation (9.41) produces a resultant vertical force P. The resultant vertical

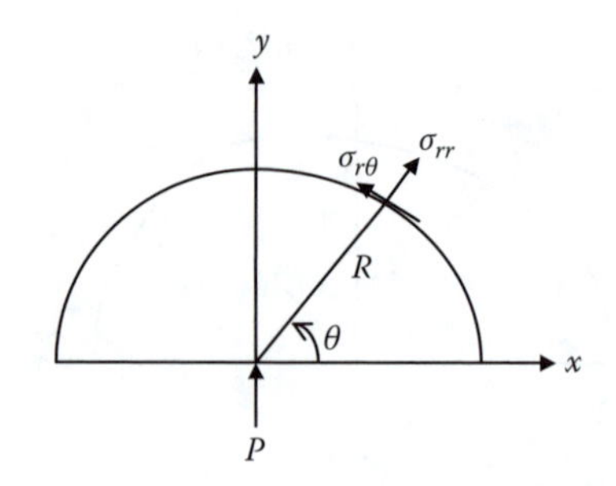


**FIGURE 9.6**
Stress directions in the semicircular region and the applied concentrated load.

force on the curved surface (see Figure 9.6) can be computed in the following manner:

$$F_y = \int_0^{\pi} (\sigma_{rr} \sin\theta + \sigma_{r\theta} \cos\theta) R d\theta \tag{9.42}$$

From the stress transformation law,

$$\begin{aligned}
\sigma_{rr} &= \ell_{rx}\ell_{rx}\sigma_{xx} + \ell_{rx}\ell_{ry}\sigma_{xy} + \ell_{ry}\ell_{rx}\sigma_{yx} + \ell_{ry}\ell_{ry}\sigma_{yy} \\
&= \sigma_{xx}\cos^2\theta + \sigma_{yy}\sin^2\theta + 2\sin\theta\cos\theta\sigma_{xy} \\
\sigma_{r\theta} &= \ell_{rx}\ell_{\theta x}\sigma_{xx} + \ell_{rx}\ell_{\theta y}\sigma_{xy} + \ell_{ry}\ell_{\theta x}\sigma_{yx} + \ell_{ry}\ell_{\theta y}\sigma_{yy} \\
&= -\sigma_{xx}\sin\theta\cos\theta + \sigma_{yy}\sin\theta\cos\theta + \sigma_{xy}(\cos^2\theta - \sin^2\theta)
\end{aligned} \tag{9.43}$$

Substituting equations (9.43) and (9.41) into equation (9.42),

$$\begin{aligned}
F_y &= \int_0^{\pi} [\sigma_{rr}\sin\theta + \sigma_{r\theta}\cos\theta] R d\theta \\
&= \int_0^{\pi} \Big[(\sigma_{xx}\cos^2\theta + \sigma_{yy}\sin^2\theta + 2\sigma_{xy}\sin\theta\cos\theta)\sin\theta \\
&\quad + (-\sigma_{xx}\sin\theta\cos\theta + \sigma_{yy}\sin\theta\cos\theta + \sigma_{xy}\{\cos^2\theta - \sin^2\theta\})\cos\theta\Big] R d\theta \\
&= \int_0^{\pi} [\sigma_{yy}(\sin^3\theta + \sin\theta\cos^2\theta) + \sigma_{xy}(2\sin^2\theta\cos\theta + \cos^3\theta - \sin^2\theta\cos\theta)] R d\theta \\
&= \int_0^{\pi} [\sigma_{yy}\sin\theta(\sin^2\theta + \cos^2\theta) + \sigma_{xy}\cos\theta(\sin^2\theta + \cos^2\theta)] R d\theta
\end{aligned}$$

$$= \int_0^{\pi} [\sigma_{yy} \sin\theta + \sigma_{xy} \cos\theta] R d\theta$$

$$= -\frac{2P}{\pi R} \int_0^{\pi} [\sin^3\theta \sin\theta + \sin^2\theta \cos^2\theta] R d\theta = -\frac{2P}{\pi R} \int_0^{\pi} \sin^2\theta [\sin^2\theta + \cos^2\theta] R d\theta$$

$$= -\frac{2P}{\pi} \int_0^{\pi} \sin^2\theta d\theta = -\frac{2P}{\pi} \int_0^{\pi} \frac{1}{2}(1 - \cos 2\theta) d\theta = -\frac{P}{\pi} \left[ \theta - \frac{1}{2} \sin 2\theta \right]_0^{\pi} = -P \tag{9.44}$$

Vertical force $-P$ computed in equation (9.44) is independent of the radius $R$ of the semicircle of Figure 9.6. Therefore, for any value of $R$ (small, medium, or large), the resultant force acting on the semicircle should always be $-P$. Thus, the semicircular region should always be in equilibrium when the external applied force $P$ acts at the origin, as shown in Figure 9.6.

## 9.5 Griffith Crack Subjected to Concentrated Crack Opening Loads *P*

Westergaard stress function for the problem geometry shown in Figure 9.7 is obtained in this section. Based on the Westergaard stress functions presented in sections 9.3 and 9.4, the stress function for this problem geometry can be postulated in the following form:

$$Z(z) = \frac{C_0}{(z - x_0)\sqrt{z^2 - a^2}} \tag{9.45}$$

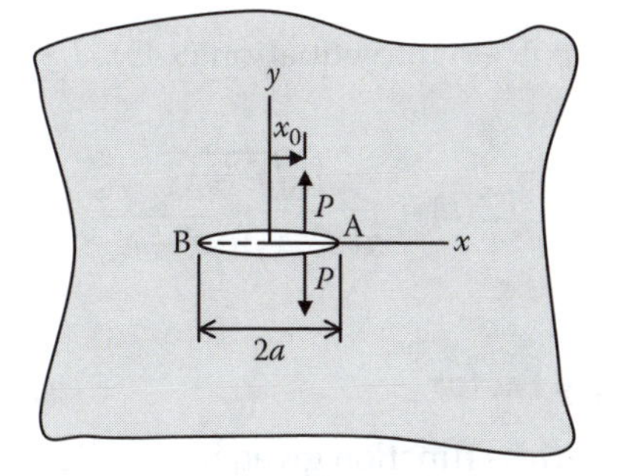


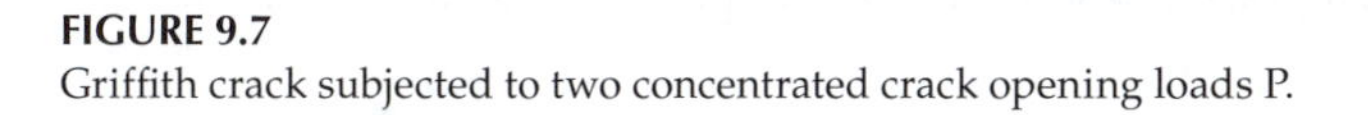

**FIGURE 9.7**
Griffith crack subjected to two concentrated crack opening loads P.

For a concentrated load at the origin acting on a half space (see section 9.4), the stress function has the $z$ dependence in the form $\frac{1}{z}$. In this problem geometry, since the concentrated force is acting at $x_0$, in the denominator of the stress function of equation (9.45), $(z - x_0)$ is taken instead of $z$. In section 9.3 (equation 9.38), we also learned that if the crack tip is present at the origin, then the Westergaard stress function should have the $z$ dependence in the form $\frac{1}{\sqrt{z}}$. Since the Griffith crack has its two crack tips at $z = \pm a$, the stress function is taken in the form $\frac{1}{\sqrt{z-a}} \times \frac{1}{\sqrt{z+a}} = \frac{1}{\sqrt{z^2-a^2}}$. Therefore, equation (9.45) takes care of both crack tips and concentrated force application conditions. The constant $C_0$ in the numerator of equation (9.45) is obtained in the following manner.

Take the limit as $z \to x_0$; in other words, $z = x_0 + \xi$ where $\xi$ is small. Then,

$$\begin{aligned} Z &= \frac{C_0}{(z - x_0)\sqrt{z^2 - a^2}} = \frac{C_0}{\xi\sqrt{z-a}\sqrt{z+a}} \\ &= \frac{C_0}{\xi\sqrt{\xi + x_0 - a}\sqrt{\xi + x_0 + a}} \approx \frac{C_0}{\xi\sqrt{x_0 - a}\sqrt{x_0 + a}} \end{aligned} \tag{9.46}$$

Simplification given in equation (9.46) is possible when $\xi$ is small compared to $(x_0 \pm a)$. Further manipulation gives

$$Z = \frac{C_0}{\xi\sqrt{x_0 - a}\sqrt{x_0 + a}} = \frac{C_0}{\xi\sqrt{x_0^2 - a^2}} = \frac{C_0}{i\xi\sqrt{a^2 - x_0^2}} \tag{9.47}$$

Comparing equations (9.47) and (9.39), one can write

$$\begin{aligned} \frac{C_0}{i\xi\sqrt{a^2 - x_0^2}} &= \frac{P}{i\pi\xi} \\ \therefore C_0 &= \frac{P\sqrt{a^2 - x_0^2}}{\pi} \end{aligned} \tag{9.48}$$

Substituting equation (9.48) into equation (9.45),

$$Z(z) = \frac{P\sqrt{a^2 - x_0^2}}{\pi(z - x_0)\sqrt{z^2 - a^2}} \tag{9.49}$$

### 9.5.1 Stress Intensity Factor

From the Westergaard stress function given in equation (9.49), the normal stress can be computed at a point ahead of the crack tip. Note that, for $z = x$, $(|x| > a)$,

$$Z(z) = \frac{P\sqrt{a^2 - x_0^2}}{\pi(z - x_0)\sqrt{z^2 - a^2}} = \frac{P\sqrt{a^2 - x_0^2}}{\pi(x - x_0)\sqrt{x^2 - a^2}} \tag{9.50}$$

Therefore, for $z = x$, ($|x| > a$), $Z'(z)$ is real:

$$\therefore \sigma_{yy}(x,0) = \text{Re}(Z) + y\,\text{Im}(Z') = \frac{P\sqrt{a^2 - x_0^2}}{\pi(x - x_0)\sqrt{x^2 - a^2}} \tag{9.51}$$

Substituting $x = a + r$, where $\frac{r}{a} << 1$, we get

$$\sqrt{x^2 - a^2} = \sqrt{x - a}\sqrt{x + a} = \sqrt{r}\sqrt{2a + r} \approx \sqrt{2ar} \tag{9.52}$$

Substitution of equation (9.52) into equation (9.51) gives

$$\begin{aligned}\therefore \sigma_{yy}(x,0) &= \frac{P\sqrt{a^2 - x_0^2}}{\pi(x - x_0)\sqrt{x^2 - a^2}} = \frac{P\sqrt{a^2 - x_0^2}}{\pi(a + r - x_0)\sqrt{2ar}} \approx \frac{P\sqrt{a^2 - x_0^2}}{\pi(a - x_0)\sqrt{2ar}} \\ &= \frac{P}{\sqrt{\pi a}}\sqrt{\left(\frac{a + x_0}{a - x_0}\right)}\frac{1}{\sqrt{2\pi r}} = \frac{K}{\sqrt{2\pi r}}\end{aligned} \tag{9.53}$$

From equation (9.53) the stress intensity factor for the problem geometry is obtained:

$$K = \frac{P}{\sqrt{\pi a}}\sqrt{\frac{a + x_0}{a - x_0}} \tag{9.54}$$

## 9.6 Griffith Crack Subjected to Nonuniform Internal Pressure

In this section the Westergaard stress function for the problem geometry shown in Figure 9.8 is obtained. Note that over an elemental length $d\alpha$ at a distance $\alpha$ from the origin, the crack opening force acting on the crack surface is given by $f(\alpha)d\alpha$. From equation (9.49) one can postulate the Westergaard stress function for this elemental force $f(\alpha)d\alpha$ as

$$dZ = \frac{\sqrt{a^2 - \alpha^2}\, f(\alpha)d\alpha}{\pi(z - \alpha)\sqrt{z^2 - a^2}} \tag{9.55}$$

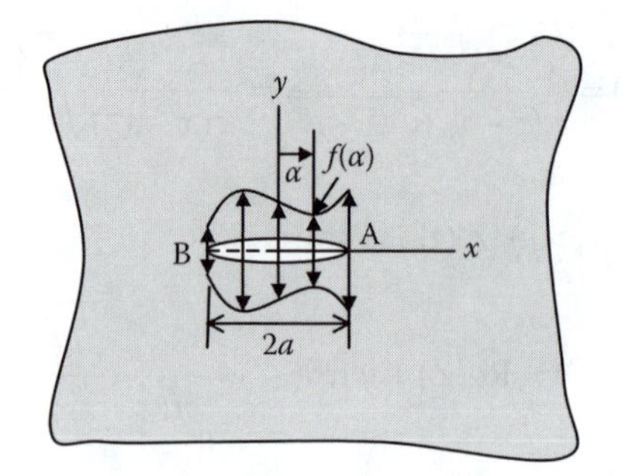


**FIGURE 9.8**
Griffith crack subjected to nonuniform distribution of normal traction.

Integrating equation (9.55), the complete Westergaard stress function is obtained:

$$Z = \int_{-a}^{a} \frac{\sqrt{a^2 - \alpha^2}\, f(\alpha)}{\pi(z-\alpha)\sqrt{z^2 - a^2}} d\alpha \tag{9.56}$$

Following similar steps as given in section 9.5.1, the SIF for this stress function is obtained:

$$K = \frac{P}{\sqrt{\pi a}} \int_{-a}^{a} \sqrt{\frac{a+\alpha}{a-\alpha}} f(\alpha)\, d\alpha \tag{9.57}$$

For the special case of $f(\alpha) = \sigma_0$, equation (9.57) gives

$$K = \frac{\sigma_0}{\sqrt{\pi a}} \int_{-a}^{a} \sqrt{\frac{a+\alpha}{a-\alpha}} d\alpha = \sigma_0 \sqrt{\pi a} \tag{9.58}$$

## 9.7 Infinite Number of Equal Length, Equally Spaced Coplanar Cracks

It can be shown that the Westergaard stress function for the problem geometry shown in Figure 9.9 is given by

$$Z(z) = \sigma \left\{ 1 - \left[ \frac{\sin(\frac{\pi a}{W})}{\sin(\frac{\pi z}{W})} \right]^2 \right\}^{-\frac{1}{2}} \tag{9.59}$$

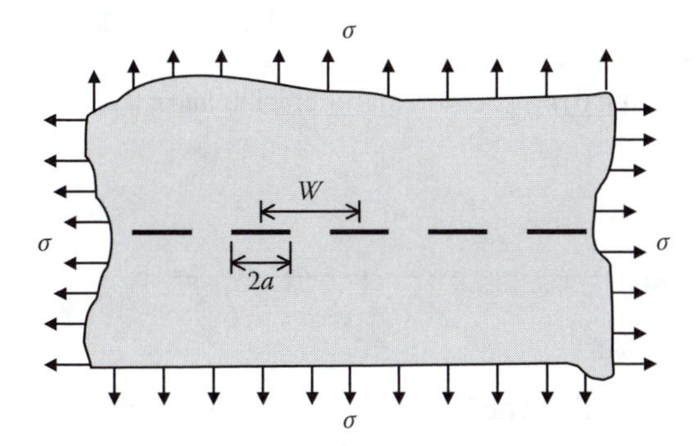


**FIGURE 9.9**
Infinite number of equal length ($2a$) equally spaced (W) coplanar cracks subjected to biaxial state of stress.

The SIF from this stress function is obtained as

$$K = \sigma \left\{ W \tan\left( \frac{\pi a}{W} \right) \right\}^{\frac{1}{2}} \tag{9.60}$$

## 9.8 Concluding Remarks

A few basic problems of fracture mechanics are solved in this chapter. For analytical solutions of more complex problems, readers are referred to the references cited in the reference list.

## References

Eftis, J. and Liebowitz, H. On the modified Westergaard equations for certain plane crack problems. *International Journal of Fracture Mechanics,* 8, 383–392, 1972.

Goodier, J. N. Mathematical theory of equilibrium of cracks. In *Fracture II,* pp. 2–67, ed. Liebowitz, New York: Academic Press, 1969.

Muskhelishvili, N. I. *Some basic problems of the mathematical theory of elasticity* (1933), English translation, Groningen, the Netherlands: Noordhoff, 1953.

Paris, P. C. and Sih, G. C. Stress analysis of cracks. *ASTM STP* 391, 30–81, 1965.

Rice, J. R. Mathematical analysis in mechanics of fracture. In *Fracture II,* pp. 192–308, ed. Liebowitz, New York: Academic Press, 1969.

Sih, G. C. On the Westergaard method of crack analysis. *International Journal of Fracture Mechanics,* 2, 628–631, 1966.

———, ed. *Methods of analysis and solutions of crack problems.* Groningen, the Netherlands: Noordhoff, 1973.

Westergaard, H. M. Bearing pressures and cracks. *Journal of Applied Mechanics,* 61, A49–53, 1939.

## Exercise Problems

Problem 9.1: Prove that the Westergaard stress function for the problem geometry shown in Figure 9.10 should be

$$Z(z) = \frac{2Pz\sqrt{a^2 - b^2}}{\pi(z^2 - b^2)\sqrt{z^2 - a^2}}$$

Problem 9.2: For the problem geometry shown in Figure 9.10, compute the stress field along lines (a) $x = 0$, $y > 0$; and (b) $y = 0$, $x > a$.

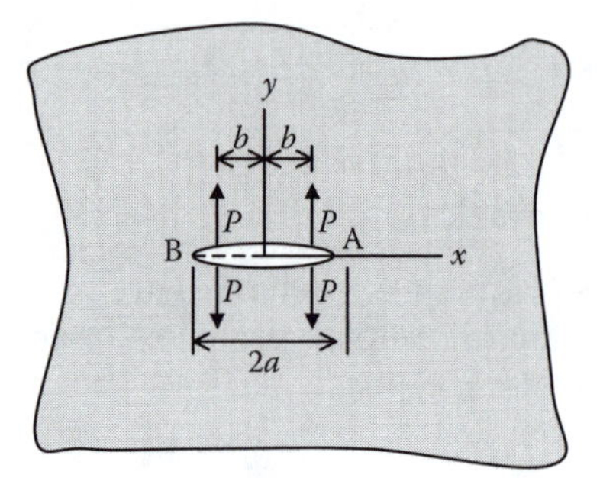


FIGURE 9.10

# 10

# *Advanced Topics*

## 10.1 Introduction

Solutions of some advanced problems are discussed in this chapter and relevant references are given.

## 10.2 Stress Singularities at Crack Corners

In all crack problems considered so far the crack is assumed to have a sharp edge and a smooth front in the crack plane. Four different crack geometries are shown in plan views A, B, C, and D in Figure 10.1. Note that in the elevation view all four cracks look similar—a crack of length 2a. Out of these four cracks so far we have discussed the solutions of crack geometries A (Griffith crack) and B (elliptical crack). No discussion on the stress field variation near the crack corners as shown in problem geometries C and D has been presented in previous chapters. This problem has been solved by Xu and Kundu (1995).

Stress singularity at the crack corner of angle $2\alpha$, as shown in Figure 10.2, has been investigated by Xu and Kundu (1995). It should be noted here that at point A of this crack two tangents can be drawn. The angle between these two tangents on the crack side is the crack corner angle. At a point on the smooth crack front only one tangent can be drawn and it has been shown earlier that the stress field has a square root singularity in front of the smooth crack front. The stress singularity near the crack corner is shown in Figure 10.3. In this Figure the singularity parameter $n$ is plotted against the half-crack angle ($\alpha$). Note that when $\alpha = 90°$ (or $2\alpha = 180°$), the crack front becomes smooth. From Figure 10.3 one can see that for $\alpha = 90°$, $n = 0.5$. Therefore, the stress field should vary near $r = 0$, in the following manner:

$$\sigma_{ij}\Big|_{r\to 0} = \frac{1}{r^{1-n}} = \frac{1}{r^{1-0.5}} = \frac{1}{\sqrt{r}} \text{ for } \alpha = 90° \qquad (10.1)$$

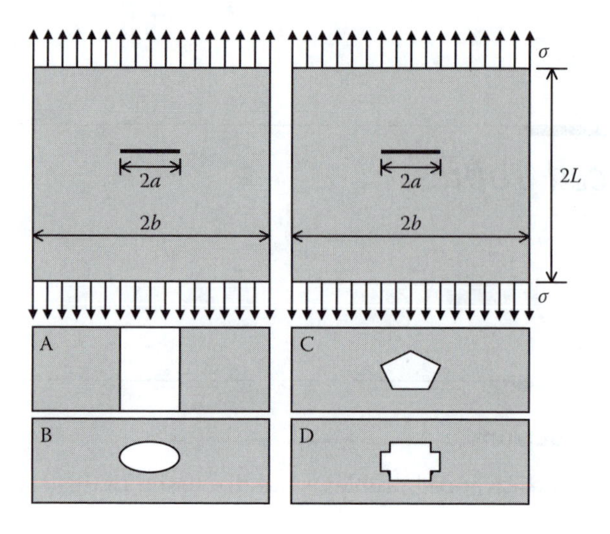


**FIGURE 10.1**
Elevation (top two figures) and plan views (A, B, C, and D) of different crack geometries.

For $\alpha < 90°$, in Figure 10.3 one can clearly see that $n > 0.5$. Therefore, the order of singularity $m = (1 - n) < 0.5$ for a crack corner angle less than 180°. Similarly, for a crack corner angle greater than 180° the order of singularity $m = (1 - n) > 0.5$. Therefore, for the heart-shaped crack shown in Figure 10.4, point A has the highest order of singularity and is most likely to fail while point B has the lowest order of singularity and is least likely to fail.

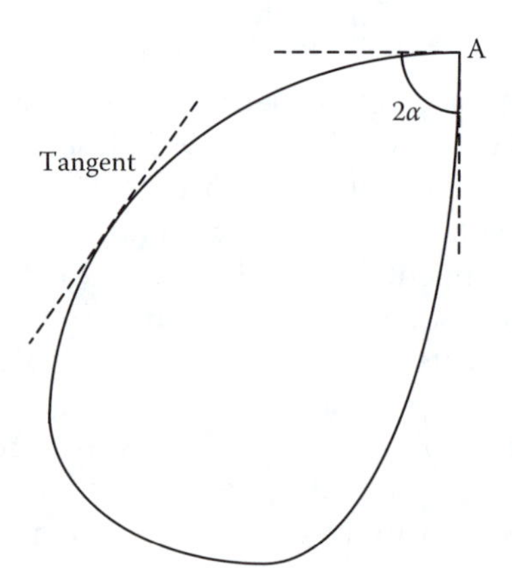


**FIGURE 10.2**
Crack geometry with a sharp corner.

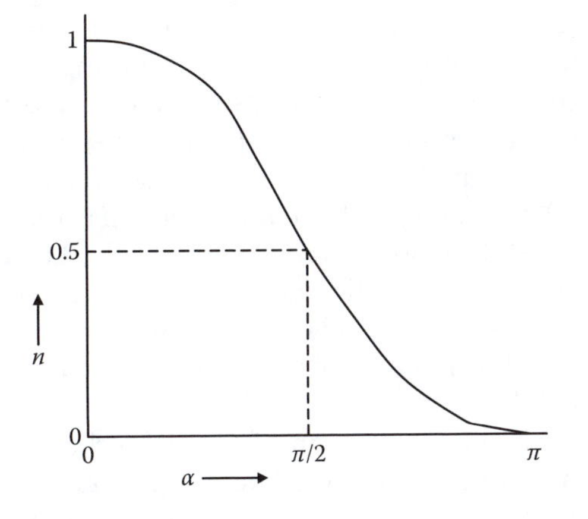


**FIGURE 10.3**
Singularity parameter $n$ versus $\alpha$. Order of singularity is $m$ ($m = 1 - n$).

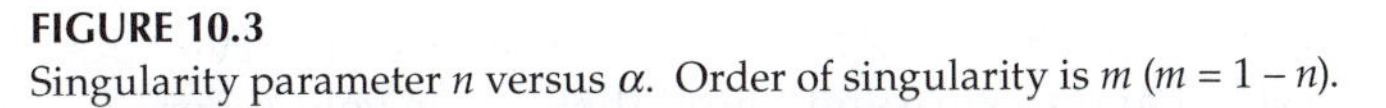

## 10.3 Fracture Toughness and Strength of Brittle Matrix Composites

Shah and Ouyang (1993) described four different mechanisms—crack shielding in fracture process zone, crack deflection, crack surface roughness induced closure, and bridging mechanism—that contribute to the increasing toughness of fiber reinforced brittle matrix composite (FRBMC) materials. Li et al. (1992) have shown that for continuous aligned fiber composites, the main mechanism that contributes to the increasing fracture toughness is multiple cracking of the matrix. However, for randomly distributed fibers in a brittle matrix, the fiber bridging force is the major contributor to the toughening mechanism of FRBMCs.

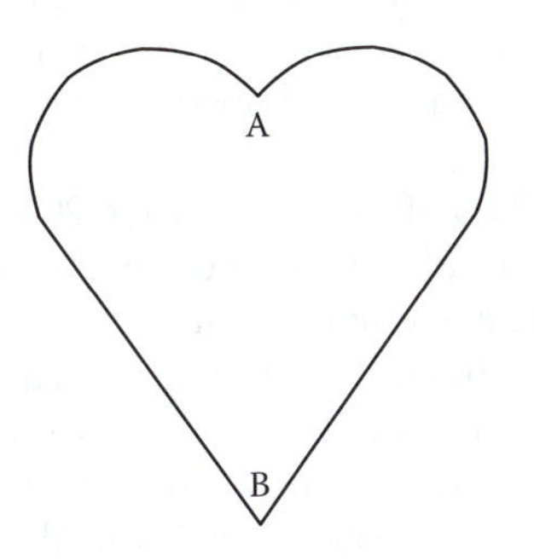


**FIGURE 10.4**
Heart-shaped crack. Point A has the highest order singularity and is most likely to fail while point B has the lowest order singularity and is least likely to fail.

Many studies (Cox and Marshall 1991; Cox 1993; Li and Ward 1989) have concentrated on the determination of the fiber bridging forces. In these studies the fiber bridging force ($p$) has been assumed to be a function of the crack opening ($u$). Different types of functional variations, $p(u)$, have been postulated and mathematical analyses based on such assumed $p(u)$ have been carried out. Cox (1993) assumed $p(u)$ to be $\beta u^{\alpha}$, where $\beta$ is the bridging stiffness and $\alpha$ can be 0.5 (for sliding fibers of infinite strength), 1.0 (for linear springs), or greater than unity for rubber ligaments. Li and Ward (1989) assumed $p(u)$ to be the first derivative of the Bessel function with respect to the crack mouth opening displacement; $p(u)$ has been also modeled as a summation series of Legendre polynomials.

Experiments by Cha et al. (1994) and Luke, Waterhouse, and Wooldridge (1974) have shown that the addition of fibers significantly increases the fracture toughness of FRBMCs, but the effect on the elastic stiffness of the composite is small. Ward et al. (1989) studied fracture resistance of acrylic fiber reinforced mortar in shear and flexure and found that, as the volume fraction of fibers is increased, the strength in shear and flexure, the fracture energy and the critical crack opening all increase, the tensile strength remains essentially constant and the compressive strength shows some reduction.

Counter examples of these observations are also available in the literature. For example, Shah (1991) has shown that addition of 10–15% volume fraction of fibers may substantially increase the tensile strength of matrices. This is probably because fibers suppress the localization of microcracks into macrocracks. Li and Hashida (1993) observed a ductile nature of the FRBMC fracturing with a large diffused microcrack region. They classified the FRBMC fracture process into three categories: brittle fracture dominated by microcracking, quasibrittle fracture dominated by fiber bridging, and ductile fracture. These different types of fracture processes observed in FRBMCs make the FRBMC strength and toughness analysis and estimation very complex. Furthermore, depending on the definition, the toughness of FRBMCs may or may not be sensitive to fiber parameters. Gopalaratnam et al. (1991) have shown that ASTM C108 toughness index is insensitive to the fiber type, fiber volume fraction, and specimen size, but the toughness as a measure of absolute energy is sensitive to these parameters.

Wang, Backer, and Li (1987) found that when 2% volume fraction of acrylic fibers of different length is added to concrete, the fracture toughness monotonically increases with the fiber length (variation from 6.4 to 19.1 mm) for one definition of fracture toughness ($T_0$) and this variation is nonmonotonic for another definition of fracture toughness ($DTI_{20}$). On the other hand, for constant fiber length (6.4 mm), when fiber volume fraction is increased (from 2 to 6.5%), the $DTI_{20}$ toughness increases monotonically but $T_0$ toughness shows nonmonotonic variation. Wang, Li, and Backer (1990) found that Aramid B fiber reinforced concrete shows nonmonotonic variation of tensile strength and fracture energy with the increase of fiber volume fraction from 1 to 3%. In their study they reported that 100% increase in fiber length while keeping other parameters unchanged increases the tensile strength by 5.4–34.1% for different fiber volume fractions.

In another study Li et al. (1996) found that steel fiber reinforced concrete shows significant increase of strength with optimum fiber–matrix interface bond strength. Importance of proper fiber–matrix bonding has been recognized and studied by Kanda and Li (1998) and Marshall et al. (1994). Different damage mechanisms for strongly and weakly bonded composites have been observed by Marshall et al. They have shown that strong interfacial bonding does not necessarily lead to optimum transverse strength of the composite.

### 10.3.1 Experimental Observation of Strength Variations of FRBMCs with Various Fiber Parameters

Luke et al. (1974) carried out a parametric study of steel fibers on the flexural strength of concrete and studied the effect of fiber length, diameter, shape, and volume fraction on the first crack strength (the point at which the load-deflection curve deviates from linearity) and ultimate strength of concrete. They concluded from their experimental investigation that the fiber length, fiber diameter, fiber shape, and fiber volume fraction affect the flexural strength of the concrete. Longer fibers with smaller diameter and higher volume fraction increase the ultimate strength of fiber reinforced concrete.

Luke et al. (1974) have shown that the ultimate flexural strength of fiber reinforced concrete increases from 910 psi (6.27 MPa, 1 psi = 6.89 kPa) to 1880 psi (12.95 MPa, 107% increase) as the fiber volume fraction increases from 0.3 to 2.5% (733% increase). For 1% volume fraction of fibers, with the increase of the fiber length the flexural strength increases from 1190 psi (8.2 MPa) to 1500 psi (10.34 MPa, 26% increase) for 0.01 in. (0.254 mm) diameter fibers; from 1155 psi (7.96 MPa) to 1455 psi (10.02 MPa, 26% increase) for 0.016 in. (0.406 mm) diameter fibers; and 1050 psi (7.23 MPa) to 1580 psi (10.89 MPa, 50% increase) for 0.02 in. (0.508 mm) diameter fibers. These three increases correspond to the fiber length variations of 0.5–1.25 in. (12.7–31.75 mm, 150% increase), 0.75–2 in. (19.05–50.8 mm, 167% increase), and 1.5–2.5 in. (38.1–63.5 mm, 67% increase), respectively. Luke et al. have also shown that for 1% volume fraction and 1 in. (25.4 mm) fiber length, as the fiber diameter decreases from 0.016 to 0.006 in. (0.406 to 0.152 mm, 62.5% decrease), the ultimate strength increases from 1115 psi (7.68 MPa) to 1795 psi (12.37 MPa, 61% increase). For 2% volume fraction and 1 in. (25.4 mm) long fibers, a decrease of fiber diameter from 0.016 to 0.012 in. (0.406 mm to 0.305 mm, 25% decrease) results in an increase in the ultimate strength from 1470 psi (10.13 MPa) to 1645 psi (11.33 MPa, 11.9% increase).

Luke and colleagues (1974) also studied the effect of the fiber shape and found that flat fibers increase the strength more than round fibers. The increase in the ultimate strength with flat fibers, in comparison to round fibers, was quite significant; this increase varied from 30 to 100% for different lengths and volume fractions of fibers. There was no attempt in their paper to quantitatively predict the change in the ultimate strength or fracture toughness of the

**TABLE 10.1**

Effect of Steel Fibers on Ultimate Strength ($\sigma_U$) of Reinforced Concrete

| Fiber Parameter that Changes | Amount of Change | Percentage (%) Change of Parameters | $\Delta\sigma_U$ = Ultimate Strength Change (in psi, 1 psi = 6.89 kPa) | Percentage Change of $\sigma_U$ (%) |
|---|---|---|---|---|
| Volume fraction | 0.3–0.5% | 66.7 | 910–1055 | 15.9 |
| Volume fraction | 0.3–1% | 233 | 910–1115 | 22.5 |
| Volume fraction | 0.3–1.5% | 400 | 910–1325 | 45.6 |
| Volume fraction | 0.3–2.0% | 567 | 910–1470 | 61.5 |
| Volume fraction | 0.3–2.5% | 733 | 910–1880 | 107 |
| Length (v.f. = 1%, dia. = 0.01 in. = 0.254 mm) | 0.5–1.25 in. | 150 | 1190–1500 | 26 |
| Length (v.f. = 1%, dia. = 0.016 in. = 0.406 mm) | 0.75–1.5 in. | 100 | 1155–1180 | 2.2 |
| Length (v.f. = 1%, dia. = 0.016 in. = 0.406 mm) | 0.75–2 in. | 167 | 1155–1455 | 26 |
| Length (v.f. = 1%, dia. 0.02 in. = 0.508 mm) | 1.5–2.5 in. | 67 | 1050–1580 | 50 |
| Length (v.f. = 2%, dia. 0.016 in. = 0.406 mm) | 0.75–1.25 in. | 67 | 1295–1905 | 47 |
| Dia. (v.f. = 1%, length = 1 in. = 25.4 mm) | 0.016–0.014 in. | –12.5 | 1115–1215 | 9 |
| Dia. (v.f. = 1%, length = 1 in. = 25.4 mm) | 0.016–0.012 in. | –25 | 1115–1230 | 10.3 |
| Dia. (v.f. = 1%, length = 1 in. = 25.4 mm) | 0.016–0.010 in. | –37.5 | 1115–1320 | 18.4 |
| Dia. (v.f. = 1%, length = 1 in. = 25.4 mm) | 0.016–0.006 in. | –62.5 | 1115–1795 | 61 |
| Dia. (v.f. = 2%, length = 1 in. = 25.4 mm) | 0.016–0.014 in. | –12.5 | 1470–1615 | 9.9 |
| Dia. (v.f. = 2%, length = 1 in. = 25.4 mm) | 0.016–0.012 in. | –25 | 1470–1645 | 11.9 |

*Source:* Luke, C. E., Waterhouse, B. L., and Wooldridge, J. F. *Steel fiber reinforced concrete optimization and applications, an international symposium on fiber reinforced concrete.* Detroit, MI: American Concrete Institute, SP-44, pp. 393–413, 1974.

composite with the change in the fiber diameter, length, shape, or volume fraction. Details of the experimental observations of Luke et al. are shown in Table 10.1. Their experimental observations were later quantitatively justified by Kundu et al. (2000) using a simple model of linear elastic fracture mechanics. Kundu et al. (2000) predicted strength variations in FRBMCs with the variations of fiber length, diameter, and volume fraction. This model predicts that a composite material's tensile and flexural strength should increase nonlinearly with the fiber volume fraction. They also predicted that similar nonlinear behavior should be observed with the reduction of the fiber diameter when other parameters are kept constant. They showed in their paper how the variation of the FRBMC strength and toughness can be quantitatively related to the variations in fiber parameters: volume fraction, fiber length, and diameter.

### 10.3.2 Analysis for Predicting Strength Variations of FRBMCs with Various Fiber Parameters

From the preceding discussion it is clear that fracture toughness and strength properties of FRBMCs depend on many factors (e.g., geometries and properties of fibers, matrices, and interfaces). As mentioned before, several investigators have carried out rigorous studies on different aspects of fracture and failure processes of FRBMCs. These detailed analyses are important for understanding the intricate mechanics of the fracture process of FRBMCs. However, sometimes too many variables, too many unknowns, and complex mathematical analyses create obstacles to the fundamental or conceptual understanding of the basic process that is responsible for increasing the strength of FRBMCs. A simplified analysis based on the linear elastic fracture mechanics (LEFM) knowledge presented by Kundu et al. (2000) provides qualitative and quantitative explanations of some observed phenomena of strength variations in FRBMCs with the volume fraction, length, and diameter of reinforcing fibers.

Addition of fibers to a brittle material increases both its ultimate strength and fracture toughness. The mechanisms responsible for these two increases are related. Brittle materials, like concrete and ceramic, contain a large number of randomly distributed microcracks. Size and shape of these microcracks vary from material to material. When fibers are added, the randomly distributed fibers intersect these microcracks and bridge the gap between two surfaces of the crack, as shown in Figure 10.5. When the material is loaded and the cracks want to propagate, the fibers apply a restraining force that makes it harder for the crack to propagate. As a result, in the presence of fibers the strength of the material increases. The restraining force applied by the fibers comes from the friction and cohesive force (due to chemical bonding) between the fiber and the matrix material. Since the cohesive and friction forces increase with the fiber surface area and flat fibers have greater surface area than the round fibers for the same volume of fibers, the flat fibers

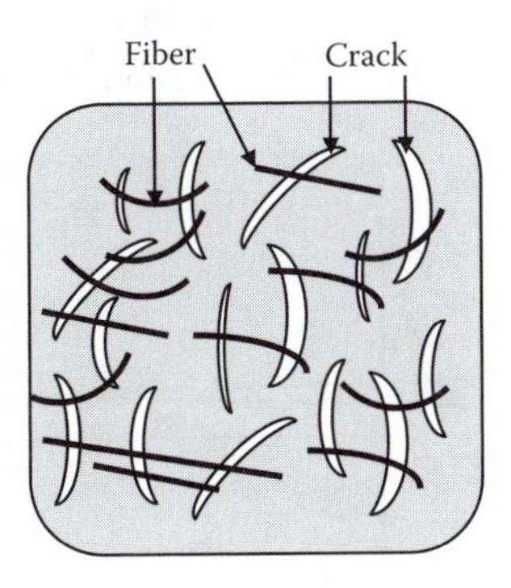


**FIGURE 10.5**
Fibers intersecting cracks increase the strength of the fiber-reinforced brittle matrix composite materials.

increase the fracture toughness more in comparison to the round fibers. This has been experimentally observed by Luke et al. (1974).

#### *10.3.2.1 Effect of Fiber Volume Fraction*

Table 10.1 shows that an increase in the fiber volume fraction increases the FRBMC strength when fiber diameter and length are kept unchanged. However, this increase is nonlinear. Note that 733% increase in the fiber volume fraction increases the ultimate strength of concrete by 107%, while 567% increase in fiber volume fraction increases the ultimate strength by only 61.5%. Can the nonlinear increase of the ultimate strength of FRBMCs be explained using the LEFM?

Following the approach suggested by Kundu et al. (2000), let us assume that in the absence of fibers, the matrix material has a fracture toughness of $K_c$. The fibers bridge the gap between the two surfaces of a crack as shown in Figure 10.5 and generate restoring forces when the crack tries to open more and propagate. The stress intensity factor (SIF), $k_1$, for a semi-infinite crack subjected to two opposing forces as shown in Figure 10.6 is given by (see equation 4.23)

$$k_1 = P\sqrt{\frac{2}{\pi\alpha}} \tag{10.2}$$

where $P$ is the force magnitude and $\alpha$ is the distance of the applied force from the crack tip.

Let the SIF of a cracked specimen in absence of the fibers be $K$; then for the same problem geometry, the stress intensity factor in presence of the fibers should be $(K–k_1)$ since the fibers produce a restoring force $P$ that tries to close the crack opening. Note that $K$ and $k_1$ are independent of each other; $K$ depends on the specimen geometry and applied loads while $k_1$ depends on the restoring force magnitude $P$ and its point of application relative to the crack tip. If one considers a large number of fibers and microcracks being

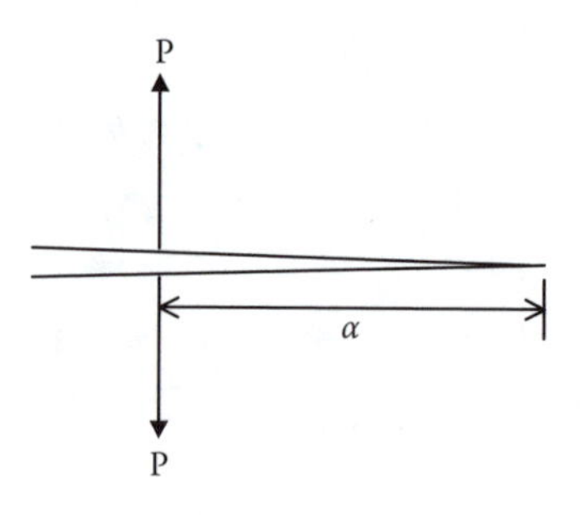


**FIGURE 10.6**
Two concentrated loads of magnitude $P$ are acting at a distance $\alpha$ from the crack tip.

randomly distributed in a material, as shown in Figure 10.5, then it can be assumed that the restoring force $P$ is proportional to the fiber volume fraction when the fiber volume fraction is changed without altering the number of microcracks in the material. As a result, for $Y$% increase of the fiber volume fraction, the restoring force ($P$) will increase by $Y$%. Then, from equation 10.2, it is easy to see that $k_1$ should also increase by about $Y$%, while $K$ remains unchanged. Then the resulting stress intensity factor becomes $[K - (1 + Y\%)k_1] = [K - (1 + 0.01Y)\, k_1]$. If we assume that the matrix fails when the resulting SIF reaches the critical stress intensity factor $K_c$, then the failure criterion in absence and in presence of fibers would be

$$\begin{aligned} &K_c = K \text{ (in absence of any fiber)} \\ &K_c = K - k_1 \text{ (in the presence of fibers)} \\ &K_c = K - k_1 - 0.01Y k_1 \text{ (for } Y\% \text{ increase of the fiber volume fraction)} \end{aligned} \tag{10.3}$$

If it is assumed that $k_1$ is $n$ times $K$, then from the relation (10.3) it can be shown that the percentage increase ($X$) in the failure load for a $Y$% increase of the fiber volume fraction is given by

$$X = \frac{\frac{K_c}{1-(1+0.01Y)n} - \frac{K_c}{1-n}}{\frac{K_c}{1-n}} = \frac{Yn}{1-(1+0.01Y)n} \tag{10.4}$$

From equation (10.4), $n$ can be expressed in the following form:

$$n = \frac{X}{X+Y+0.01XY} \tag{10.5}$$

where $X$ and $Y$ of equation (10.5) can be obtained experimentally.

From Table 10.1, we get $X = 107\%$ for $Y = 733\%$; then equation (10.5) gives $n = 0.066$. If one substitutes this value of $n$ in equation (10.4) and calculates $X$ for different values of $Y$ (varying from 0 to 733%), then a theoretical prediction of $X$ for different fiber volume fraction increases can be obtained. This plot is shown in Figure 10.7; the experimental values from Table 10.1 are shown by square markers on the same plot. The theoretical curve (equation 10.4) obtained from this simple model shows good agreement with the experimental values. Only one experimental data point that corresponds to the 0.5% volume fraction (66.7% increase in the fiber volume) shows comparatively large deviation from the theoretical curve in Figure 10.7. This may be due to some experimental error associated with that point or with the very first point (0.3% fiber volume fraction). Note that if the ultimate strength associated with the 0.3% fiber volume fraction (the first experimental point of Figure 10.7) is increased from 910 psi (6.27 MPa) to, say, close to 1000 psi (6.89 MPa), then the matching between the theoretical curve and the experimental values becomes even better.

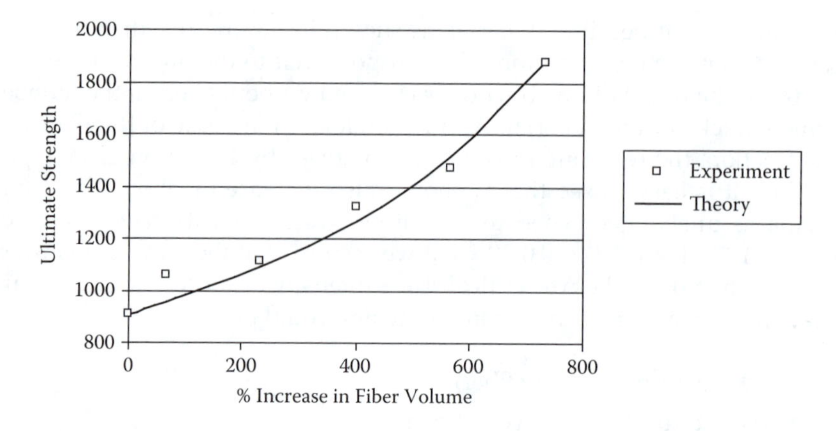


**FIGURE 10.7**
Theoretical prediction (continuous line) and experimental results (square markers) showing the ultimate strength (in psi) variation in FRBMCs with the increase of the fiber volume fraction (1 psi = 6.89 kPa).

It should be noted here that the increase in the ultimate strength has been so far explained considering only the bridging effects of fibers. However, presence of fibers in front of the crack tip, as shown in Figure 10.8, also increases the value of $K_c$; the exact amount of increase depends on the fiber distribution (spacing, etc.) ahead of the crack tip (Evans 1988, 1989, 1991; Evans and Zok 1994). Thus, if one accounts for both the bridging effect and the effect of fibers ahead of the crack tip, then the failure criterion should be

$$
\begin{aligned}
&K_c = K \text{ (in absence of any fiber)}\\
&K^*_c = K - k_1 \text{ (in the presence of fibers)}\\
&K^{**}_c = K - k_1 - 0.01Y\,k_1 \text{ (for } Y\% \text{ increase of the fiber volume fraction)}
\end{aligned}
\tag{10.6}
$$

where $K^*_c$ and $K^{**}_c$ are critical stress intensity factors of the composite in presence of fibers in front of the crack tip. However, although an analysis based on equation (10.6) will be more accurate than the one presented here (equations 10.3–10.5), the improved analysis also involves a greater number of unknowns, such as $K^*_c$ and $K^{**}_c$, and is not considered.

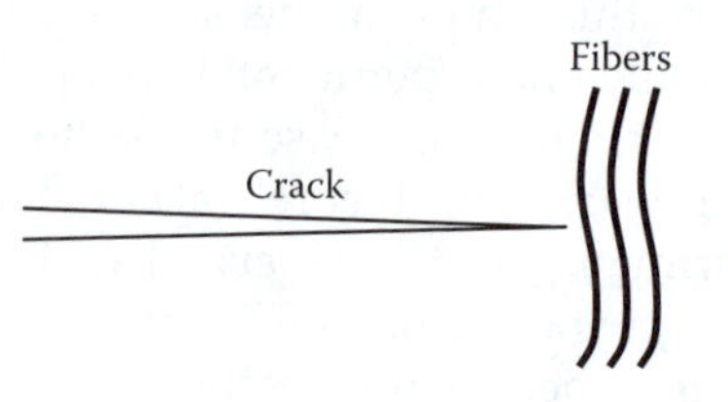


**FIGURE 10.8**
Presence of fibers in front of the crack tip increases composite strength by creating barriers to the crack propagation.

#### *10.3.2.2 Effect of Fiber Length*

Fiber forces are generated due to cohesion (chemical bonding) and friction between the fiber and the matrix. As a result, one can logically assume that the fiber force is proportional to the interface area between the fiber and the matrix. A simplifying assumption of our analysis is that this bridging force mechanism is mainly responsible for the fracture toughness increase. If this assumption is correct, then we should observe that when the fiber volume fraction and diameter are kept unchanged but the fiber length is increased, then, because the fiber surface area does not change, the ultimate strength should not change. Experimental results sometimes support this prediction, but often they do not (see Table 10.1).

Table 10.1 shows that, without altering the fiber volume fraction, if the fiber length is increased by 100% (for 1% fiber volume fraction and 0.016 in. fiber diameter), the ultimate strength increases by only 2.2%. However, for the same volume fraction and fiber diameter, when the fiber length is increased by 167%, a 26% rise in the ultimate strength is observed. For different volume fractions and fiber diameters, when the fiber length increases by 150% and 67%, the ultimate strength increases by 26% and 50%, respectively. From these experimental results, it can be clearly seen that the strength increase of FRBMCs with the change in the fiber length is not consistent. Sometimes it is almost independent of the fiber length (100% increase in the fiber length showing only 2.2% increase in the ultimate strength) and can be explained by the simplified theory presented here.

However, at other times, the ultimate strength increases with the fiber length but the rate of increase is not consistent; while a 67% length increase provides 50% increase in the ultimate strength, a 150% length increase (for a different fiber diameter) shows only 26% increase in the ultimate strength. A quantitative theoretical prediction like in Figure 10.7 is not possible in this case since the ultimate strength variation is not consistent with the fiber length change. To explain this phenomenon one may have to take into account other mechanisms (such as large-scale bridging [LSB], small-scale bridging [SSB], etc.) in addition to the fiber–matrix interface cohesive force.

In addition to the total fiber surface area, distribution of fibers relative to the crack geometry may also play an important role in deciding the level of resistance that should be generated to oppose the crack propagation. To investigate the effect of the fiber distribution on the SIF of a cracked material, the SIF of a semi-infinite crack in an infinite medium is computed with fiber forces of different distributions. As mentioned earlier, the SIF for the geometry shown in Figures 10.6 and 10.9a is given by equation (10.2). The SIF for the problem geometry shown in Figure 10.9b can be obtained from equation (10.2) after applying the superposition principle:

$$k_1 = \frac{P}{2}\sqrt{\frac{2}{\pi(\alpha-\varepsilon)}} - \frac{P}{2}\sqrt{\frac{2}{\pi(\alpha+\varepsilon)}} = P\frac{\sqrt{\alpha+\varepsilon}+\sqrt{\alpha-\varepsilon}}{\sqrt{2\pi(\alpha^2-\varepsilon^2)}} \tag{10.7}$$

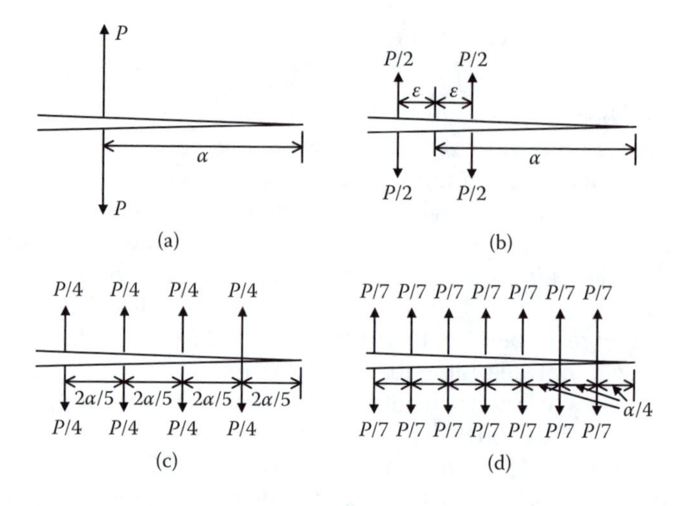


**FIGURE 10.9**
Semi-infinite cracks subjected to a resultant opening load $P$ at a distance $\alpha$ from the crack tip when the loads act at (a) one point, (b) two points, (c) four points, and (d) seven points.

where $2\varepsilon$ is the distance between the two crack opening forces of magnitude $P/2$ and $\alpha$ is the distance between the crack tip and the line of action of the resultant force $P$ (see Figure 10.9b).

For $\varepsilon = \alpha/3$, $k_1$ in equation (10.7) becomes $1.045P(2/\pi\alpha)^{0.5}$; compared to equation (10.2) this represents an increase of 4.5% in the stress intensity factor. Hence, if a cracked infinite plate has a stress intensity factor of $K$ in absence of any fiber, then addition of fibers will reduce the stress intensity factor to $(K - k_1)$ if one fiber applies a closing force $P$ (in this case the fiber force will be opposite to the force direction, shown in Figure 10.9a). The SIF will be reduced to $(K - 1.045k_1)$ if two fibers apply a total load of $P$ as shown in Figure 10.9b.

Distributing the total force $P$ over a larger number of fibers changes the stress intensity factor. When the load is distributed over four fibers, each carrying $P/4$ as shown in Figure 10.9c, the SIF becomes $1.101P(2/\pi\alpha)^{0.5}$—an increase of 10.1% compared to the single fiber case. If the load is distributed over seven fibers as shown in Figure 10.9d, the SIF becomes $1.15P(2/\pi\alpha)^{0.5}$—a 15% increase when compared to equation (10.2). For the uniform distribution of fiber forces as the load is distributed over a greater number of fibers, the SIF ($k_1$) gradually increases, causing a larger resistance to failure or increase of the ultimate strength. According to this calculation, if the fiber length is decreased keeping its diameter and the volume fraction constant, then the ultimate strength should increase because shorter fibers will give a greater number of fibers with smaller force per fiber. But note that the experimental observation (Table 10.1) contradicts this prediction.

A major drawback of the preceding analysis is that the fiber loads are assumed to be constant over the crack length. It should be noted here that

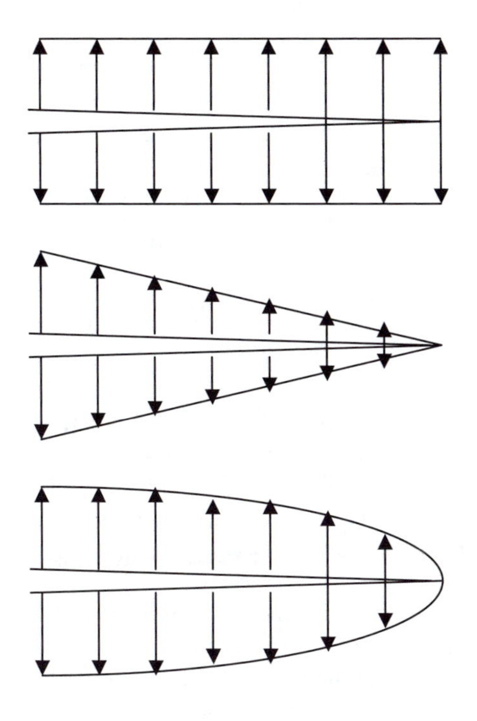

**FIGURE 10.10**
Uniform, linear, and quadratic variations of the fiber bridging force along the crack length, considered in the analysis.

although the fiber pull-out force or the fiber failure force is independent of the crack opening, the fiber bridging force is not. The fiber failure force depends on either the cohesive force at the fiber–matrix interface for short fibers or the tensile strength of the fiber for long fibers. However, the fiber bridging force generated by the tensile strain of a fiber depends on the crack opening displacement. As a result, the fiber bridging force is a function of the crack opening displacement and it increases from the crack tip to the center of the crack.

Linear and quadratic variations of the fiber force distribution (see Figure 10.10), in addition to the uniform distribution that has been already considered, are analyzed next. It should be noted that the nonlinear fiber force distribution is more realistic. Let the stress intensity factor $k_1$ be defined by the relation $k_1 = \gamma P\sqrt{\frac{2}{\pi\alpha}}$. Then $\gamma$ is computed numerically for different numbers of fibers and various types of fiber force distribution (quadratic, linear, and constant). Table 10.2 shows the $\gamma$ values computed for these various cases. After knowing $\gamma$, one can compute the SIF ($k_1$). Note that greater $\gamma$ means greater resistance to failure.

Fiber positions for all three distribution functions are the same and the total fiber force for all these cases is $P$. It is interesting to note here that for the uniform fiber force distribution, $\gamma$ increases with the number of fibers, but

**TABLE 10.2**

Variation of $\gamma$ Value as Number of Fibers and Fiber-Force Distribution Functions (see Figure 10.10) Change on a Semi-infinite Crack

| Number of Fibers | Uniform Distribution | Linear Distribution | Quadratic Distribution |
|---|---|---|---|
| 2 | 1.045 | 0.986 | 1.004 |
| 4 | 1.101 | 0.972 | 1.004 |
| 7 | 1.15 | 0.958 | 0.975 |

an opposite trend is observed for both linear and quadratic distributions. This is because, as the total load is distributed over a larger number of fibers, for linear and quadratic distributions the resultant force moves away from the crack tip. It is easy to see from equation (10.2) that if the resultant force moves away from the crack tip, the $k_1$ value will be reduced. This movement is larger for the linear case, and hence the reduction in the $\gamma$ value is also greater. When the resultant force does not move (for the uniform fiber force distribution), $\gamma$ increases slightly, but for linear and quadratic distributions this increase is offset by the decrease of the $\gamma$ value due to the movement of the line of action of the resultant force away from the crack tip. This gives a qualitative justification why the ultimate strength increases with the increase of fiber length (or decrease of the number of fibers) when other parameters are kept constant.

Another important mechanism that contributes to the increasing fracture toughness with the fiber length is that long fibers intersect multiple cracks contributing to the bridging force in more than one crack. These two important mechanisms increase the fracture toughness of FRBMCs when the fiber length is increased.

#### *10.3.2.3 Effect of Fiber Diameter*

If the volume fraction and the fiber length are kept unchanged and the fiber diameter is reduced, the number of fibers increases, which results in an increase in the ultimate strength. It is easy to see that $n\%$ reduction in the fiber diameter causes $n\%$ reduction in the surface area of one fiber. If the cohesive force between the fiber surface and the matrix does not change, the force produced by individual fibers will be reduced by $n\%$. Then the individual fiber's cross-sectional area and volume are reduced by $m\%$, where

$$m = 100\left[1 - \left(1 - \frac{n}{100}\right)^2\right] \tag{10.8}$$

For the same volume fraction and fiber length, the number of fibers is increased by $\frac{100m}{100-m}\%$ for $n\%$ reduction of the fiber diameter. Hence, the

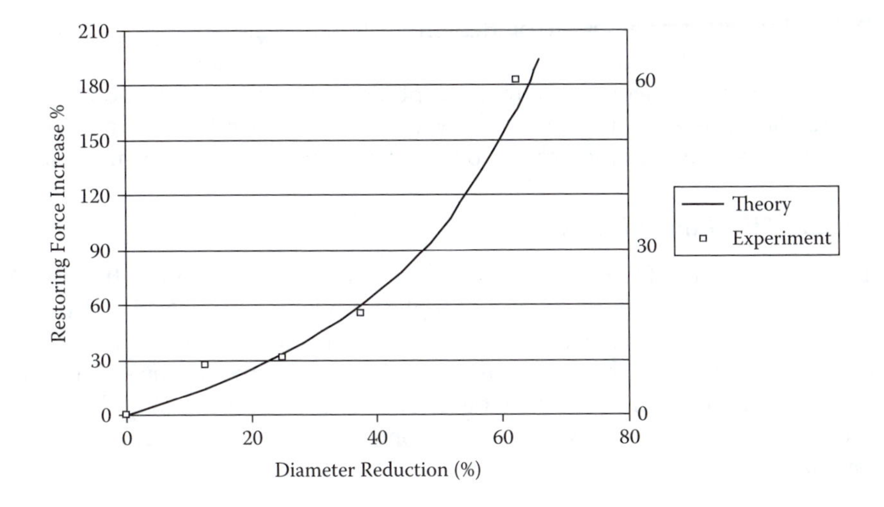


**FIGURE 10.11**
Theoretical prediction of the percentage increase of the fracture toughness (scale on the left side) as a function of the percentage reduction of the fiber diameter (horizontal axis). Experimental points (square markers) show the percentage increase of the ultimate strength (scale on the right side).

percentage increase in the restoring force due to $n$% reduction in fiber diameter is given by

$$X = m\left(\frac{100-n}{100-m}\right) - n \tag{10.9}$$

A plot of $n$ versus $X$ is shown in Figure 10.11. Since restoring force is directly proportional to the stress intensity factor $k_1$ (equation 10.2), Figure 10.11 shows the variation of $k_1$ (the part of SIF that depends on the fibers; see equation 10.3) with $n$. However, the total SIF has two parts, $K$ and $k_1$, as shown in equation 10.3. An increase in $k_1$ reduces the total stress intensity factor $(K - k_1)$, since $K$ depends only on the loading and the problem geometry and not on the fiber geometry. Thus, as $k_1$ increases, net SIF decreases and, as a result, the critical load for failure as well as the ultimate strength increases. From Figure 10.11 it can be seen that as $n$ (plotted along the horizontal axis) increases, $X$ of equation (10.9) (plotted along the vertical axis) also increases monotonically with increasing slope. Experimental results show that for 1% volume fraction of fibers as $n$ increases by 12.5, 25, 37.5, and 62.5%, the ultimate strength is increased by 9, 10.3, 18.4, and 61%, respectively, which represents a nonlinear monotonic increase.

Since $X$ of equation (10.9) represents the variation of the stress intensity factor $k_1$ and values listed in Table 10.1 correspond to the variation of the ultimate strength $\sigma_U$, it is not expected that these two sets of values be numerically equal; however, they should follow the same trend that is observed here.

If the experimental values are plotted on a different scale (shown on the right side of Figure 10.11) on the same Figure, then the experimental values correlate very well with the theoretical curve except for the first point corresponding to the 12.5% diameter reduction. Thus, the experimental results associated with the fiber diameter variation can also be justified from this simple analysis.

### 10.3.3 Effect on Stiffness

The mechanism of fiber force resisting the crack propagation has the direct effect on the increase of the fracture toughness, which increases the ultimate strength. However, increase in the fracture toughness should not have much influence on the stiffness of the material. Therefore, the Young's modulus should remain essentially unchanged or should vary only slightly when short fibers are added to strengthen the composite. This was observed experimentally by Cha et al. (1994) and others.

### 10.3.4 Experimental Observation of Fracture Toughness Increase in FRBMCs with Fiber Addition

In Table 10.1 the variations in ultimate strength of FRBMCs with fiber parameters have been listed. Although the ultimate strength and the fracture toughness of FRBMCs are related, no fracture toughness information is given in Table 10.1. Kundu et al. (2000) measured fracture toughness of FRBMCs taking three-point bending specimens. They carried out tests on steel fiber reinforced ceramic matrix composite specimens. Steel fibers of size $15.24 \times 0.76 \times 0.25$ mm$^3$ ($0.6 \times 0.03 \times 0.01$ in.$^3$) were uniformly mixed with a ceramic powder material and then the mixture was placed in a mold and heated in an oven to produce the FRBMC. Fiber amounts in the five specimens tested were 0, 2, 5, 10, and 15% of the specimen weight. Table 10.3 shows the recorded fracture toughness for these five specimens.

Note that an increase of fiber weight fraction from 2% in specimen 2 to 15% in specimen 5 corresponds to an increase of fiber weight fraction by 650%. It increased the fracture toughness from 1181 psi-in.$^{1/2}$ to 3954 psi-in.$^{1/2}$.

**TABLE 10.3**

Fiber Weight Fraction and Fracture Toughness for Five Ceramic Specimens

| Specimen No. | Fiber Weight Fraction (%) | Fracture Toughness ($K_c$) in psi-in.$^{1/2}$ (1 psi-in.$^{1/2}$ = 1.1 N-cm$^{-3/2}$) |
|---|---|---|
| 1 | 0 | 1043 |
| 2 | 2 | 1181 |
| 3 | 5 | 1524 |
| 4 | 10 | 2238 |
| 5 | 15 | 3954 |

*Source:* Kundu, T. et al. *International Journal for Numerical and Analytical Methods in Geomechanics*, 24, 655–673, 2000.

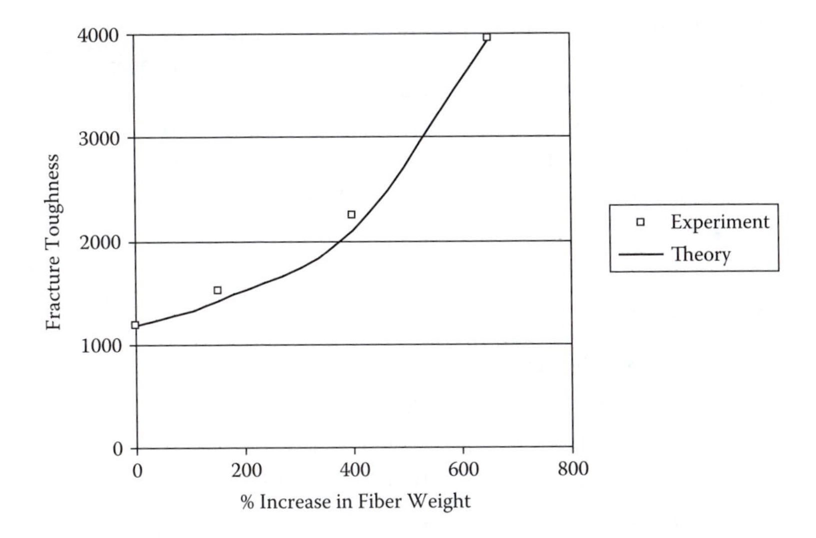


**FIGURE 10.12**
Variation of fracture toughness (psi-in.$^{1/2}$) with increase in fiber volume in FRBMCs.

The theoretical curve for this set of experimental data points can be obtained from equations (10.4) and (10.5). This curve, along with the experimental data points, is shown in Figure 10.12. Note that Figures 10.7 and 10.12 show similar variations of ultimate strength and fracture toughness with the fiber volume fraction, as expected.

## 10.4 Dynamic Effect

This book analyzed cracked problem geometries under static loads only. If the cracked structures are subjected to dynamic (or time-dependent) loads, then both the SIF and the crack opening displacement are amplified due to the dynamic effect. Therefore, a cracked plate that does not fail under the static loading condition may fail when the load is applied suddenly. Transient response of cracked bodies subjected to time-dependent loads showing amplification of the SIF and the crack opening displacement has been studied by Kundu (1986, 1987, 1988), Kundu and Hassan (1987), Karim and Kundu (1988, 1989, 1991), Awal, Kundu, and Joshi (1989), and others.

Elastodynamic analysis of cracked bodies is also necessary to study the interaction between elastic waves and cracks. This study is important for the nondestructive inspection of internal cracks by ultrasonic waves. One needs to know how ultrasonic waves are scattered by cracks if the location, orientation and shape of the crack are to be determined from the scattering pattern of the ultrasonic waves. Many publications on this topic are available in the literature. A good number of these publications are referred to in various papers published by the author and his colleagues over the last two decades (Kundu and Mal 1981; Karim and Kundu 1990; Kundu 1990; Kundu and Boström 1991, 1992; Karim, Awal, and Kundu 1992a, 1992b; Banerjee and Kundu 2007a, 2007b) and in many other similar papers. Interested readers are referred to these publications.

## 10.5 Concluding Remarks

Two advanced topics that have been discussed in detail in this chapter are on the stress singularity near crack corners and the fracture toughness of FRBMCs. In section 10.2 it is shown that, for cracks with sharp corners, the strength of the stress singularity increases or decreases from the square root singularity, observed for the smooth crack front. Section 10.3 discusses how the length, diameter, and volume fraction of fibers affect the fracture toughness, ultimate strength, and stiffness of short fiber reinforced brittle matrix composites (SFRBMC). Importance of solving dynamic problems for predicting the dynamic SIF and for detecting cracks by ultrasonic nondestructive techniques is briefly discussed in section 10.4 and adequate references are provided for interested readers.

The three topics discussed in sections 10.2, 10.3, and 10.4 are too advanced for an introductory fracture mechanics book. These materials are introduced for readers who are interested in exploring this subject beyond the basic knowledge of linear elastic fracture mechanics. These readers can acquire adequate knowledge on these advanced topics from the references provided in this chapter.

## References

Awal, M. A., Kundu, T., and Joshi, S. P. Dynamic behavior of delamination and transverse cracks in fiber reinforced laminated composites. *Engineering Fracture Mechanics, an International Journal*, 33(5), 753–764, 1989.

Banerjee, S. and Kundu, T. Semi-analytical modeling of ultrasonic fields in solids with internal anomalies immersed in a fluid. *Wave Motion*, in press, 2007a.

———. Scattering of ultrasonic waves by internal anomalies in plates immersed in a fluid. *Optical Engineering*, 46(5), 053601-1–053601-9, 2007b.

Cha, Y.-H., Kundu, T., Stewart, W., and Desai, C. S. Effects of steel fibers on the fracture toughness of ceramic composites from a lunar simulant. *Proceedings of the First International Conference on Composites Engineering*, New Orleans, LA, Aug. 1994, ed. D. Hui, pp. 271–272, 1994.

Cox, B. N. Scaling for bridged cracks. *Mechanics of Materials*, 15(2), 87–98, 1993.

Cox, B. N. and Marshall, D. B. Determination of crack bridging forces. *International Journal of Fracture*, 49(3), 159–176, 1991.

Evans, A. G. Mechanical performance of fiber reinforced ceramic matrix composites. *Proceedings of the 9th Riso International Symposium on Metallurgy and Materials Science*, Riso National Laboratory, Riso Library, Roskilde, Denmark, 13–34, 1988.

———. High toughness ceramics and ceramic composites. *Proceedings of the 10th Riso International Symposium on Metallurgy and Materials Science*, Riso National Laboratory, Riso Library, Roskilde, Denmark, pp. 51–91, 1989.

———. Mechanical properties of reinforced ceramic, metal and intermetallic matrix composites. *Materials Science and Engineering A: Structural Materials: Properties, Microstructure and Processing*, A143, 63–76, 1991.

Evans, A. G. and Zok, F. W. Physics and mechanics of fiber-reinforced brittle matrix composites. *Journal of Materials Science*, 29, 3857–3896, 1994.

Gopalaratnam, V. S., Shah, S. P., Batson, G. B., Criswell, M. E., Ramakrishnam, V., and Wecharatna, M. Fracture toughness of fiber reinforced concrete. *ACI Materials Journal*, 88(4), 339–353, 1991.

Kanda, T. and Li, V. C. Interface property and apparent strength of a high strength hydrophilic fiber in cement matrix. *ASCE Journal of Materials in Civil Engineering*, 10(1), 5–13, 1998.

Karim, M. R., Awal, M. R., and Kundu, T. Numerical analysis of guided wave scattering by multiple cracks in plates: SH-case. *Engineering Fracture Mechanics, an International Journal*, 42(2), 371–380, 1992a.

———. Elastic wave scattering by cracks and inclusions in plates: In-plane case. *International Journal of Solids and Structures*, 29(19), 2355–2367, 1992b.

Karim, M. R. and Kundu, T. Transient surface response of layered isotropic and anisotropic half-spaces with interface cracks: SH case. *International Journal of Fracture*, 37, 245–262, 1988.

———. Transient response of three layered composites with two interface cracks due to a line load. *Acta Mechanica*, 76, 53–72, 1989.

———. Scattering of acoustic beams by cracked composites. *ASCE Journal of Engineering Mechanics*, 116, 1812–1827, 1990.

———. Dynamic response of an orthotropic half-space with a subsurface crack: In-plane case. *ASME Journal of Applied Mechanics*, 58(4), 988–995, 1991.

Kundu, T. Transient response of an interface crack in a layered plate. *ASME Journal of Applied Mechanics*, 53(3), 579–586, 1986.

———. The transient response of two cracks at the interface of a layered half space. *International Journal of Engineering Science*, 25(11/12), 1427–1439, 1987.

———. Dynamic interaction between two interface cracks in a three layered plate. *International Journal of Solids and Structures*, 24(1), 27–39, 1988.

———. Scattering of torsional waves by a circular crack in a transversely isotropic solid. *Journal of the Acoustical Society of America*, 88(4), 1975–1980, 1990.

Kundu, T. and Boström, A. Axisymmetric scattering of a plane longitudinal wave by a circular crack in a transversely isotropic solid. *ASME Journal of Applied Mechanics*, 58(3), 695–702, 1991.

———. Elastic wave scattering by a circular crack in a transversely isotropic solid. *Wave Motion*, 15(3), 285–300, 1992.

Kundu, T. and Hassan, T. A Numerical study of the transient behavior of an interfacial crack in a bimaterial plate. *International Journal of Fracture*, 35, 55–69, 1987.

Kundu, T., Jang, H. S., Cha, Y. H., and Desai, C. S. A simple model to predict the effect of volume fraction, diameter and length of fibers on strength of fiber reinforced brittle matrix composites. *International Journal for Numerical and Analytical Methods in Geomechanics,* 24, 655–673, 2000.

Kundu, T. and Mal, A. K. Diffraction of elastic waves by a surface crack on a plate. *ASME Journal of Applied Mechanics,* 48(3), 570–576, 1981.

Li, S. H., Li, Z., Mura, T., and Shah, S. P. Multiple fracture of fiber-reinforced brittle matrix composites based on micromechanics on micromechanics. *Engineering Fracture Mechanics,* 43(4), 561–579, 1992.

Li, V. C. and Hashida, T. Engineering ductile fracture in brittle matrix composites. *Journal of Materials Science Letters,* 12, 898–901, 1993.

Li, V. C., Wu, H. C., Maalej, M., Mishra, D. K., and Hashida, T. Tensile behavior of cement based composites with random discontinuous steel fibers. *Journal of American Ceramic Society,* 79(1), 74–78, 1996.

Li, V. C. and Ward, R. A novel testing technique for post-peak tensile behavior of cementitious materials. In *Fracture toughness and fracture energy—Test methods for concrete and rock,* ed. A. A. Mihashi. Rotterdam, the Netherlands: A. A. Balhema, pp. 183–195, 1989.

Luke, C. E., Waterhouse, B. L., and Wooldridge, J. F. Steel fiber reinforced concrete optimization and applications. *An International Symposium on Fiber Reinforced Concrete.* Detroit, MI: American Concrete Institute, SP-44, pp. 393–413, 1974.

Marshall, D. B., Morris, W. L., Cox, B. N., Graves, J., Porter, J. R., Kouris, D., and Everett, R. K. Transverse strengths and failure mechanisms in $Ti_3Al$ matrix composites. *Acta Metallurgica et Materialia,* 42(8), 2657–2673, 1994.

Shah, S. P. Do fibers increase the tensile strength of cement-based matrixes? *ACI Materials Journal,* 88(6), 595–602, 1991.

Shah, S. P. and Ouyang, C. Toughening mechanisms in quasibrittle materials. *ASME Journal of Engineering Materials and Technology,* 115(3), 300–307, 1993.

Wang, Y., Backer, S., and Li, V. C. An experimental study of synthetic fiber reinforced cementitious composites. *Journal of Materials Science,* 22, 4281–4291, 1987.

Wang, Y., Li, V. C., and Backer, S. Experimental determination of tensile behavior of fiber reinforced concrete. *ACI Materials Journal,* 87(5), 461–468, 1990.

Ward, R. J., Yamanobe, K., Li, V. C., and Backer, S. Fracture resistance of acrylic fiber reinforced mortar in shear and flexure. In *Fracture mechanics: Application to concrete,* eds. V. C. Li and Z. P. Bazant. ACI, SP 118-2, 17–68, 1989.

Xu, L. and Kundu, T. Stress singularities at crack corners. *Journal of Elasticity,* 39, 1–16, 1995.

## Exercise Problems

Problem 10.1: For different crack geometries shown in Figure 10.13, identify the point(s) where the crack will start to propagate under mode I loading.

Problem 10.2: Derive equation (10.4).

Problem 10.3: Derive equation (10.8).

Problem 10.4: Derive equation (10.9).

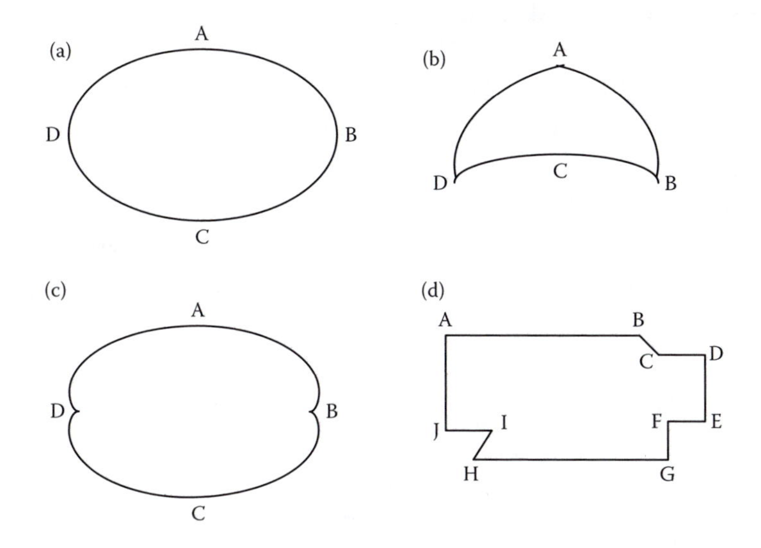

**FIGURE 10.13**

Problem 10.5:

(a) For SFRBMCs, if the fiber diameter is reduced, keeping the fiber length and volume fraction unchanged, do you expect an increase or decrease in the fracture toughness of the composite?

(b) Calculate the percentage change in the fiber bridging force for 25% reduction in fiber diameter. Show all steps of your calculation.

# Index

## D



## E



## F



## G



## H



## I